Applications in LabVIEW

Leonard Sokoloff

PEARSON

Prentice
Hall

Upper Saddle River, New Jersey

Columbus, Ohio

Library of Congress Cataloging in Publication Data

Editor in Chief: Stephen Helba
Editor: Charles E. Stewart, Jr.
Editorial Assistant: Maria Rego
Production Editor: Kevin Happell
Design Coordinator: Diane Ernsberger
Cover Designer: Kristina D. Holmes
Cover art: Digital Images
Production Manager: Matthew Ottenweller
Marketing Manager: Ben Leonard

This book was printed and bound by Courier-Kendallville, Inc. The cover was printed by Phoenix Color Corp.

LabVIEW images of objects are reproduced for use in this book with permission of National Instruments Corp.

Pearson Education Ltd.
Pearson Education Singapore Pte. Ltd.
Pearson Education Canada, Ltd.
Pearson Education—Japan

Pearson Education Australia Pty. Limited
Pearson Education North Asia Ltd.
Pearson Educación de Mexico, S.A. de C.V.
Pearson Education Malaysia Pte. Ltd.

10 9 8 7 6 5 4 3 2 1
ISBN 0-13-016194-2

This book is dedicated to my wife Elena and my daughter Lara.

Contents

Preface

High-performance computation and display capabilities offered by personal computers (PCs) have dramatically changed the field of instrumentation. Modern instruments use a PC and a graphical software unit with a variety of interfaces to communicate with and to control the hardware. These instruments use the computer's extensive processing power to offer a very high level of performance.

Over the years, graphical programming language has gained ground as a programming tool. Although it has not displaced traditional programming such as C or C++, it offers an environment that is fast and easy to use. In applying the language, users do not have to remember the code; they simply manipulate the objects on the computer's screen.

LabVIEW®, a product of National Instruments Corporation, uses graphical language in creating a program called a virtual instrument (VI). Virtual instruments can acquire and process data, display results on a graph, control another instrument and/or an external system, and perform simulation and many other tasks. Because VI is a software file, it can be easily reconfigured to meet the requirements of a new specification. The ability to alter the functionality of an instrument is an advantage that was never before available to the user. In the past, the vendor or the manufacturer controlled the instrument's functionality.

ORGANIZATION OF THE TEXT

A wide range of data acquisition, analysis, and simulation experiments using LabVIEW software have been designed and included in this book. In order to better understand LabVIEW, the user is often required to modify the existing software in order to achieve a specific measurement, as in the case of the position control servo where the user is required to design the settling time measurement software.

Chapter 1 introduces the reader to some of the basic tools and operating features of LabVIEW, including loops arrays and graphs. Data acquisition, data processing, and GPIB instrument control are introduced. Chapter 2 presents the five structures in LabVIEW: sequence, cases that can be configured as numeric or Boolean, for and the while loops, and the formula node. Most structures, except formula node and sequence, have their counterpart in C language. The sequence structure is not required in C language because all commands are executed in the order that they are written. Execution in LabVIEW VI, however, is based on the flow of data. An object in the block diagram executes only if data is available at all inputs. The formula node gives the designer an option of either wiring objects or writing formulas inside the node. Exercises in this chapter illustrate practical application of the structures.

Chapter 3 introduces the user to one-dimensional and two-dimensional arrays that are created inside the for loop or while loop. It also includes array functions in LabVIEW, which are essential in processing array-based data. Chapter 4 presents applications where the raw or processed data must be displayed. The chart, waveform, and X-Y graphs are used throughout the text. Each graph has a specific input requirement. For example, the waveform graph requires an array input for the X-axis, while the X-Y graph requires arrays for the X-axis and the Y-axis.

Strings offer the most versatile format because they can represent any character on the keyboard. Typically a string is an array of ASCCI characters, with each character is assigned an ASCCI code. A character is a number, an alphabetical character, or any other character on the keyboard. Strings are used to communicate information over a network. TCP/IP protocol uses strings. GPIB uses strings in communicating with remote instruments. LabVIEW includes string palettes in its analysis library providing string controls, indicators, and string functions for building VIs. Experiments in Chapter 5 introduce the application of various string functions.

The data generated by VI can be saved to a file, which can later be opened, appended, or modified. In LabVIEW special rules and procedures must be followed to save data to a file. These rules include creating a new file or reading from and writing to an existing file. Chapter 6 illustrate this process.

Mathematical analysis software is given special attention in Chapter 7. Fourier analysis software plots the square wave that corresponds to any number of Fourier terms. The Fourier spectrum is generated and displayed for a pulse train with adjustable duty cycle. Although it may take a few seconds, the software can sum and display as many as one million terms, a task that would have been unthinkable in the not too distant past.

Chapter 8 introduces basic electronic communication concepts and LabVIEW simulation exercises. Amplitude modulation and frequency modulation are covered extensively. DSBFC, DSBSC, and SSB are discussed in detail. Theoretical concepts are supported by the simulation software. Transmission line concepts in microwaves are presented using a software design project. Extensive background material is provided to refresh the reader's comprehension.

Chapter 9 introduces the fundamentals of data acquisition. When an analog signal is brought inside the LabVIEW environment via the A/D converter, it must first be digitized or converted to an array of samples. An interface called the data acquisition board (DAQ board) is required to acquire data. LabVIEW supports the data acquisition process with the Data Acquisition subpalette in the Functions floating palette. The VIs used in data acquisition fall into one of three categories: Easy, Intermediate, and Advanced. Easy VIs have the highest software overhead but are easiest to use; Advanced VIs have the lowest software overhead but in order to use them, the designer must be very familiar with LabVIEW.

Chapter 10 introduces the data acquisition process through very simple experiments. Real data is acquired and displayed, stored to a spreadsheet, or otherwise processed. Acquire Waveforms VIs, which reduce software overhead, are used. They employ buffered and hardware timed types of data acquisition.

A broad area of physics is explored in Chapters 11 to 13 with numerous experiments that support and illustrate various theoretical principles. Uncommon experiments include the diffraction pattern measurement in optics and thermodynamic experiments, including the

Seebeck and Peltier effects. Classical concepts in optics and the inverse square law are investigated in Chapter13 using commercial equipment and LabVIEW.

Chapters 14 and 15 present an extensive theoretical and experimental coverage of motors and generators, AC and DC. For example, the motor generator assembly used in Chapter 14 is open so that the armature and other details are easily seen. Motion control experiments in Chapter 15 illustrate the speed control of a DC motor. Of special interest is the control system presented in Chapter 16. A commercial trainer is used to demonstrate the characteristics of speed control and position control servos. Velocity feedback to control system damping is explored in a separate experiment. Ramp waveform generated in LabVIEW drives the position control servo in order to measure the steady state error.

TCP/IP is a protocol that is supported by virtually all operating systems, including LabVIEW. Chapter 17 explains and uses the server-client model to communicate data over a local area network such as the Ethernet. The chapter also illustrates how a client can operate a VI at a remote server location. CGI, HTTP servers, and the use of hyperlinks and HTML are also covered as they play an important role in communication and control over the Internet.

Chapter 18 covers the GPIB instrument control. GPIB commands sent to a remote instrument with GPIB capability can control its operation. In a typical setup, the GPIB interface on the controlling station is connected by the GPIB cable to the GPIB connectors on the remote instruments. Experiments in this chapter involve controlling the operation of a power supply, multimeter, and the function generator; the GPIB control of oscilloscope; the Bode Plotter; and the Spectrum Analyzer.

ACKNOWLEDGMENTS

I am grateful to Amin Karim, the director of curriculum development for all DeVry campuses. He has made possible my development work in implementing physics experiments in LabVIEW.

Many thanks to Bhupinder Sran, Dean of Academic Affairs, and John Abdellatif, Dean of Electronics and TCM at the North Brunswick campus, for their support and encouragement in the course of writing this book.

I wish to thank my student assistants who provided time and effort in developing the laboratory experiments for this book. They are as follows:

Allan Kennedy Jr.: LabVIEW on the Internet

Gianfranco Comune: Servo Control

Raul Lasluisa: TCP/IP, Distributed Process Control

Chaz Byrne: GPIB

Justin Berhang, Nathan Rathmell: GPIB

Salvatore Bonello, Paul James: GPIB

Marc Deltoro: Physics II, Physics III

Marc Deltoro, Herve Yoerg: PID

Kurt Jack, Joe Carrer: PID

Annette Navarro, Mina Makram: TCP/ IP

Manuel Marrero: Physics I

Ryan Lewis, Edward Barrett: PID

Lloyd Gossler, Michael Ronaldo: PID

Nicolas Michael, Hernan Alvarez: Distributed Process Control

Robert Doinov, Joseph Buttry, Ruben Beltran, David Good: Spread Spectrum Communication

And last but not least I am grateful to my wife Lena for her patience and support throughout the writing of this book.

Chapter 1
About LabVIEW...

LabVIEW® (Laboratory Virtual Instrument Engineering Workbench) is a graphical programming software used in developing programs for simulation, data acquisition, control, and communication applications. In the LabVIEW environment icons are interconnected to create a program generally referred to as a VI (Virtual Instrument), and that is why all LabVIEW programs have the extension .vi. There are rules for creating VIs; however, the user does not have to know programming languages such as C^{++}.

Front Panel and Block Diagram

As shown in Fig. 1-1, all VIs must have two components: the Front Panel and the Block Diagram. The Front Panel generally includes controls, indictors, switches, graphs, and so forth, and the Block Diagram contains function icons that are connected by wires.

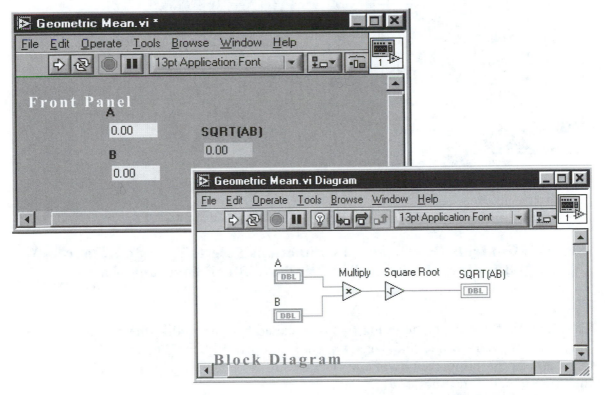

Fig. 1-1 The Front Panel and the Block Diagram of a Typical VI

The program called Geometric Mean.vi, shown in Fig. 1-1, produces a geometric mean of numbers A and B and displays the result on the digital indicator SQRT(AB). The user inputs the values of A and B in the Front Panel. These values are processed in the Block Diagram; they are first multiplied, and then a square root operation is performed on their product.

Controls and Functions Palettes

The Front Panel and the Block Diagram concept resemble an oscilloscope, one of the instruments most often used by scientists and engineers, providing the user with an environment that is familiar and intuitive. An oscilloscope has a control panel that includes switches and various indicators resembling the Front Panel of a VI. The functionality of the oscilloscope reflects the internal wiring of chips much as the wiring of the function icons in the Block Diagram reflects the functionality of the VI.

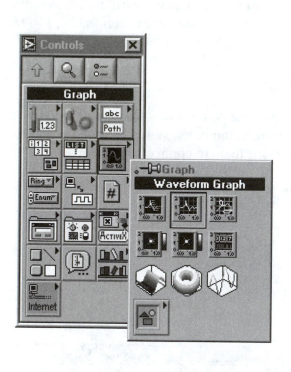
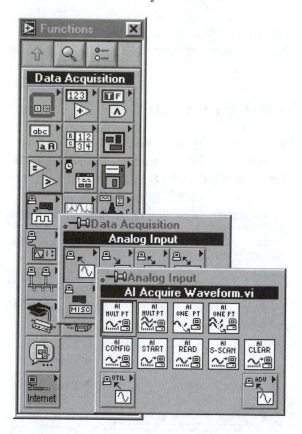

Fig. 1-2 Controls Palette and the Functions Palette Provide Resources for Building The Front Panel and the Block Diagram of a VI

The Controls floating palette in Fig. 1-2 is so called because, unlike the pull-down menu, it can be moved around the window, once it opens. It can be opened by clicking anywhere in the Front Panel window. The Controls palette contains objects for building the Front Panel of the VI.

Included in the Controls palette are subpalettes for different categories of Front Panel objects. Fig. 1-2 shows the selection of the Waveform Graph form the Graph subpalette.

A comparison of LabVIEW versions 5.x and 6i shows that the latter adds three buttons at the top of the palette. The middle button called "Search" opens a list of objects from which to choose in the Controls palette. The Search list for the Controls palette is shown in Fig. 1-3.

The More button at the bottom of the palette provides options to choose objects from the Controls palette, Functions palette, or both.

The function of the first button at the top of the palette, the Back button, is the same as that of the Back button in an Internet browser.

The Functions palette (also floating) provides various objects and functions for building the Block Diagram of a VI. A category button in the Functions Palette may have a subpalette that in turn may also have a subpalette.

The illustration in Fig. 1-2 shows the selection of AI Acquire Waveform.vi from the Analog Input subpalette that is one of the options in the Data Acquisition subpalette of the Functions Palette. An alternative method of making this selection is to choose AI Acquire Waveform.vi from the Search list and open the Analog Input subpalette.

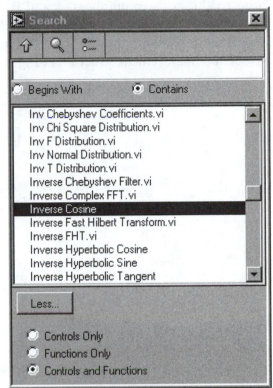

Fig. 1-3 The Search List of Objects for the Front Panel and the Block Diagram

SubVIs

The Block Diagram of a typical VI includes various operational tasks. Each task can be incorporated into a subVI that has an icon and a connector, as shown below.

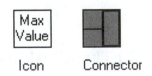

Icon   Connector

Each task is thus represented by a module. Modularization or breaking up the code into modules simplifies the Block Diagram and makes it easier to follow. Modularization also simplifies the task of troubleshooting or modifying a module.

Fig. 1-4 shows an example of modularization. The three tasks in Three Tasks.vi, shown in Fig. 1-4, include the determination of the arithmetic mean, the geometric mean, and the largest value for two numbers, A and B. Each task is incorporated into a subVI that has an icon and a connector. The face of the icon can be edited by the user to reflect an important feature of the subVI. The modularized VI is shown in Fig. 1-4b and its associated hierarchical structure in Fig. 1-4c.

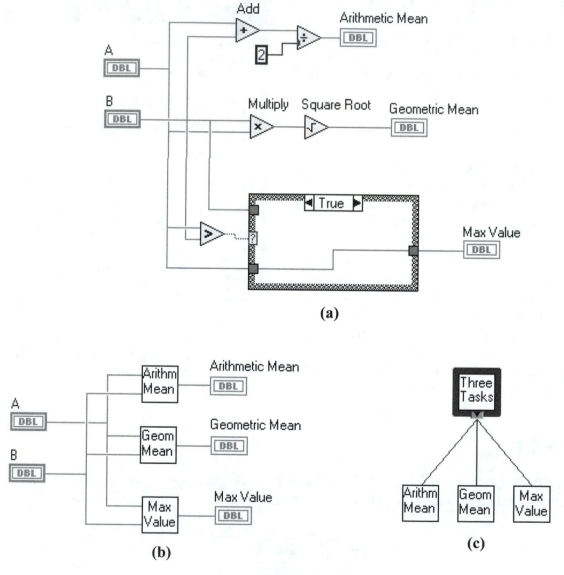

Fig. 1-4   (a) Three Tasks.vi , (b) Three Tasks.vi Modularized, (c) the Hierarchy Tree for Three Tasks.vi

Loops and Arrays

Loops provide a repetitive execution of the code. The While Loop shown in Fig. 1-5 has its "Do While" equivalent in C and other high-level languages. It has two terminals: the Iteration terminal that increments each time that the code inside the loop has been executed once, and the Condition terminal, which requires a Boolean input, a TRUE or a FALSE. As long as the input is TRUE, the loop will continue to execute indefinitely, and a FALSE input terminates the execution. If the input is FALSE at the beginning of execution, the loop will execute once. The While Loop is used in operations such as search, where the number of loop iterations is not known.

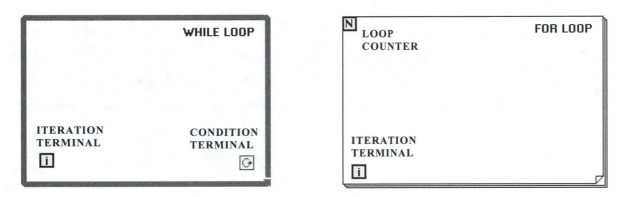

Fig. 1-5 The While Loop and the For Loop

The For Loop, also shown in Fig. 1-5, is more suitable for applications in which the total number of iterations is known. In that case the Loop Counter input N requires a numerical input representing the number of iterations.

Loops are used to create arrays. In Fig. 1-6, a For Loop generates five elements of a one-dimensional array. Data can be passed into and out of the loop only when the loop completes its iterations. As shown in the illustration, a thick line designates an array. The "indexing" feature, a default setting in For Loops allows the formation of the array. It may be disabled by right-clicking on the tunnel and then choosing "Disable Indexing"

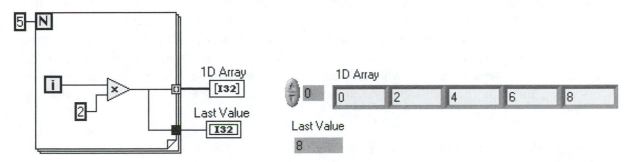

Fig. 1-6 Creating a One-Dimensional Array and the Result of a Typical Run

from the popup menu. In Fig. 1-6 this has been done for the data passed to the Last Value digital indicator, with the result of only the last value of the array being passed out of the loop.

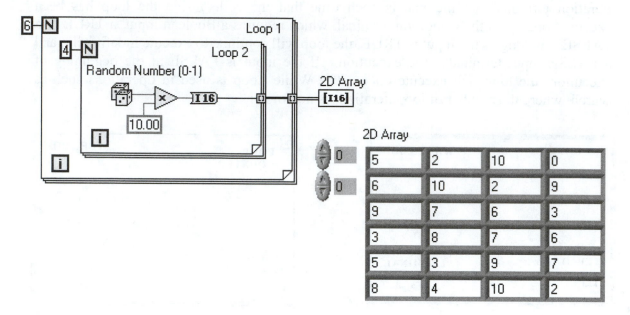

Fig. 1-7 Creating a Two-Dimensional Array and the Result of a Typical Run

As shown in Fig. 1-7, two nested loops are necessary to create a two-dimensional array. Loop 1 creates rows and Loop 2 creates columns. Thus, for each iteration of Loop 1, Loop 2 produces four random numbers between 0 and 10. As shown in the illustration, the result is a 6x4 array.

Charts and Graphs

Charts and graphs are Front Panel objects that are used for displaying waveforms. Despite the analog-like continuous appearance, the displayed waveforms are actually represented by points or samples. One must be aware that in a digital environment continuous waveforms are not possible. Once interpolation is removed, the illusion of continuity is gone and one can clearly see the samples that make up the waveform. Consequently, regardless of whether the wave is generated within the computer or imported through the data acquisition process, it must necessarily be represented by samples.

Two graphs and a chart are shown in Fig. 1-8. Despite their striking similarity in the Front Panel, their implementation in the Block Diagram requires special care.

Waveform Chart can display in an ongoing fashion the data points as they are generated. One or more waveforms can be displayed on the same chart. As shown in the illustration below, the Bundle function must be used when displaying more than one waveform.

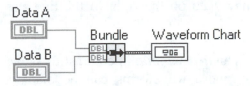

The Waveform Chart has three update modes: Strip Chart, Scope Chart, and Sweep Chart.

Strip Chart resembles the old paper strip chart with a scrolling display. As the visible display reaches 1024 data points, the limit of the data buffer, the display scrolls off the screen as new data is displayed.

Scope Chart has a retracing display that resembles the oscilloscope. As the display reaches the right edge of the screen, the screen is cleared and the new display scrolls from the left edge of the screen.

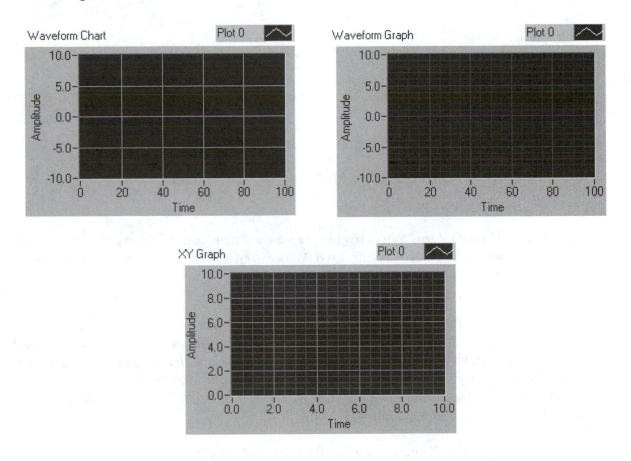

Fig. 1-8 The Two Graphs and the Chart are Front Panel Indicator Objects

Sweep Chart has also a retracing display, except that the screen is not cleared as the display reaches the right edge of the screen. Instead, a vertical line moves with the scrolling display, showing new data on the right of the line and the old data on its left side.

Waveform Graph requires an *array input*, in contrast to Waveform Chart, which plots data point by point. One or more waveforms can be displayed on the same graph. The arrangement shown in Fig. 1-9a displays one waveform as a function of samples. In Fig. 1-9b the X-axis is formatted by specifying the waveform start point (usually 0) and the spacing between points; the display here consists of two waveforms, Array 1 and Array 2.

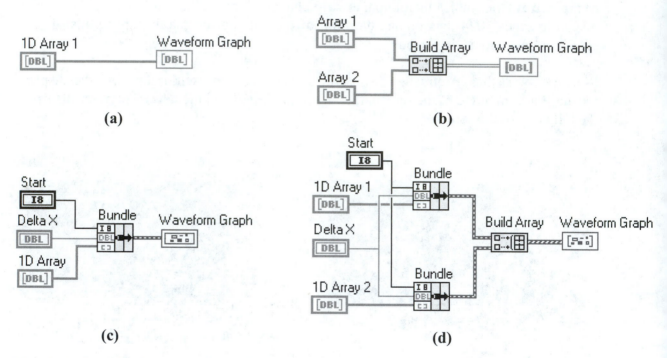

Fig. 1-9 Displaying One Waveform: (a) as a Function of Samples, (b) X-axis Formatted. Displaying Two Waveforms: (c) as a Function of Samples, (d) X-axis Formatted

The Build Array function is used in displaying multiple waveforms, as shown in Fig. 1-9b and d. Here the X-axis may also be formatted; otherwise, the data points are plotted versus sample points. The Build Array function has the **Concatenate Inputs** feature that plots both inputs on the same trace, the upper input first, followed by the lower input. The default setting for the Build Array function is a two-dimensional array output that plots two waveforms on the same graph.

The formatting of the X-axis is especially useful in data acquisition applications. AI Acquire Waveforms.vi performs timed data acquisition on multiple analog inputs. One of the outputs in this VI is actual scan period, the reciprocal of the scan rate input. This value may be wired to the delta X input of the Bundle function in Fig. 1-9, resulting in the real-time scale in the acquired waveforms.

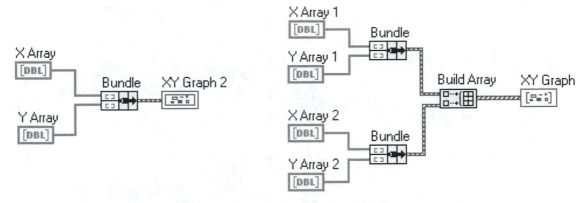

Fig. 1-10   Displaying One Waveform and Multiple Waveforms on the X-Y Graph

X-Y Graph requires array inputs for the X-axis and the Y-axis. Fig. 1-10 shows the Block Diagram arrangements for displaying a single waveform, or multiple waveforms. The X-Y graph is useful in applications where the generated data must have a well defined abscissa scale, such as time scale. A small inconvenience in this case is the generation of two one-dimensional arrays.

Data Acquisition

Hardware

The hardware required to perform computerized data acquisition is shown in Fig. 1-11. The data acquisition (DAQ) board converts the applied analog waveforms to digital form. National Instrument Corporation offers a variety of DAQ boards. Speed is generally the determining criterion of their cost. The LabPC$^+$ DAQ board shown in the illustration has a speed of 80 kbps on one channel and 40 kbps on the two analog input channels used in the illustration. As shown, the input and the response are acquired on channels 0 and 1, respectively. Because the DAQ board is almost entirely inside the computer housing, an extender board is used to provide access to various I/O pins on the DAQ board.

Software

As shown in Fig. 1-11, the pre-loaded software, transparent to the user, includes Windows, LabVIEW software (version 5.1 and higher), and NI DAQ, the driver for the DAQ board. The data acquisition software is the responsibility of the user.

All programs written in LabVIEW must have a Front Panel and a Block Diagram, and their file names must have extension .vi. The Front Panel contains controls and indicators, as does the Front Panel of an oscilloscope. The Block Diagram contains function icons interconnected by wires. LabVIEW provides the user with the graphical language (G language) for writing a custom LabVIEW program (called a VI) that meets the user's needs. As with all high-level programming languages, LabVIEW programs must be compiled before execution because the computer understands only the machine language.

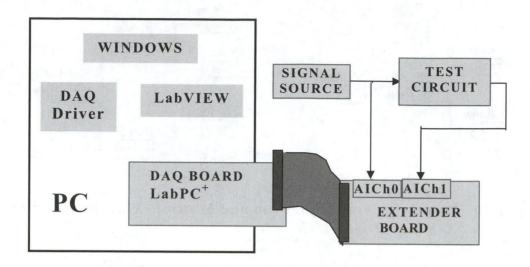

Fig. 1-11   Data Acquisition Hardware

There is one important difference between LabVIEW and any high-level language program during their execution. The execution process in any program written in a high-level language such as C progresses step by step in the order in which its instructions were written by the programmer.

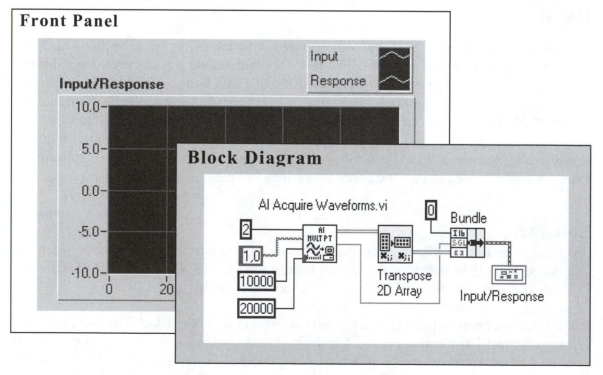

Fig. 1-12   Front Panel and Block Diagram of Experiment2.vi, the Program that Acquires Data

In LabVIEW, on the other hand, a function in the Block Diagram executes only if all input values are present. In that respect the execution of a LabVIEW VI is data driven.

Fig. 1-12 shows the Front Panel and the Block Diagram of Experiment2.vi, the program that acquires data.

AI Acquire Waveforms.vi is the function that acquires data, input squarewave on Ch. 0 and response on Ch.1. As shown, 10,000 scans at 20,000 scans/s result in 1/2 sec worth of acquired data. The data is formatted into a two-dimensional array. Each line in the array represents one scan that includes one data point from Ch. 0 and one data point from Ch. 1. The array thus contains 10,000 lines or rows.

The array must be transposed, as shown, in order to display its contents on a waveform graph. The abscissa of the waveform graph is formatted to reflect real time through the use of the Bundle function. The "0" input begins the display at t = 0, and the Δt input, equal to the reciprocal of scan rate, provides real time indication.

Fig. 1-13 shows the input step and circuit response of the acquired data.

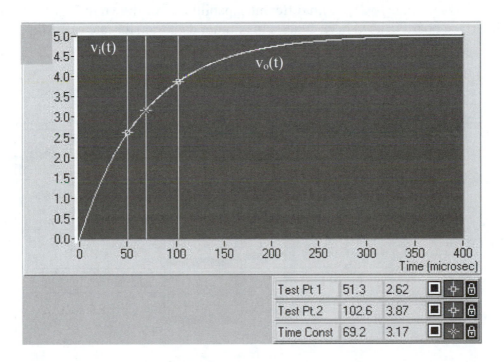

Fig. 1-13   Step Response of R-C Circuit. The Data Is Acquired and Displayed in LabVIEW in Real Time.

GPIB Instrument Control

With the proper interface cards, external instruments can be controlled by a VI built in the LabVIEW environment.

The GPIB interface is a parallel 24-pin conductor bus. It includes eight data lines for control messages that are often ASCII encoded, and various management and handshake lines. Handshaking is used to transfer messages between the PC and the instrument being controlled.

Fig. 1-14 shows a typical connection between the PC with the GPIB interface board and the software for controlling the operation of an instrument. Not every instrument has a GPIB capability. This capability is designed into the instrument by the manufacturer and a GPIB connector is provided on the instrument housing. The cable used to connect the instrument must also be GPIB compatible and meet the IEEE-488 protocol.

As shown in Fig. 1-14, the operation of several instruments may be controlled. Consequently, each instrument must have a unique address between 0 and 30 (address 0 is usually reserved for the GPIB interface board). National Instrument Corporation has an array of GPIB interface boards with different capabilities. As shown in Fig. 1-14, the AT-GPIB interface board is used here. It has a maximum data transfer rate of 1 Mbps.

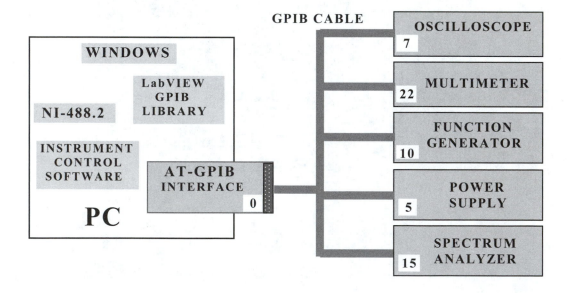

Fig. 1-14 GPIB Instrument Control Configuration. Each Instrument and the GPIB Interface Has a Unique Address.

The NI-488.2 is the driver software for the GPIB interface. GPIB devices can be Talkers, Listeners, or Controllers. The Listener receives messages while the Talker sends messages. The Controller, usually the PC where the GPIB board and NI.488.2 software are installed, manages the flow of commands or messages on the GPIB bus. The communication between the PC and the instrument is message based. ASCII-encoded message strings are transferred using handshaking.

Functions>Instrument I/O>GPIB is the location of available GPIB VIs for building an instrument control VI, the VI that is custom tailored to meet a user's exact needs. On a small scale this is a driver.

The most used GPIB VIs are the GPIB Read and the GPIB Write VIs, shown below. The former reads a data string from the instrument with the specified address and the latter writes a control string to the instrument with the specified address.

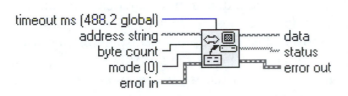

The **GPIB Read** VI reads the specified number of bytes from an instrument with the specified address string.

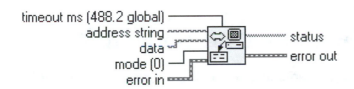

The **GPIB Write** VI writes the data string to an instrument with the specified address string.

GPIB Instrument Commands

Five instruments are used in the experiment described in this chapter: power supply, multimeter, function generator, oscilloscope, and the spectrum analyzer. If an instrument has GPIB capability then the manual for that instrument must include a list of GPIB commands. A typical instrument control command has the following format:

Command name: command value
Command name: parameter name: parameter value

For instance, to select a square wave from the function generator and to set its voltage to 5 V and its frequency to 500 Hz, the following commands must be transferred to the instrument at address 10:

```
SOUR:FUNC: SHAP SQU
SOUR:VOLT 5.0
SOUR:FREQ   500.0
```

In a similar fashion, commands are issued to control the operation of other instruments.

The functional capability of many commercially available instrument drivers is often much greater than one requires in a specific instrument control application. Custom designed instrument control software offers unique advantages. First, the software meets the immediate needs of the user. There is no overhead as there would be with commercial drivers. Second, the software can control multiple instruments. And third, the user can modify the software by implementing new requirements and deleting portions that he no longer needs. The modification process is quick and simple.

An additional benefit of the custom designed instrument driver is the inclusion in the same software package of specific data processing tasks that a commercial package will not have. The VI that integrates instrument control and data processing can be modified and expanded with relative ease to include other instruments as well as additional processing tasks. For example, the waveform that an oscilloscope is displaying in process monitoring operation can be imported into a GPIB VI over the GPIB bus. The data can then be displayed on a waveform graph or stored to a file for future reference. The data can also be applied to immediate signal processing such as displaying its frequency spectrum in order to assess process performance in real time. Commercial drivers do not have such features because the commercial software designer does not know the specific needs of individual users.

Distributed Process Control With GPIB

The ability to control an instrument at a remote location and to import data from a remote instrument presents interesting possibilities in controlling multiple processes on a network. The TCP/IP protocol is used to communicate with a workstation on a network. The integration of distributed and diverse process controls into a single software package offers a process control mechanism from a centralized location

Following is a paper that illustrates distributed process control. This paper was accepted for presentation by this author at a technical conference.

Distributed Process Control

Leonard Sokoloff
DeVry University

Introduction

Virtual Instrumentation is an important technology that is making a significant impact in today's industry, education, and research. Virtual instrument software can be used for simulation, and with appropriate interfacing, it can also be applied to data acquisition and control. A number of virtual instrument packages are available commercially. For this presentation LabVIEW, a product of National Instruments Corporation, is used.

This paper presents real-time multistation process control over a network such as Ethernet. The remote stations, referred to as servers, supply data to the Central Control. The block diagram in Fig. 1-15 illustrates this process.

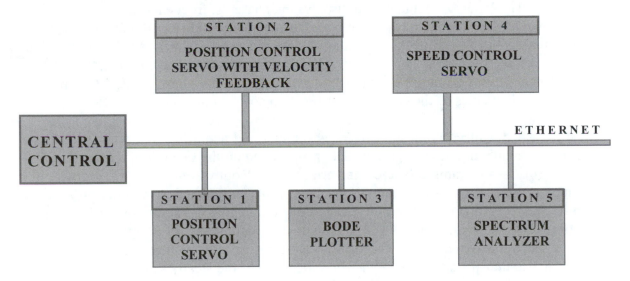

Fig. 1-15 Distributed Process Control Over a Network

At station 1 the position control servo VI is running and acquiring the motor shaft position data over a data acquisition interface. This data is displayed on the graph in the Front Panel of the VI. This is an example of a computerized positioning of a load. Central Control downloads and examines this data on the same Front Panel as that of station 1. Central Control also has access to the Front Panel control objects. For example, Central Control can issue a command to change the angular displacement of the motor shaft and then view the resulting response.

At station 2 the position control servo with velocity feedback VI is running and acquiring the motor shaft position data over a data acquisition interface. Here the operation is similar to that of the station except that in this case the derivative or the velocity feedback is also used to control step response percent overshoot. Central Control downloads and examines this data on an identical Front Panel that displays response waveforms and other response data. Central Control can select one of several tests from a vertical slide control and run the remote servo. For example, in the step response test Central Control examines response data such as settling time, percent overshoot and the associated damping ratio, and the amount of velocity feedback used; and in the ramp response test, Central Control examines the input ramp, the response, and the steady state error. Central Control can issue commands to change the input ramp amplitude or its frequency and inspect the result of such changes.

At station 3 the Bode Plotter VI is running and acquiring data for a filter over the GPIB interface. The VI controls the functions of the signal generator and the multimeter. The VI operates over a preset sweep range, issuing a series of commands to increment the frequency, measure the filter output voltage, and return this voltage for further processing. The VI uses this data to produce the magnitude vs. frequency plot. The VI also generates a theory-based magnitude plot and displays it on the same graph as the real data curve. The VI also calculates and displays the phase vs. frequency curve. In addition, the VI also displays the error (the difference between magnitude based on theory and that of the real data) vs. frequency. Central Control downloads and examines the data displayed on the Front Panel at station 3. Central Control may issue commands to station 3 to change the sweep range or number of points to be measured. Central Control can also change the values of the filter transfer function coefficients.

At station 4 the speed control servo VI is running and acquiring the motor speed data over a data acquisition interface. This data is displayed on the graph in the Front Panel of the VI. This is an example of a computerized speed control of a load. Central Control downloads and examines this data on the same Front Panel as that of station 4. Central Control has several test options. It can issue a command to station 4 to run the step response test after setting the value of the input and examine the resulting data such as the desired speed, the actual speed, the steady state error, and the response time constant; or it can choose the speed test where the servo operates in the steady state. Central Control can change the input and examine the response in real time.

At station 5 the spectrum analyzer VI is running and acquiring the spectral response data. The VI controls the operation of the function generator and the spectrum analyzer equipment over the GPIB interface. This data is displayed on the graph in the Front Panel of the VI. Central Control may download the Front Panel of the VI at station 5 and examine the data. The specific process conducted at station 5 is the acquisition and display of a tone modulated AM carrier in the frequency domain. The display shows three spectral lines: the carrier and the two sidebands. Upon examination of the sideband power and the AM bandwidth response data, Central Control may issue commands to alter the start, stop, and center frequencies on the spectrum analyzer as well as commands to change the carrier and the modulating tone parameters.

The real time process control at remote stations over the network involves integration of several methods that include data acquisition, GPIB, and the TCP/IP protocol that is used to send commands and receive data over the network. LabVIEW plays the central role in acquiring data and in controlling the GPIB operation at the remote server station as well as at the Central Control station. LabVIEW supports the TCP/IP protocol as do most operating systems.

LabVIEW Software

LabVIEW[TM] (Laboratory Virtual Instrument Engineering Workbench), a product of National Instruments[TM], is a powerful software system that accommodates data acquisition, instrument control, data processing, and data presentation. LabVIEW can run on PCs under Windows, Sun SPAR stations, and Apple Macintosh computers; and uses graphical programming language (G language), departing from the traditional high-level languages such as C, Basic, or Pascal.

All LabVIEW graphical programs, called Virtual Instruments or simply VIs, consist of a Front Panel and a Block Diagram. The Front Panel contains various controls and indicators while the Block Diagram includes a variety of functions. The functions (icons) are wired inside the Block Diagram, where the wires represent the flow of data. The execution of a VI is data dependent, which means that a node inside the Block Diagram will execute only if data is available at each input terminal of that node. By contrast, the execution of a traditional program, such as C language program, follows the order in which the instructions are written.

LabVIEW incorporates data acquisition, analysis, and presentation into one system. For acquiring data and controlling instruments, LabVIEW supports IEEE-488 (GPIB) and RS-232 protocols as well as other D/A, A/D, and digital I/O interface boards. The Analysis Library offers the user a comprehensive array of resources for signal processing, filtering, statistical analysis, linear algebra operations, and many other applications. LabVIEW also supports the TCP/IP protocol for exchanging data between the server and the client. LabVIEW version 5 and higher also supports Active X controls, allowing the user to control a web browser object.

GPIB Interface

The GPIB (General Purpose Interface Bus) interface has its origin in the HPIB (Hewlett Packard Interface Bus) developed by Hewlett Packard Corporation in the 1970s for interfacing their instruments. It gained widespread popularity and soon became a de facto industry standard. In 1975 the IEEE standards group published the standard IEEE-488 for the GPIB interface. Industry uses this standard to this day as a parallel interface for various applications in instrument control.

The GPIB interface is a parallel 24-pin conductor bus. It includes eight data lines for control messages that are often ASCII encoded, and various management and handshake lines. Handshaking is used to transfer messages between the PC and the instrument being controlled.

Fig. 1-16 shows a typical connection between the PC with the GPIB interface board and the software for controlling the operation of an instrument. Not every instrument has a GPIB capability. This capability is designed into the instrument by the manufacturer and a GPIB connector is provided on the instrument housing. The cable used to connect the instrument must also be GPIB compatible and meet the IEEE-488 protocol.

As shown in Fig. 1-16, the operation of several instruments may be controlled. Each instrument must have a unique address between 0 and 30, address 0 usually being reserved for the GPIB interface board. National Instrument Corporation has an array of GPIB interface boards with different capabilities. As shown in Fig. 1-16, the AT-GPIB interface board is used in this paper. It has a maximum data transfer rate of 1 Mbps.

The NI-488.2 is the driver software for the GPIB interface. GPIB devices can be Talkers, Listeners, or Controllers. The Listener receives messages while the Talker sends messages. The Controller, usually the PC where the GPIB board and NI.488.2 software are installed, manages the flow of commands or messages on the GPIB bus. The communication between the PC and the instrument is message based. ASCII-encoded message strings are transferred using handshaking.

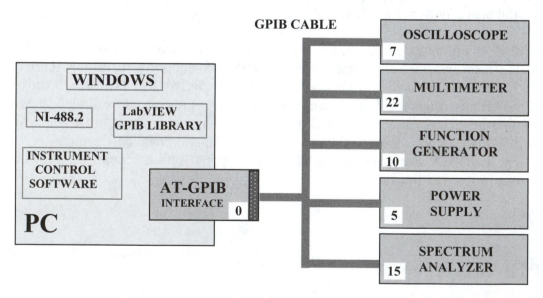

Fig. 1-16 GPIB Instrument Control Setup. Each Instrument and the GPIB Interface Has a Unique Address.

Functions>Instrument I/O>GPIB is the location of available GPIB VIs for building an instrument control VI, the VI that is custom tailored to meet a user's exact needs. On a small scale this is a driver.

The most used GPIB VIs are shown below: the GPIB Read and the GPIB Write VIs. The former reads a data string from the instrument with the specified address and the latter writes a control string to the instrument with the specified address.

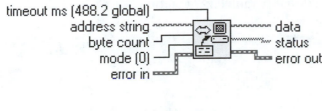

The **GPIB Read** VI reads the specified number of bytes from an instrument with the specified address string.

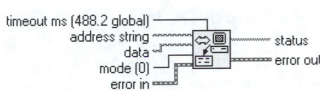

The **GPIB Write** VI writes the data string to an instrument with the specified address string

GPIB Instrument Commands

Four instruments are used in the experiment described in this paper: power supply, multimeter, function generator, and oscilloscope. If an instrument has GPIB capability then the manual for that instrument must include a list of GPIB commands. A typical format for an instrument control is as follows:

Command name: command value
Command name: parameter name: parameter value

For instance, to select a square wave from the function generator and to set its voltage to 5 V and its frequency to 500 Hz, the following commands must be transferred to the instrument at address 10:

SOUR:FUNC: SHAP SQU
SOUR:VOLT 5.0
SOUR:FREQ 500.0

In a similar fashion, we can issue commands that will control the operation of other instruments.

Central Control Software

Fig. 1-17 shows the Front Panel of Central Control.vi, a LabVIEW program. In this particular display the Central Control station has downloaded the Bode plotting operation from the remote station 3 in Fig. 1. This is a real time interaction as the operation at station 3 is ongoing. Central Control examines the data and has several options at its disposal. It can switch to another remote station if the data is within limits. However, if the data is outside acceptable limits, Central Control may choose to investigate further by sending commands that change the operational parameters at station 3.

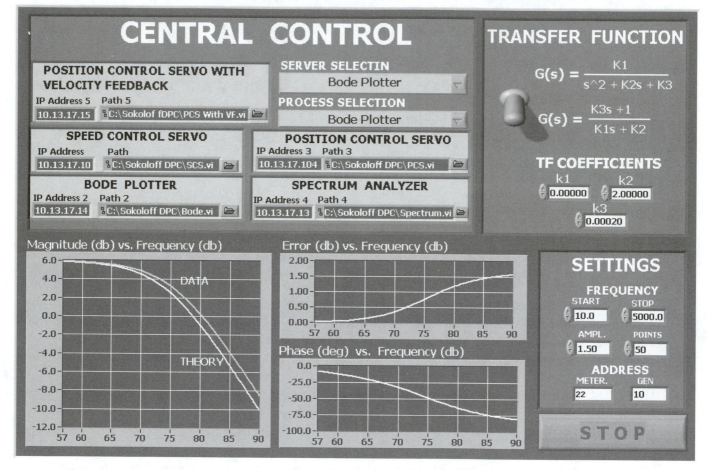

Fig. 1-17 The Front Panel of Central Control.vi. The Bode Plotting Operation Shown Here Has Been Downloaded From the Remote Station by Central Control

The Settings box includes several digital controls that are also in the Front Panel at station 3. Central Control may choose different Start and Stop frequencies in order to examine the response in a different frequency range, or it can change the amplitude of the sinewave being applied to the filter at station 3. The third option available to Central Control is to terminate the operation. It can do so by clicking on the Stop button. Central Control may also select a different filter and vary the transfer function coefficients in the Transfer Function box.

As stated earlier, Central Control can access any of the remote stations in order to view the ongoing operation. In order to do this, it must execute a sequence of steps. In the Front Panel of Fig. 1-17 the recessed box labeled Central Control includes several control objects that provide the access.

The Server Selection menu ring includes five server options including Bode Plotter and Spectrum Analyzer. By choosing one of these options, the code in the Block Diagram selects the remote station IP address and the path to the VI file that is stored at the remote station. In the case of the Bode Plotter, the IP address is 10.13.17.14 (string control object) and the path to the VI Bode.vi is C:\Sokoloff DPC\Bode.vi (file path control object).

Next, Central Control must select one of the options from the Process Selection menu ring. If, for example, Bode Plotter is selected, LabVIEW activates the appropriate code in the Block Diagram. The code includes various commands that are transmitted to the remote station. For example, if the operator at Central Control changes the Start Frequency from 10 to 1, the command that changes the frequency is dispatched to the remote   station.

When the Stop button is activated, the command to terminate the VI execution at the remote station is transmitted.

Position Control Servo Software

Fig. 1-18 shows the Front Panel of the Position Control Servo VI. This file is stored at remote server station 1 in Fig. 1. When Central Control accesses this station, it will view the same panel. Step Response is the selected test and the waveform graph shows the 35° input step and the motor's shaft response.

Central Control may examine the response waveform and detailed data in the Step Response Data box that includes percent values for overshoot (POT), damping (ζ), settling time (Tsetl), system time constant (Taus), and the natural and the damped frequencies. Also included is a table providing the coordinate values of the first four response peaks.

The Settings box provides Central Control with process control objects. Central Control may wish to change the input angle or the value of the Band. The Band value specifies the band for measuring the settling time.

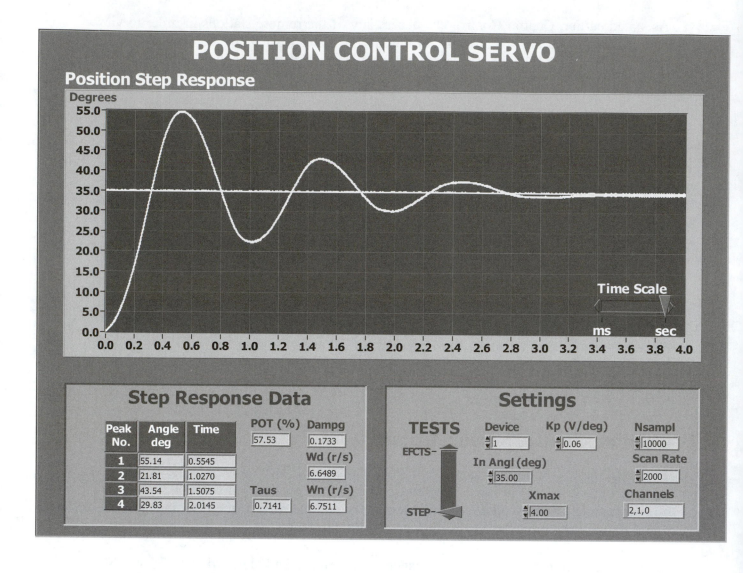

POSITION CONTROL SERVO

Position Step Response

Step Response Data				
Peak No.	Angle deg	Time	POT (%)	Dampg
			57.53	0.1733
1	55.14	0.5545		**Wd (r/s)**
2	21.81	1.0270		6.6489
3	43.54	1.5075	**Taus**	**Wn (r/s)**
4	29.83	2.0145	0.7141	6.7511

Settings

TESTS

EFCTS — STEP

Device: 1
Kp (V/deg): 0.06
Nsampl: 10000
In Angl (deg): 35.00
Scan Rate: 2000
Xmax: 4.00
Channels: 2,1,0

Fig. 1-18 The Front Panel of Position Servo.vi. The Waveform Graph Shows the Step Response of the Servo.

Position Control Servo With Velocity Feedback Software

Shown in Fig. 5 is the Front Panel of the Position Control Servo with Velocity Feedback.vi. This VI is stored and running at remote station 2 in Fig. 1. As usual Central Control can download the Front Panel and view the operation. In this instance station 2 is conducting a ramp test on the on the position control servo. The display shows the system response to a 0.5 Hz, 42° peak ramp. Since the position control system is a Type 1 system, it exhibits an error that is also included on the same graph.

Central Control examines the response and may issue commands to run the test in real time at a different frequency or different amplitude.

By switching to the Step Response test, Central Control will view the step response test and the effect of velocity feedback as shown in Fig. 1-20. The display shows the motor shaft's angular response and shaft's velocity in RPM scaled by a factor of 10. The Step Response Data

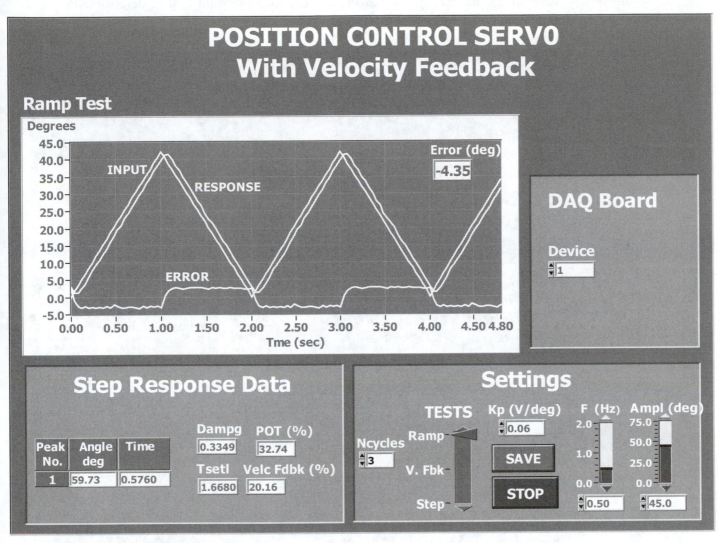

**Fig. 1-19 The Front Panel of Position Servo With Velocity Feedback.vi. The Graph
Shows the Ramp Response of the Servo, the Input and the Error**

box includes additional information on percent overshot, damping ratio, percent velocity feedback used, and so forth.

Speed Control Servo Software

Fig. 1-21 shows the Front Panel of the Speed Control Servo VI. As is the case with other remote stations, this VI is stored in station 4 in Fig. 1-15 at the IP address as shown on the Front Panel of the Central Control VI in Fig. 1-17. The path to this VI is also shown in Fig. 1-17.

Waveform Graph in Fig. 1-21 displays the step response of the speed control servo. In addition, the Step Response Data recessed box provides response values such as the final or the steady state speed, the steady state error, and the time constant that is determined from the response data by the LabVIEW software.

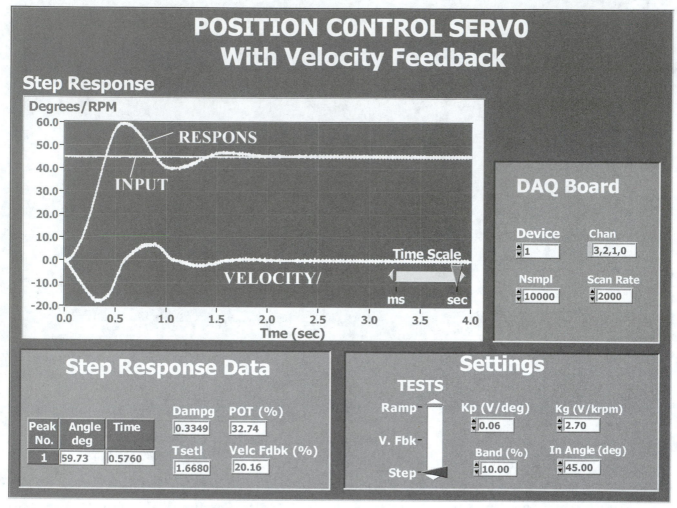

Fig. 1-20   The Front Panel of Position Servo With Velocity Feedback.vi. The Graph
Shows the Step Response of the Servo, the Input and Motor Speed

Upon examination, Central Control may issue commands to repeat the test with a different input.

There are four tests to choose from. Central Control may wish to run the Motor Speed test. The input or the desired speed, the motor speed, and the steady state error indicators and controls are included in the Motor Speed recessed box in the right side of the Front Panel in Fig. 1-21. The operator at Central Control can use the input speed vertical slide and examine the response.

Two other tests are the Time Constant test and the Steady State Error test. They are plotted as a function of loop gain as it is varied in increments. The curves show that the steady state error and the system time constant decrease as loop gain is increased.

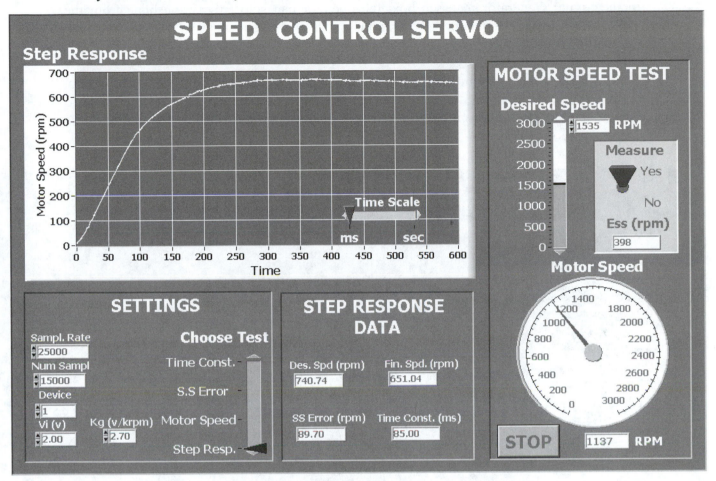

Fig. 1-21   The Front Panel of Speed Control Servo With the Velocity Feedback VI. The Graph Shows the Step Response of the Servo.

Spectrum Analyzer Software

Fig. 1-22 shows the Front Panel of the Spectrum Analyzer VI. As is the case with other remote stations, this VI is stored in station 5 in Fig. 1-15 at the IP address as shown on the Front Panel of the Central Control VI in Fig. 1-17. The path to this VI is also shown in Fig. 3.

Waveform Graph in Fig. 1-22 displays the double sideband full carrier (DSBFC) spectrum of a 10 MHz carrier amplitude modulated by a 25 kHz tone. The indicators in the recessed box below the graph display the carrier and the sideband powers as well as the bandwidth of the AM wave. The sweep width slider control allows the operator to change the start and stop frequencies. In the illustration the sweep width is set to 100 kHz, resulting in the sweep extending from 9.95 MHz to 10.05 MHz.

In the Signal Generator recessed box, the operator can select the waveshape from the menu, the carrier frequency, the modulating signal frequency and level, and the modulation index.

Central Control can download this Front Panel, examine the data; and, if it chooses to, operate the controls in the Signal Generator box. It can also choose the sweep width to view, examine the additional data, and make a determination regarding the quality of the spectral response.

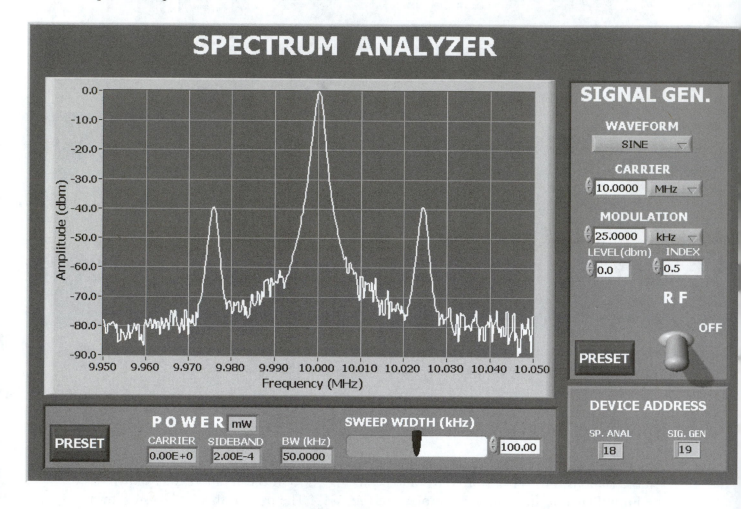

Experiment Setup

1. At each of the five remote stations do the following:

    Connect the equipment (such as the position control servo).

    Determine the IP address and record it.

    Store the process control VI (such as the position control servo) and record the path.

    Turn ON the equipment at each station.

    Run the LabVIEW software.

2. At the Central Control station, do the following:

    Open Central Control.vi.

    Enter the IP addresses for the five remote stations (from step 1).

    Enter the path for the process VI at each remote station (from step 1).

    Set Central Control TCP/IP access. To do this, choose *Tools>Options,* then select *VI Server: Configuration* from the pull-down menu. Check the TCP/IP box under Protocols in the Options window.

    Return to the pull-down menu and choose *VI Server: TCP/IP Access.* In the dialog window type the IP address of the remote station, then click on *Add.* Repeat this for each remote station. If the remote stations have the same first three numbers of the IP address, as is the case in Fig. 3, where the numbers are 10.13.17, then enter 10.13.17*. This will allow Central Control to gain access to any station whose IP address has those numbers.

3. At this time the experiment setup is complete. Central Control.vi may be executed. Any of the five remote stations may be selected, the responses examined, and the Front Panel controls operated.

Conclusion

Distributed Process Control that includes five remote stations, the position control servo, the speed control servo, the position control servo with velocity feedback, the Bode plotter, and the spectrum analyzer, has been successfully tested. Central Control downloaded the front panels of each process, one at a time, and examined the responses. It is able to operate various Front Panel control objects and thus alter or modify the operation at the remote stations.

One obvious industrial application of this paper is in testing facilities. The operator at the Control Center can make decisions to accept or reject a subsystem being tested based on limits of acceptance. The Central Control structure has the convenience of accessing many remote stations in any order and examining the test data. With additional code, the Central Control VI may be completely automated, removing the presence of the operator.

At this time, Distributed Process Control software package is used by the advanced students at DeVry College of Technology. Because it integrates the TCP/IP protocol and the GPIB and data acquisition interfaces, it provides the student with hands-on experience in meaningful computerized process control.

Chapter 2
Structures

It has been said that the entire C programming language can be broken down into three **Control Structures** and that any C program can be written in terms of these structures. These structures are:

Repetition Structure
Selection Structure
Sequence Structure

LabVIEW also has these structures. It has two repetition structures: the **While Loop,** which is equivalent to Do/While in the C language, and the **For Loop.** The **Case** structure in LabVIEW, which can do single or multiple selections, is an example of a selection structure. The sequence structure, which occurs naturally in the C language because all instructions are executed in sequential order, is given special attention in LabVIEW. As you may remember from the last chapter, the execution order in LabVIEW is based on data flow; a node can execute only if data is available at all input terminals. Therefore, LabVIEW has a special node called the **Sequence** structure that is intended for special sequence operations. LabVIEW also includes the **Formula Node** for mathematical and logical operations. (Formula Node is not in C.)

The While Loop

Consider the concept of the repetition structure shown in Fig. 2-1. The program instructions are executed once before the condition is tested. If the result of the test is *true,* the program instructions will be executed again. If it is *false,* the next statement following

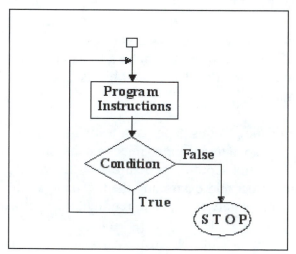

Fig. 2-1 The Do While Repetition Structure

the loop will be executed in C language and in LabVIEW the loop terminates. This type of repetitive execution of a group of program instructions is called the **Do/While** structure in C language, and in LabVIEW it is known as the **While Loop.**

To Open the While Loop, click on the While Loop option with the *positioning tool* in the *Structures* subpalette of the *Functions* menu, as illustrated in Fig. 2-2.

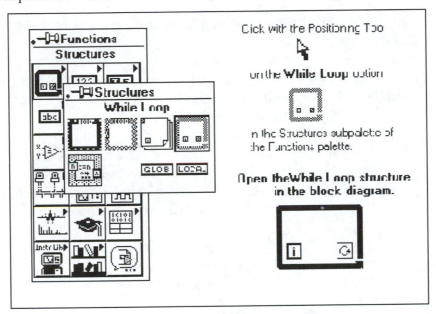

Fig. 2-2 Opening the While Loop in LabVIEW

As shown in the illustration, the While Loop has an iteration terminal **i** that counts the number of times the loop has executed. The **Condition** terminal expects a *true* or *false* input. A *true* input forces the While Loop to run indefinitely, and a *false* input terminates execution.

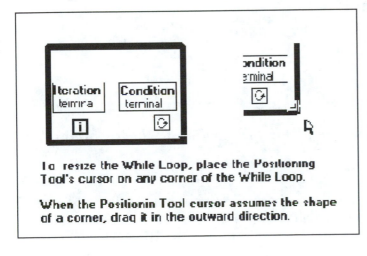

As shown in the illustration, the While Loop may be resized by placing the Positioning Tool's cursor over any corner of the loop. When the cursor assumes the shape of the corner, drag it in the outward direction.

To move the loop, catch the border of the loop with the positioning tool's cursor and drag it to a new position.

Exercise 2-1: Using the While Loop (Lottery Game)

In this example we will consider an application of the While Loop. You will create a lottery game. The VI, called Play Three, will guess the three numbers that you pick and will indicate the number of tries that it took to guess the numbers.

But first, you will build a VI, including wiring, with an icon and a connector that will be used as a subVI in our game.

1. Construct the Front Panel and the Block Diagram shown in Fig. 2-3.

   Front panel objects: ***Digital control*** is in the *Numeric* subpalette of the *Controls* palette, and the **Square LED** is in the *Boolean* subpalette of the Controls palette.

   Block diagram objects: *Numeric* subpalette of the *Functions* palette includes the following objects: ***Random Number, Multiply***, and ***Round To Nearest***. The ***Equal?*** function is in the *Comparison* subpalette of the Functions palette.

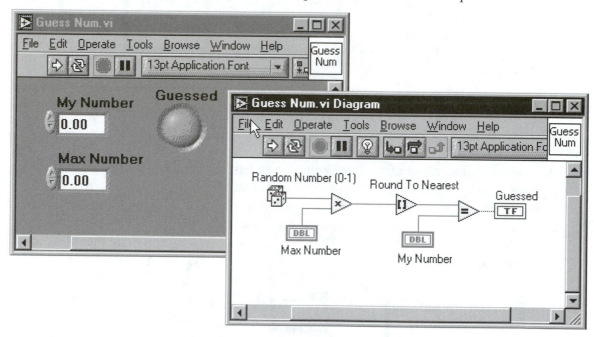

Fig. 2-3 The Front Panel and Block Diagram of Guess_Number.vi

2. Next, create the connector and the icon for this VI. You should refer to Exercise 3 of Chapter 3 (Basic Concepts of LabVIEW 4) on how to create an icon and a connector for a VI. To create a connector, select the entire Block Diagram of this VI (the dotted rectangle must include all objects) and choose ***subVI From Selection*** from the *Edit* menu. Create the icon for this VI by double-clicking on the icon that was just created and choosing ***Edit Icon*** from the ***icon pane*** in the

Front Panel. The *Icon Editor* that opens allows you to edit the icon. Edit the icon by entering text as shown in the illustration. Choose OK and close the Icon Editor.

Save this VI as **Guess_Number.vi**. Close this VI.

3. Construct the VI whose Front Panel and Block Diagram are shown in Fig. 2-4. The Front Panel includes three *digital controls* and one *digital indicator*. Label with owned labels the digital controls and the indicator, as shown in Fig. 2-4. Choose *I-16* from the popup palette for each of these by clicking with the right mouse button on the digital control and indicator object.

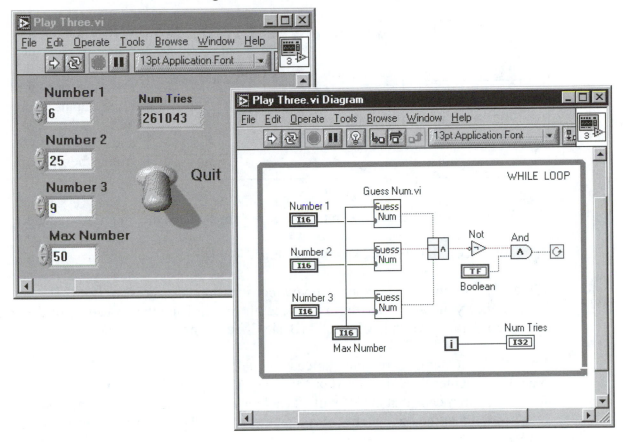

Fig. 2-4 The Front Panel and Block Diagram of Play Three

The switch shown in the Front Panel is the **Vertical Switch,** which is in the *Boolean* subpalette of the Controls palette. Click on the switch with the *Positioning* tool found in the Tools palette to move it to the upper position, as shown in Fig. 2-4. Click on the switch (the right mouse button) and choose **Latch When Pressed** from the **Mechanical Action** popup menu.

Also choose *Make Current Value Default* from the *Data Operations* option in the popup menu for this switch. Label the switch, with an owned label, as *Quit*.

4.   Turning now to the **Block Diagram**, open the icon of the Guess_Number.vi. As indicated before, you open the icon of a VI by clicking on the *Select a VI...* button in the Functions palette. This gives you access to directories and files. Open *Guess_Number.vi*. The icon appears in the Block Diagram. Duplicate this icon so that you have a total of three.

The *Compound Arithmetic* function is in the *Numeric* subpalette of the Functions palette. Configure the Compound Arithmetic function as a three-input

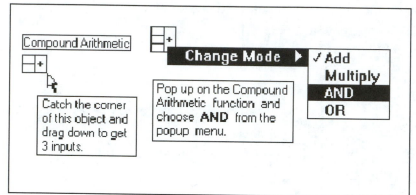

And gate. As shown in the illustration to the left, resize the Compound Arithmetic icon for three inputs, pop up on the icon, and choose *AND* from the popup menu.

The *Not* and *And* functions are in the *Boolean* subpalette of the Functions palette.

Wire all objects in the Block Diagram of Fig. 2-4.

5.   If the *Run* button is not broken, then your construction is successful. If it is broken, then you have syntax errors. Correct them following the procedure of the Troubleshooting section in Chapter 3 (Basic Concepts of LabVIEW 4).

Enter the three numbers of your choice in digital controls Num1, Num2, and Num3. Also enter the value for Max Num. Try Max Num of 10. A larger value for Max Num means that the VI will take a longer time to guess your numbers. Don't forget that the three numbers you pick cannot be larger than Max Num. Run the VI.

Play the game also with larger values of Max Num.

Save this VI as **Play_Three.vi** and close it.

The For Loop

Another instance of the repetition control structure is the For Loop. As in the case of the While Loop, the For Loop can execute a group of instructions repetitively. The While Loop checks the state of the condition with each iteration and as long as the condition is true, it continues the repetitive execution of the instructions. It stops execution as soon as the condition becomes false.

The For Loop, on the other hand, executes a group of instructions a *fixed number* of times. It initializes a counter to N=0, as shown in Fig. 2-5, checks the counter against a user-supplied value C, and then executes a group of instructions that are inside the For Loop. The counter is then incremented and the procedure repeats. As soon as the value of the counter is equal to C, the loop stops.

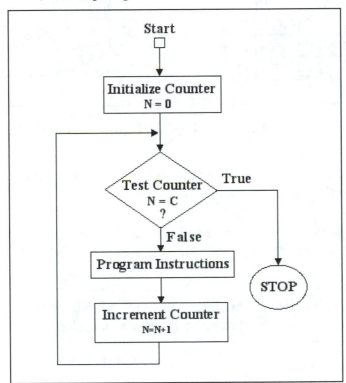

Fig. 2-5 The For Loop Repetition

Note: *The For Loop outputs values only after it completes executing N times the program that is inside the boundary of the loop. The loop stops execution when the loop counter N reaches the value supplied by the user. Thus, if you wired a value of 50 to the loop counter, the loop will begin with N=0 and continue through N=49. On the count of N=50, the loop stops execution and only now will be able to pass values to objects outside the loop.*

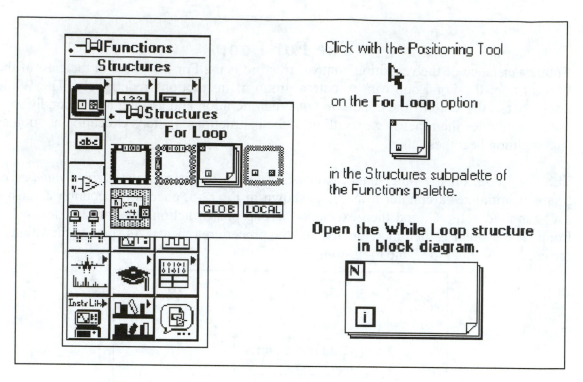

Fig. 2-6 Opening the For Loop in LabVIEW

To Open the For Loop, click with the *Positioning Tool* in the *Structures* subpalette of the *Functions* menu on the For Loop option, as illustrated in Fig. 2-6.

As shown in the illustration below, the For Loop has the iteration terminal **i**, which counts the number of times that the loop has executed; **i** can be used as an output to a digital indicator.

It also has the Loop Counter **N**. This is an input that the user must provide to specify the number of times that the loop must execute. N must be an integer number.

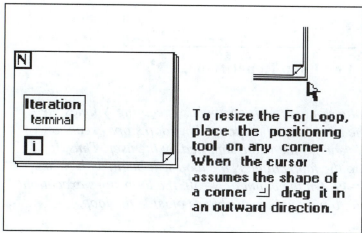

To Move the For Loop to a new location in the window, place the cursor of the positioning tool on the border of the loop and drag the loop to the new location.

To Copy the For Loop, select the loop, hold down the ***Ctrl*** key, and drag the loop.

The Replace Option. If you completed all the code inside the For Loop and then realized that a While Loop might do better, you can make the switch easily using the *Replace* option. Pop up with the positioning tool cursor on the border of the loop and choose Replace from the popup menu. While still holding down the right mouse button, navigate to the Functions palette and make another choice from the Structures subpalette. If you choose ***Remove the For Loop*** from the popup menu, you will delete the For Loop in your Block Diagram. Incidentally, the *Replace* option can be applied to any object in the Front Panel or in the Block Diagram.

The Shift Register

A shift register is a device that can store data that has occurred in the past. This concept is similar to a shift register moving or shifting binary data from one stage to the next in a digital circuit. In a digital circuit the shifted data is restricted to the binary type only. In the LabVIEW shift register, however, the data may be of any type including strings, floating point, and so on. The shift register in LabVIEW may be implemented in a For Loop or a While Loop.

To Create a Shift Register, open the ***For Loop*** or the ***While Loop***; the shift register can be created in either the For Loop or the While Loop. To open the For Loop in the Block Diagram, choose *For Loop* from the *Structures* subpalette of the Functions palette.

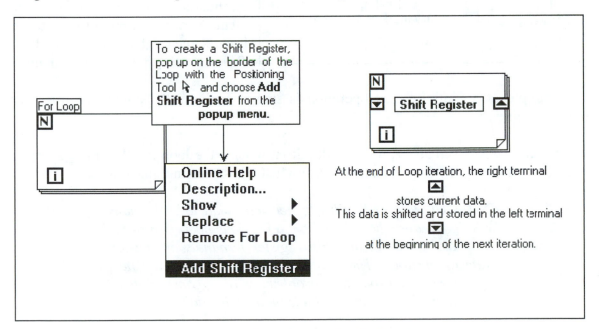

Fig. 2-7 Creating a Shift Register

As shown in Fig. 2-7, click on the border of the For Loop (either the right side or the left side, but not the top or the bottom) and choose ***Add Shift Register*** from the popup menu. As soon as you release the mouse button, two shift register terminals appear, one on the right side and the other on the left side.

Data is applied to the left terminal. At the end of the iteration (in one iteration all nodes inside the loop border are executed), this data is stored in the left terminal and is available at the left terminal at the beginning of the next iteration.

A single terminal on the left side of the loop is capable of storing only one value from the previous iteration. If you want to store values from two iterations ago, you need two terminals on the left side of the loop and three terminals to accommodate values from the three previous iterations. You can have as many terminals as the room on the left side of the loop allows. The more past history that you want to store, the more terminals you will need.

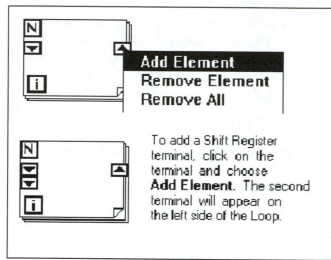

To add a Shift Register terminal, click on the terminal and choose **Add Element**. The second terminal will appear on the left side of the Loop.

To Add a Shift Register Terminal, pop up on the terminal, as shown in the illustration to the left, and choose ***Add Element*** from the popup menu. The second terminal will appear on the left side of the For Loop. Repeat this process if you want to add more elements to the loop.

To Delete Terminals, pop up on the terminal and choose from the popup menu ***Remove Element*** to delete the last element that was added or ***Remove All*** to delete all terminals.

An example of a Shift Register operation is shown in Fig. 2-8. Notice that this shift register is initialized.

To Initialize a Shift Register, wire a value from outside the loop to all terminals on the left side of the loop. In this example the shift register has been initialized to 10.

> **Note:** *Uninitialized shift registers can lead to ambiguous results because the values from a previous operation of the loop are stored in the shift register terminals on the left side of the loop. These values will be used for the next operation. To avoid this problem, wire a constant from outside the loop to all terminals on the left side of the loop.*

In the example of Fig. 2-8, the iteration terminal **i** is wired to the right terminal of the For Loop and the digital indicators SR1, SR2, SR3, and SR4 are wired to the four shift register terminals on the left side of the loop. At the beginning of loop operation, **i=0**, 10 is moved in the four left terminals, and a 0 is stored in the right shift register terminal. When i is incremented to 1, 0 is shifted to the SR1 terminal, and 1 is shifted to the right terminal. This process continues until i=4, and by that time SR1=3, SR2=2, SR3=1, and SR4=0. As i is incremented to 5, the value of 4 is not stored in the shift register because the loop counter N=5 and therefore the loop must terminate execution. Fig. 2-8 shows the values stored at the end of loop operation.

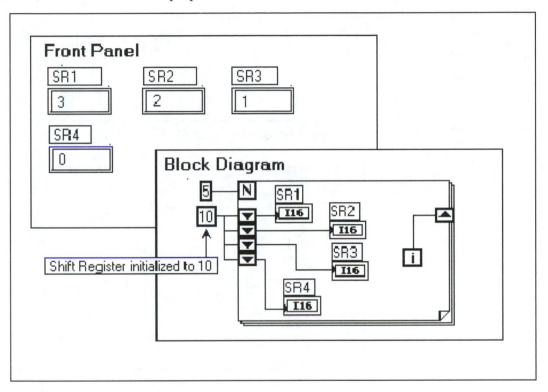

Fig. 2-8   A Shift Register Example

Exercise 2-2: Using the Shift Register

In this exercise the For Loop is used as a shift register. This VI generates four random numbers between 0 and 101 (floating point), finds and displays their sum and their average value.

1. **Build** the Front Panel and the Block Diagram of the VI shown in Fig. 2-9. The Front Panel has two digital indicators, which are found in the *Numeric* subpalette of the Controls palette. Label these with owned labels as **Sum** and **Average**.

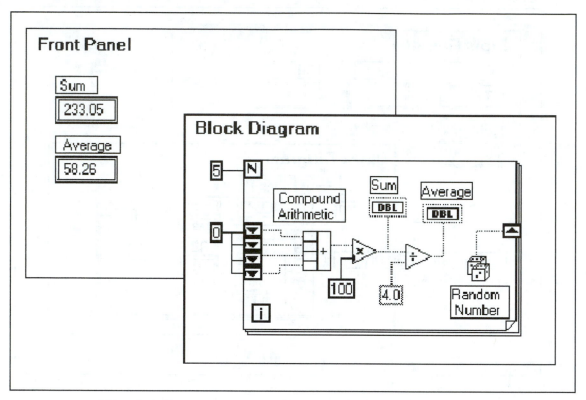

Fig. 2-9 Front Panel and Block Diagram of Exercise 2-2

Block Diagram

For Loop is in the *Structures* subpalette of the Functions palette. Open the For Loop in the Block Diagram and configure it as a *Shift Register* with four terminals on the left side of the loop. Follow the procedure described in the Shift Register section preceding this Exercise.

Initialize the shift register to **0** by wiring a **0** *numeric constant* to each shift register terminal on the left side of the loop. Also wire a constant value of **5** to the loop counter **N**. The *Numeric Constant* can be found in the *Numeric* subpalette of the Functions palette.

The **Compound Arithmetic Function** is in the *Numeric* subpalette of the Functions palette. Resize this icon to accommodate four inputs. To resize the Compound Arithmetic function, catch the lower corner with its icon and drag in the downward direction until you get four inputs.

Multiply $\boxed{\gg}$, **Divide** $\boxed{\gg}$, **and the Random Number Generator** $\boxed{}$ are found in the *Numeric* subpalette of the Functions palette.

2. **Run this VI.** You may want to practice your VI single stepping skills and single step this VI. Single stepping will reveal in slow motion how the values are shifting through the terminals on the left side of the For Loop.
Save this VI as **Shift Register.vi** and close it.

Selection Structures

The If/Else Selection Structure

The double selection structure in a high-level programming language such as C has the configuration shown in Fig. 2-10. This is the If/Else selection structure. The condition is first tested and if the result is *true*, all program statements that are included in task 1 are executed. If the result of the test is *false*, all program statements that are included in task 2 are executed.

The If/Else selection structure offers the programmer a branching option. The outcome of the condition test determines which of the two program segments is to be done next.

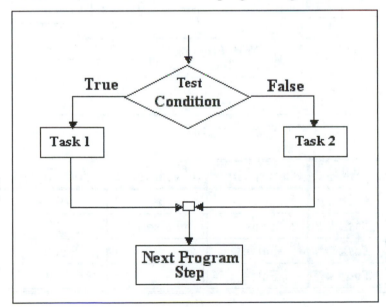

Fig. 2-10 The If/Else Selection

The Nested If/Else Selection Structure

If/Else structures may also be *nested*. This means that one If/Else structure is inside the other. The flowchart shown in Fig. 2-11 illustrates the nested If/Else structure, and the table lists the tasks to be selected on the basis of the condition tests.

As you can see either from the table or from the flowchart, task 1 will be executed if both conditions 1 and 2 test as true. If, on the other hand, both conditions 1 and 3 test as false, then task 3 will be executed.

A single If/Else statement offers a choice of two tasks to be executed. Notice that by nesting If/Else statements, a greater range of tasks is possible. By nesting two If/Else statements, shown in Fig. 2-11, we made four tasks available. Eight tasks are possible by nesting three If/Else structures, and so on.

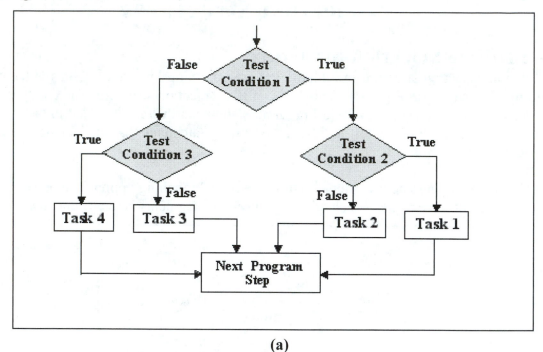

(a)

Task	Condition 1	Condition 2	Condition 3
Task 1	True	True	
Task 2	True	False	
Task 3	False		False
Task 4	False		True

(b)

Fig. 2-11 (a) If/Else Nested Selection Structure, (b) Condition/Task Table

The Boolean Case Structure in LabVIEW

In the previous section we considered If/Else, as well as the nested If/Else selection structures, through the perspective of a high-level programming language such as the C language. LabVIEW also has this structure, and it is called the **Case Structure.** Case structure can be of the Boolean type or the Numeric type. Let's consider first the Boolean type.

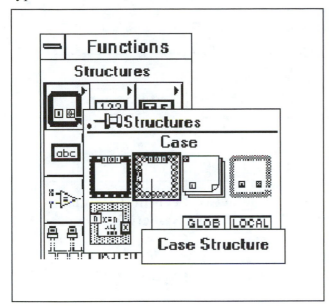

To Open the Boolean Case Structure, choose *Case* from the *Structures* subpalette of the Functions palette, as shown in this illustration.

When you open the **Case** structure in the Block Diagram, it has the appearance shown in the illustration below.

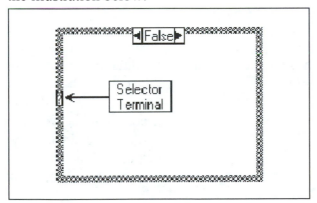

Actually, there are two overlapping frames: the **true** frame and the **false** frame. To switch between them, click on the button to the left ⌐x⌐ or on the button to the right ⌐x⌐ of the False/True window at the top of the frame.

The Boolean input to the **Selector Terminal** determines which of the two frames will be executed. A *true* input will cause the *true* frame to be executed, and a *false* input will force execution of the *false* frame.

Fig. 2-12 compares the If/Else structure that can be implemented in a high-level language such as the C language with the Case structure in LabVIEW. A closer inspection of the two structures shows that both accomplish the same objective. In LabVIEW the test condition is implemented in a Boolean control such as a switch tested by the Case structure for a *true* or a *false*. The result of the test determines which of the two Case frames will be executed.

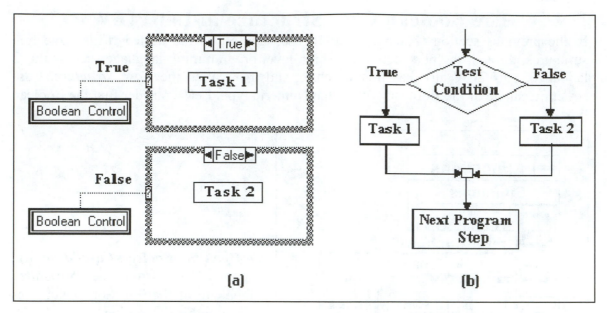

Fig. 2-12 (a) A True or False Input to the Selector Terminal Determines Which Case Will Be Executed, (b) If/Else Selection Structure

The Numeric Case Structure in LabVIEW

Conditional branching need not be limited to a choice of two tasks. When a Boolean control is wired to the selector terminal, you are led to only two Case structure frames. But suppose that you need more than two Case frames because you have more than two tasks to be done. The solution is a *Numeric Case Structure*.

To Open a Numeric Case Structure, follow the procedure for opening a Boolean Case structure. As shown in this illustration, the Boolean format is the default for the Case

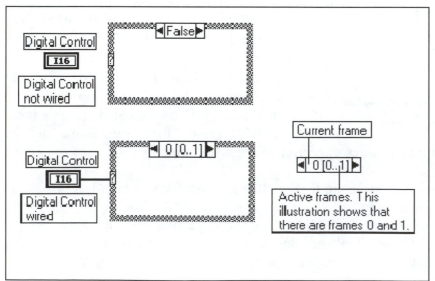

structure. But as soon as you wire a digital control to the *Selector (?)* Terminal, the Case structure becomes Numeric.

Note the window at the top of the frame. It indicates the current frame and the range of active frames that are below the current frame.

To Advance to the next or to the preceding frame, click in the frame window either on the right arrow button ▶ or the left arrow button ◀ .

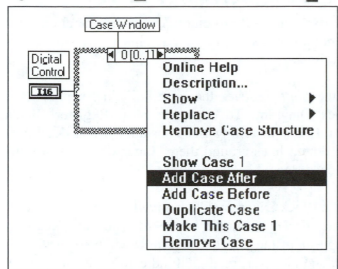

To Add a Case Frame, click anywhere inside the Case window and choose *Add Case After* from the popup menu. This will add a Case frame after the current frame. Choosing *Add Case Before* will add a Case frame before the current frame.

Note the other options in the popup menu. *Make This Case 1* will change the current frame to frame number 1. *Remove Case* will delete the current frame. *Duplicate Case* will copy the current Case frame.

Note: All Case structure frames are stacked one on top of the other. They cannot be separated and placed next to each other. To access a particular frame for the purpose of viewing it or for building a VI, you must use the arrow buttons inside the Case Window.

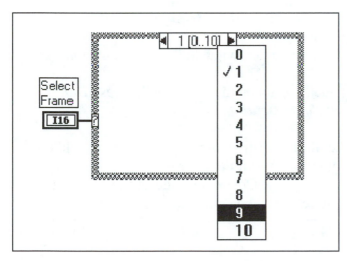

Suppose that there are a large number of active frames. In this illustration the Case structure has 10 active frames, with number 1 as the current frame, and you wish to open and view frame number 9.

Instead of clicking many times on the arrow button, click (with the left mouse button) inside the Case window and choose 9 from the popup menu, as shown in this illustration.

Exercise 2-3: Using Boolean Case Structure

This Exercise illustrates the use of the Boolean Case structure. The Front Panel and the Block Diagram of the VI that you will build is shown in Fig. 2-13. This VI will make use of the For Loop shift register from Exercise 2-2.

1. Open the Block Diagram of the *Shift Register* that you built in Exercise 2-2. Copy the objects in the Block Diagram. To copy the Shift Register Block Diagram, *select* the entire For Loop, including the constants **0** and **5**. When selected, the objects will have dotted lines around them. Enter ***Ctrl+C*** from the keyboard and close the Shift Register.vi.

2. Open a new VI and switch to the Block Diagram. Click anywhere in the center of the block diagram and then enter ***Ctrl+V*** from the keyboard. The Shift Register Block Diagram should now appear in the Block Diagram. Switch to the Front Panel and observe that the *Sum* and the *Average* digital indicators have also been copied.

3. Add to the Front Panel two **Vertical Switches**. Vertical switches can be found in the *Boolean* subpalette of the Controls palette. With *owned* labels, label one switch as the ***Function*** and the other as ***QUIT***. As shown in Fig. 2-13, add free labels ***Sum*** and ***Average*** to the Function switch. Also configure *both* switches as follows:

    Pop up on the switch and choose ***Mechanical Action>Latch When Pressed*** from the popup menu.
    Use the Operating Tool🖑 to move the switch to the upper position so

    that it appears like this ▮ then pop up on the switch and choose ***Data Operations>Make Current Value Default***.

4. ***Switch to the Block Diagram***. The Shift Register that you just copied should be inside the Block Diagram at this time. Open next the ***Case*** structure. The Case structure is in the *Structures* subpalette of the Controls palette. As soon as you click on the Case structure in the Structures subpalette, the cursor changes to

    Place this dashed rectangle cursor above and to the left of the For Loop shift register (slightly above and to the left of constant 5). Click and *hold down* the left mouse button. Drag the cursor down and to the right so that the dashed outline completely encloses the entire shift register. When you release the mouse button, the shift register will be inside the *False* frame of the Case structure.

    Remove the *Sum* terminal and place it outside the Case structure. Delete any bad wires (Ctrl+B).

5. Switch to the *true* frame of the Case structure by clicking on the arrow button in the Case window.

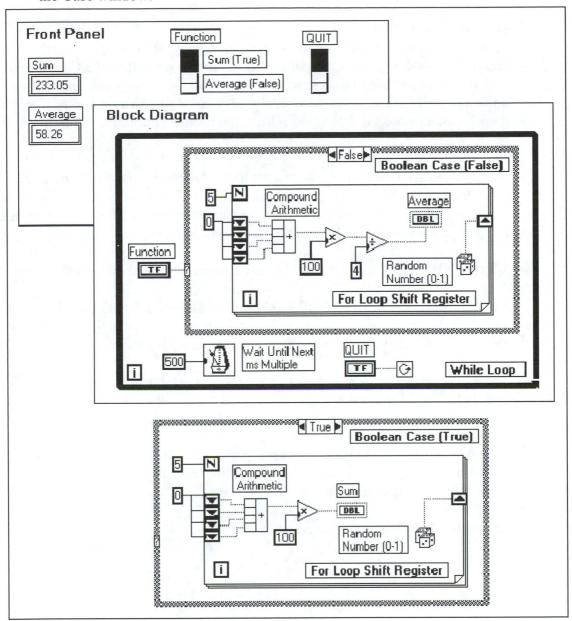

Fig. 2-13 Front Panel and Block Diagram of Exercise 2-3

Copy the For Loop shift register to the *True* frame of the Case structure. At this point, the best way to accomplish this is to pop up in the *False Case* window and choose **Duplicate Case**. Notice that the *True Case frame* now contains the same thing as the *False* frame minus the digital indicator *Average* terminal.

Delete the *Divide* function and the **4** numeric constant, and wire the *Sum* digital indicator terminal to the output of the *Multiply* function, as shown in Fig. 2-13 for the True frame.

6.   Wire the **Function** Boolean terminal to the *Selector Terminal* of the Case structure.

7.   Open the **While Loop**. You will find the While Loop in the *Structures* subpalette of the Functions palette. As described in step 4 of this Exercise, place the rectangular dashed line cursor outline above and to the left of the Case structure, then drag it (continue to hold down the mouse button) downward and to the right until the entire Case structure is enclosed. Leave some room inside the While loop for other objects. When you release the mouse button, the While Loop will be in place.

8.   Open the **Wait Until the Next ms Multiple** function from the *Time & Dialog* subpalette of the Functions palette and wire to it a numeric constant of 500. You can either copy the numeric constant from the current Block Diagram or get it from the *Numeric* subpalette of the Functions palette.

9.   Wire the **QUIT** Boolean terminal to the *Condition* terminal of the While Loop.

10.   Switch to the Front Panel and ***Run*** the VI. Make sure that the *Quit* vertical switch in the Front Panel is in the *up* or *True* position. Switch the *Function* to *Sum* and *Average* positions and observe the indicator displays.

The Function switch decides which of the two Boolean Case frames will be executed. The While Loop provides a continuous operation, allowing you to change the function switch setting while the VI is running.

The *Wait Until the Next ms Multiple* provides a 500 ms time delay between iterations, giving you an opportunity to view the display. Increasing the 500 value will make the delay longer, and vice versa.

Save this VI as **Boolean_Case.vi** and close it.

Exercise 2-4: Using the Numeric Case Structure

This VI illustrates the use of the Numeric Case structure. In the previous Exercise we used the Boolean Case structure, which has only two frames, *True* and *False*. When there are more than two tasks that must be done, the Boolean Case structure cannot be used. The Numeric Case structure has as many case frames as you need. In this Exercise we have three tasks to be executed.

Two random integer numbers are generated inside the While Loop and applied to the Case structure. The setting of the Function switch (menu ring) determines which frame will be executed.

1. The Front Panel includes five digital indicators. Digital indicators are in the *Numeric* subpalette of the Controls palette. Label them with owned labels, as shown in Fig. 2-14.

    The **Vertical Switch** is in the *Boolean* subpalette of the Controls palette. Configure the switch according to the procedure described in step 3 of Exercise 2-3 and label it *QUIT* using an owned label.

    The **Menu Ring** that we are using in the Front Panel under the owned label of *Function* is a new control that you haven't used before It is located in the *List & Ring* subpalette of the Controls palette.

Menu Ring

To enter an item, click with to *Operating Tool* on the Menu Ring and choose **Add Item After** from the popup menu.

Add Item After
Add Item Before
Remove Item

To Enter an Item into the Menu Ring, click with the Operating Tool, as shown in this illustration, on the Menu Ring button and choose *Add Item After* from the popup menu. The face of the button becomes white. Type the item name.

Type *Sum* as the first item. Repeat the above process, and type *Math Average* as the second item. Repeat this process once more and type *Geom Average*.

Note other options in the popup menu that allow you to *Add Item Before* and to *Remove Item.*

To Operate the Menu Ring, click with the Operating Tool on the Menu Ring button and choose from the menu one of the items that you typed.

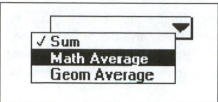

When you select *Sum,* the first item, the Menu Ring (being a control object) outputs a **0**. It outputs a **1** when you select *Math Average* and a **2** when you select *Geom Average.*

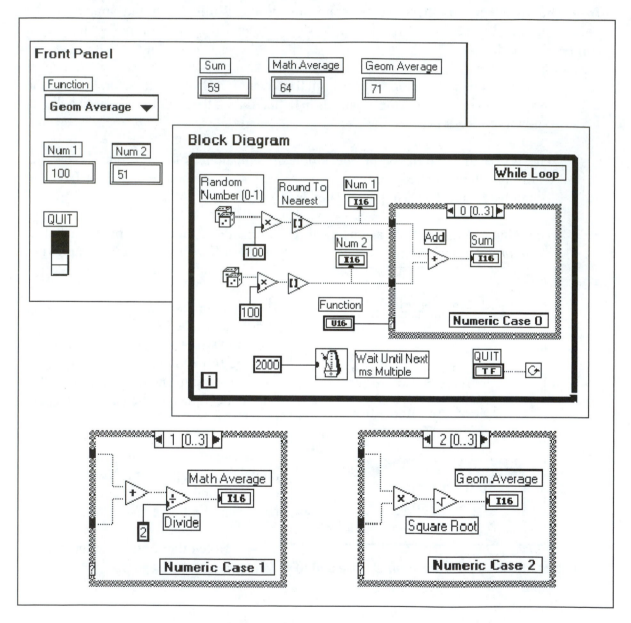

Fig. 2-14 Using the Numeric Case Structure of Example 2-4

2. In the **block diagram:**
Case Structure is in the *Structures* subpalette of the Controls palette. When you wire the *Function* terminal to the *Selector* terminal, the Boolean Case structure will change to the Numeric Case structure. Notice that Fig. 2-14 shows all three Case frames. To access frames 1 and 2, use the arrow button in the Case window.

The *Numeric* subpalette of the Functions palette contains most of the remaining items in the Block Diagram: **Random Number** generator, **Add**, **Multiply**, **Divide**, **Square Root**, **Round to Nearest**, **Numeric Constant**. **Wait Until the Next ms Multiple** is in the *Time & Dialog* subpalette of the Functions palette.

3. *Wire* all objects inside the Block Diagram as shown in Fig. 2-14. Notice that when you extend a wire into the Case structure, a **tunnel** in the shape of a black rectangle appears. The tunnel is the means of passing data to and from a structure. This was true also for the While and the For loops.

4. This VI begins its operation by generating two random integers inside the While Loop. Actually the *Random Number* generator outputs random values between 0 and 1. But after being multiplied by 100 and after passing through the *Round To Nearest* function, these values fall into the range of 0 to 100.

The *Wait Until Next ms Multiple* function provides a delay between iterations. In this example the delay is set to 2000 milliseconds, or 2 seconds. The delay pauses the execution and gives you an opportunity to view the data. You can change the delay to another value. In fact, instead of using the numeric constant, you can create a digital control and thus change the delay time from the front panel.

The *Function* terminal in the Block Diagram outputs an integer value corresponding to your choice in the Menu Ring. If you select the *Sum*, the Menu Ring outputs a **0** and applies it to the Case structure, causing the execution of frame **0**. Similarly, when you choose *Math Average,* the Menu Ring outputs a **1** and Numeric frame **1** and Geom Average will result in execution of frame 2.

The use of the While Loop will execute all code repeatedly until you stop execution with the QUIT switch.

Run the VI. Choose different options from the Function menu and note the results.

Save this VI as **Numeric_Case.vi** and close it.

Sequence Structure

Thus far we have spoken of programming tools that provide branching and a repetitive execution of a group of instructions. The Sequence structure is different. In most high-level languages the sequence or the order in which instructions are executed is simply the order in which the instructions are written. Therefore in the high-level languages there is no concern about sequence.

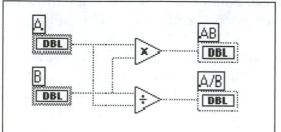

In LabVIEW the order in which nodes in the Block Diagram are executed depends on the availability of data at the input terminals of that node. A particular node will execute only if data is available at each terminal of that node. In this respect LabVIEW differs from the traditional high-level languages.

In this illustration, for example, it would be difficult to predict which operation, multiplication or division, will be done first. The Sequence structure can easily resolve this problem.

To Open the Sequence Structure, click with the *Positioning Tool* on the *Sequence* option in the *Structures* subpalette of the Functions palette, as shown in Fig. 2-15.

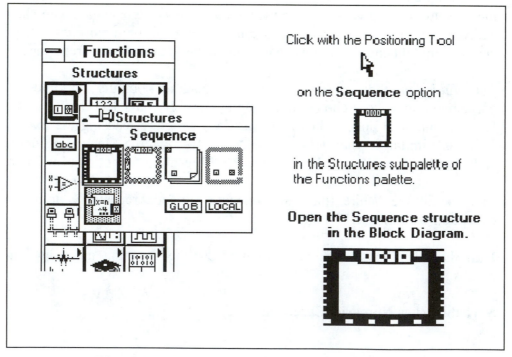

Fig. 2-15 Opening the Sequence Structure

Once opened in the Block Diagram, the Sequence structure can be *moved* or *resized* in the usual way.

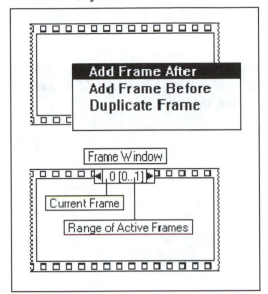

To Add a Frame, click (with the right mouse button) on the Sequence structure border and choose **Add Frame After** from the popup menu. This popup menu also allows you to **Add Frame Before** the current frame as well as to copy a frame.

This creates the **Frame Window**, as shown in the illustration. Notice that the frame window is similar to that of the Numeric Case structure. As shown in this illustration, it indicates the current frame and, in the brackets, [0..1], it shows the range of all active frames. The 0..1 indication means that there are frames 0 and 1.

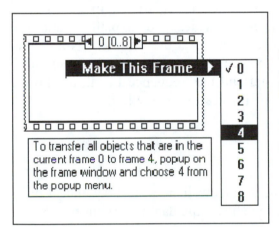

To switch to another frame, click with the left mouse button inside the frame window and choose the frame number that you wish to open from the popup menu.

In the illustration to the left, we are switching from frame 1 to frame 5.

If for some reason you wish to *transfer* all objects in the current frame to another frame, there is an easy way of doing so.

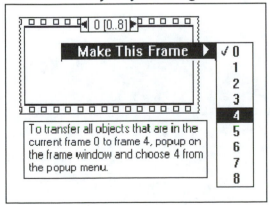

Pop up on the window and choose **Make This Frame** from the popup menu. Then choose the frame number from the submenu. In this illustration all code will be transferred from frame 0 to frame 4. All other frames will adjust accordingly.

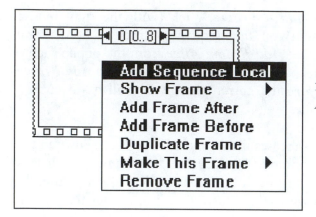

To Add a Sequence Local, pop up on the frame window and choose **Add Sequence Local** from the popup menu.

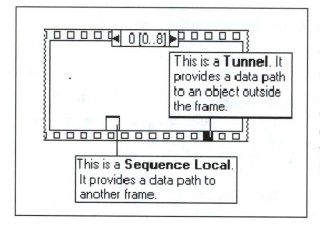

The **Sequence Local** provides a data path to any other active frame. In contrast, a **Tunnel** provides a data path to another object outside the frame. Note the difference in appearance between the tunnel and the Sequence Local, as shown in the illustration on the left.

When an object is wired to the Sequence Local, an arrow inside the Sequence Local indicates the direction of data flow. In the illustration above, the date is leaving frame and is, consequently, available in all frames higher than 3. This date is not available in frames below frame 3. The above illustration also shows the passing of data from frame 3 to an object outside the frame.

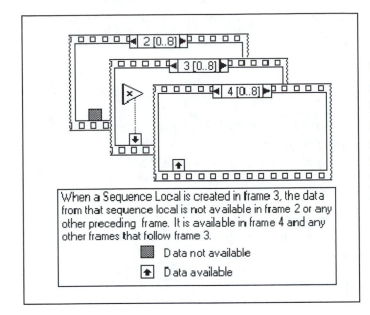

When a Sequence Local is created in frame 3, the data from that sequence local is not available in frame 2 or any other preceding frame. It is available in frame 4 and any other frames that follow frame 3.

▦ Data not available

⬆ Data available

When you create a Sequence Local in a frame, the data from that Sequence Local is not available in any of the preceding frames. It is available, however, in all other frames that follow the frame where you created the Sequence Local, as shown in this illustration.

The Formula Node

A Formula Node is a rectangular structure into which you enter formulas following specific syntax rules. These formulas are done at execution time. It's true that an equation such as $y = 2 + x^3$ can be synthesized using add and multiply function blocks, as shown in this illustration.

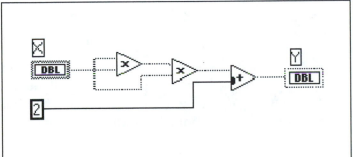

However, there is another option: type the formula inside the Formula Node rectangular box and the Formula Node will execute it for you. It is an option for the user and very often a convenient one.

To Open the Formula Node, choose ***Formula Node*** from the *Structures* subpalette of the Functions subpalette. When you open the Formula Node inside the Block Diagram, it looks like a simple rectangular box, as shown below.

But you also need input and output terminals to pass parameter values into the *Formula Node* and to pass calculated values to objects outside the *Formula Node.*

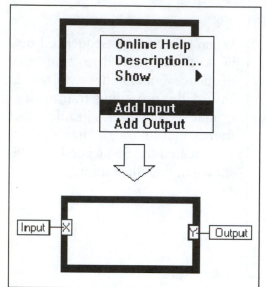

To Create Input Terminal, pop up (with the right mouse button) on the *border* of the Formula Node and choose the ***Add Input*** option from the popup menu.

To Create Output Terminal, repeat the above procedure and choose ***Add Output.***

To Enter Equation into the Formula Node, click on the Labeling Tool [A] in the Tools palette, then click inside the Formula Node with the labeling tool's cursor and start typing.

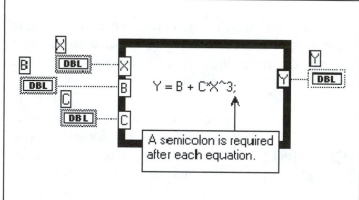

Don't forget the semicolon, because one is required at the end of each equation.

As shown in this illustration, the Formula node will output the value of the Y variable when you supply the X variable value.

In this illustration, the Front Panel digital control X supplies the X value and the digital controls B and C provide the parameter values B and C. The Front Panel indicator Y displays the values that the Formula Node outputs.

```
Formula Node operators, lowest precedence first:
assignment   =
conditional   ? :
logical OR    ||    logical AND    &&
relational    == != > < >= <=
arithmetic    + - * / ^
unary         + - !

Formula Node functions:
abs acos acosh asin asinh atan atanh ceil
cos cosh cot csc exp expm1 floor getexp getman
int intrz ln lnp1 log log2 max min mod rand
rem sec sign sin sinc sinh sqrt tan tanh
```

Fig. 2-16 Formula Node Syntax Table

The syntax or the rules that you must follow are shown in Fig. 2-16. As you can see, the set includes the hierarchy of the operators, from the assignment with the lowest precedence to the unary operator having the highest precedence. The set also includes an array of functions that you can use.

To Open Formula Node Syntax Table, enter *Ctrl+H* from the keyboard and then click with the positioning tool on the Formula Node.

Exercise 2-5: Using the Formula Node

This Exercise illustrates the use of the Formula Node. You will build a VI that solves a quadratic equation when the roots are real. When the roots are imaginary, the VI will display a message to that effect.

1   The Front Panel includes three digital controls, two digital indicators and a vertical switch. You will find the digital controls and indicators in the *Numeric* subpalette of the Functions palette, and the vertical switch in the *Boolean* subpalette of the Functions palette. Set the vertical switch to the *up* or *true* position, then pop up on the switch and choose *Mechanical Action>Latch When Pressed*.

2.   Switch to Block Diagram and open the **While Loop**. The While Loop is in the *Structures* subpalette of the Functions palette. Enlarge it so it takes up most of the window.

3.   Inside the While Loop open the first **Formula Node**. As shown in Fig. 2-17, add three *Input* terminals along the left edge and label them A, B and C. Add also the *Output* terminal and label it X. Wire the digital control terminals A, B, and C to match their corresponding inputs on the Formula Node.

Using the Labeling tool, enter the equation $X = B^2 - 4*A*C$; inside the Formula Node.

4. Open the **Case structure** inside the While Loop. The Case structure is in the *Structures* subpalette of the Functions palette. Enlarge the Case structure so that you can fit another Formula Node inside it. Make sure that the *false* frame is open and if not, click on the arrow button in the case window.

5. Open a **Formula Node** inside the Case structure. Create three inputs along the left edge of the Formula Node, as shown in Fig. 2-17, and label them as **A, B,** and **X**. Also create two outputs and label them as **RT1** and **RT2**.

    Enter into the Formula Node the following equations:

    $$RT1 = (-B + sqrt(X))/(2*A);$$
    $$RT1 = (-B - sqrt(X))/(2*A);$$

6. *Wire* the digital control terminal A to the A input of the Formula Node inside the Case structure. As you pass the wire across the border of the Case structure, a tunnel (black rectangle) will be formed in the wall of the Case structure.

    In a similar fashion wire the digital control terminal B to the B input of the Formula Node inside the Case structure. Also wire the X output of the first Formula Node to the X input of the Formula Node inside the Case structure.

    Wire the RT1 and RT2 digital control terminals to the outputs RT1 and RT2.

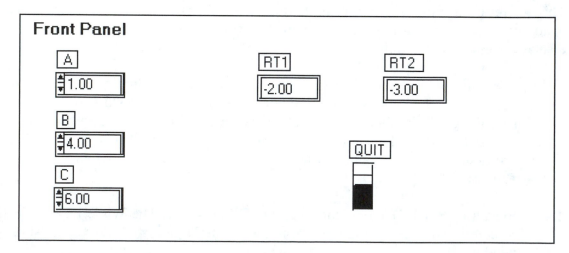

Fig. 2-17 Front Panel and Block Diagram of Exercise 2-5

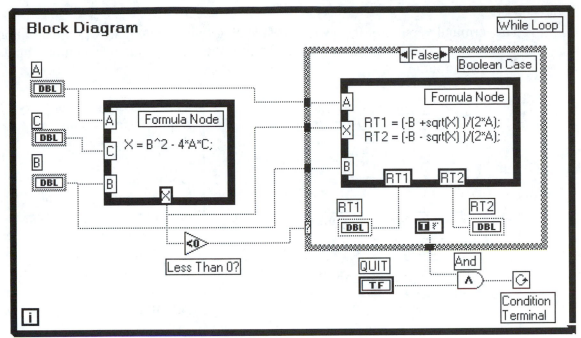

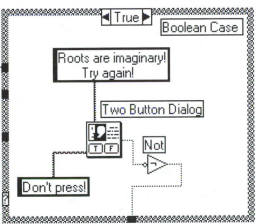

Fig. 2-17 Front Panel and Block Diagram of Exercise 2-5 (continued)

7. Open the **Less than 0? function** <0> comparison function. It is in the *Comparison* subpalette of the Functions palette. As shown in Fig. 2-17, wire the output X of the Formula Node to the Selector Terminal (**?**) of the Case structure.

8. Switch to the ***True*** frame of the Case structure by clicking on the arrow button in the Case window at the top of the Case structure.

   Open the **Two Button Dialog** . It is in the *Time & Dialog* subpalette of the Functions palette. As shown in this illustration, the *Two Button Dialog* icon has two buttons. One button is labeled as *T* and the other as *F*.

The terminal version of the Two Button Dialog (which you get by popping up on the icon and choosing *Show Terminals* from the pop-up menu) has three inputs and one output.

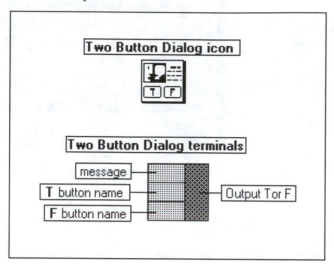

The **message** input is a required input and must be provided by the user.

Both the *T* and *F* button inputs are optional. The default name for the former is *OK* and for the latter, *Cancel*. You can, however, provide your own T and F button names if you don't like the default ones.

When the *Two Button Dialog* function executes, it outputs to the screen a dialog box with your message and two buttons whose names are either default or the ones you specified.

To create a message you must first open the **String Constant** ▯ , which looks like a purple rectangular box. It is located in the *Strings* subpalette of the Functions palette.

Open the **String Constant** from the *Strings* subpalette of the Functions palette in he Block Diagram. Immediately start typing (do not click with the mouse button anywhere).
Type the following: **Roots Are Imaginary!**
                         Try again!
It will look like this when you are done:

Roots Are Imaginary!
Try again!

Next, wire this string constant to the *message input* of the Two Button Dialog function. Notice the color of the wire; it's purple, which is appropriate for strings. Get another string constant, type ***Don't Press*** and wire it to the *F* input of the Two Button Dialog. Leave the T input unwired unless you don't like the *OK* name.

Select the two strings and the Two Button Dialog objects using multiple object selection rules. When selected, each object will be surrounded by dashed lines. Move the strings and the objects in the **True** frame of the Case structure.

9.   ***Open*** the **Not** inverter inside the True frame of the Case structure and an **AND** gate inside the While Loop.

Wire the output of the Two Button Dialog to the input of the Not and the output of the Not to the one input of the AND gate. Notice the tunnel that was formed when you moved the wire across the boundary of the Case structure to make the connection to the AND gate.

Wire the **QUIT** Boolean terminal to the other input of the AND gate.
Wire the output of the AND gate to the *Condition Terminal*.

10.   Assuming that you have followed the above directions and wired everything correctly, you should have at this time a broken Run button⬚, indicating that you have syntax errors. You can click on the Run button and find out where the errors are. The system will tell you that you are "missing assignment to tunnel" in the Case structure.

After a closer look, you will observe that the other three tunnels in the Case structure appear as **black rectangles**, but the tunnel wired to the output of the **Not** inverter appears as a **white rectangle**. The white rectangle is an indication that one input is missing. Certainly the True Case frame provides an input, but the False frame doesn't.

The best way to handle this problem without affecting anything else is to use a **Boolean Constant**. It can be found in the *Boolean* subpalette of the Functions palette. Its default form will appear as ⬚ . Click with the operating tool ⬚ on the Boolean Constant and it will change to ⬚ .

Place ⬚ in the False frame of the Case structure and wire it to the white tunnel of the Case structure. Notice that the tunnel changed to black and the broken Run arrow became solid. You are now ready to run the VI.

11.   The first Formula Node calculates the value of $B^2 - 4*A*C$ and equates it to X. The sign of X is tested by the *Less Than 0?* comparator. If X is positive, the *False* frame of the Case structure will be executed, and the roots RT1 and RT2 will be calculated and displayed on their respective Front Panel digital indicators.

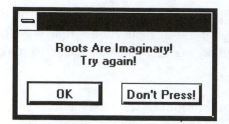

But if X is negative, the *True* Case frame will be executed, and the Two Button Dialog function will be executed, displaying a dialog box as shown in this illustration. When you click on OK button, the Two Button Dialog function will output a *TRUE,* which is inverted, and the *FALSE* from the inverter is applied to the AND gate. The AND gate in turn will apply a *FALSE* to the Condition Terminal. This will force the While Loop to stop execution.

Should you click on the *Don't Press,* nothing will happen because the Two Button Dialog will output a *FALSE,* which is inverted, resulting in a *TRUE* input to the AND gate.

You can also stop the execution by clicking on the QUIT vertical switch in the Front Panel with the operating tool.

Run the VI. Try different values of A, B, and C.
Save this VI as **Root.vi** and close it.

Exercise 2-6: Craps Game

This Exercise combines the use of the While Loop, For Loop, shift register, and a Sequence structure in a game of chance, a simulation of a craps game. In this Exercise you will explore on a larger scale the interactive aspects of structures and functions.

1.   Before building the craps game VI, you must first build another VI that will be used later as a subVI. The Front Panel and the Block Diagram for this VI are shown in Fig. 2-18.

Build the Front Panel and the Block Diagram shown in Fig. 2-18. You will find the *Random Number* generator in the *Numeric* subpalette of the Functions palette and the *To Unsigned Word Integer* in the *Conversion* subpalette of the Numeric subpalette of the Functions palette.

The Front Panel includes three digital indicators, as shown in Fig. 2-18.

A suggested icon may look like this | Roll 2 Dice | .

Save this VI as **Roll 2 Dice.vi** and close it.

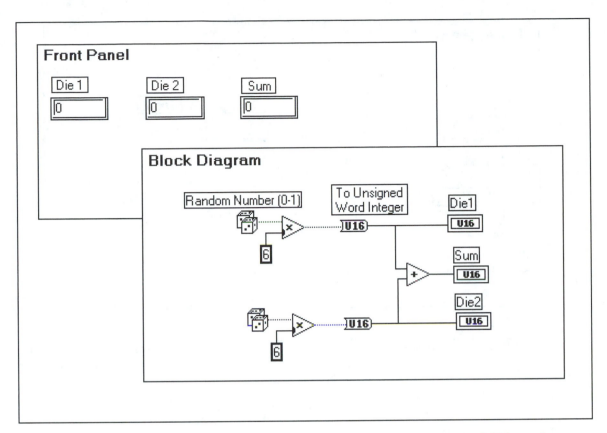

Fig. 2-18 The Front Panel and Block Diagram of Roll 2 Dice.vi

2.   ***Build*** the Front Panel of the Craps game VI as shown in Fig. 2-19. As you can see, it includes five digital indicators and a Boolean vertical switch.

3.   ***Build*** the Block Diagram as shown in Fig. 2-19. The Block Diagram consists of Sequence Frames 0, 1 and 2. Notice that the subVI Roll 2 Dice is used in frames 0 and 2.

4.   The operation begins in Frame 0 when the Roll 2 Dice subVI simulates the rolling of two dice. If the sum is 7 or 11, the One Dialog button in the Boolean case will display the message that you won. When you click on the OK button in the dialog box, the sum will be tested in frames 1 and 2, where the empty *False* case will be executed. The VI terminates its execution when all nodes in Frame 2 have been executed.

If the sum is equal to 2, 3 or 12, it will be tested in Frame 0 forcing the execution of the empty *False* case; but in the sequence of Frame 1, the *True* case will be executed, displaying the message that you lost. After you click on the OK button in the dialog box, sequence Frame 3 will be tested, forcing the execution of the empty *False* case. At this point the VI will terminate its execution because all frames of the Sequence structure have been executed.

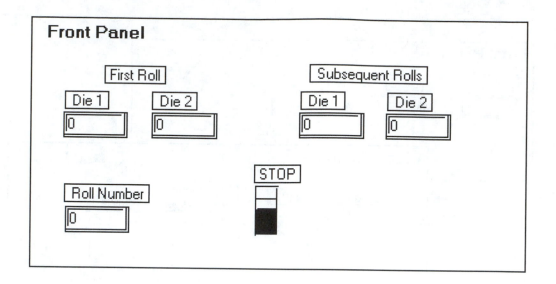

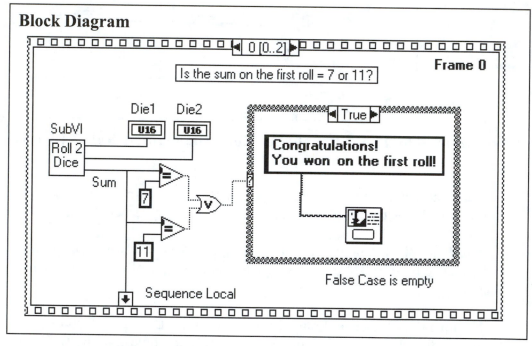

Fig. 2-19 The Front Panel and Block Diagram of Exercise 2-6

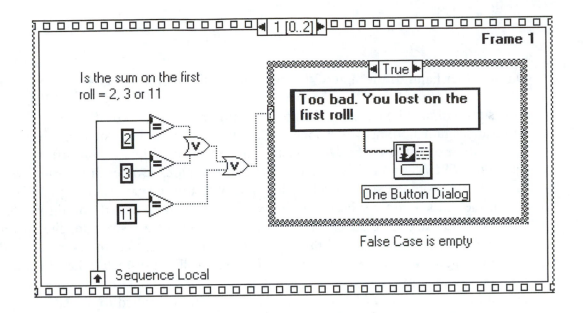

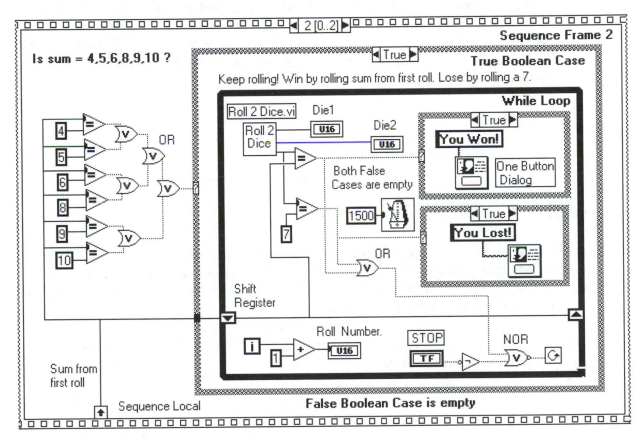

Fig. 2-19 The Front Panel and Block Diagram of Exercise 2-6 (continued)

On the other hand, if the sum of the two dice on the first throw is 4, 5, 6, 8, 9, or 10, the empty *FALSE* case will be executed in Frame 0 and nothing will happen. This sum will also be tested in Frame 1, forcing the execution of the empty *False* case, and again nothing will happen. The sum will next be tested in Frame 2, forcing the execution of the *TRUE* case. Notice that the While Loop shift register inside the *TRUE* case recirculates the sum and compares it on each iteration with the new sum from the subVI Roll 2 Dice. The object of the game at this point is to match the old sum on a subsequent throw of the dice. You lose if you throw a 7. One Boolean Case structure checks for a 7 and the other Case structure checks for the match between the old and the new sums. Both of these comparisons are fed to the OR gate. When one of them is *TRUE* (you roll a 7 or match the old sum), the appropriate message will be displayed and the While Loop will terminate execution because the *TRUE* from the OR gate will be applied to the NOR gate, which in turn will apply a *FALSE* to the condition terminal of the While Loop. This will complete execution of the sequence Frame 2 and thus terminate VI execution. Notice the 1.5 second delay inside the While Loop of Frame 2. This time delay allows you to watch the subsequent throw of the dice on the Front Panel in slow motion.

Run this VI.

Save this VI as **Craps.vi** and close it.

Additional Challenge Exercise

Suppose that we wish to make the Craps VI more user friendly so that after the game is finished, whether you win or lose, a message will be displayed that reads: *Would You Like to Play Again?* The user must then click on a YES or a NO button. Make changes to Craps.vi to accommodate this requirement.

[Hint: Enclose the entire sequence structure in a While Loop and use a Two Button dialog function in sequence Frame 2.]

Summary

1. The *While Loop* is a repetition structure. The loop executes all nodes inside the loop repeatedly as long as the Boolean input to the Condition terminal is *true*. When it becomes *false*, the loop terminates execution. Usually a Boolean control such as a switch in the Front Panel is used to provide the input to the Condition terminal in the Block Diagram.

2. The *For Loop* is also a repetition structure. It executes all code inside the loop border a fixed number of times, as dictated by the user-supplied value to the loop counter N.

3. One property that applies to the While Loop as well as to the For Loop is that the loop does not accept inputs or output data during execution time. At the beginning of its execution, the loop reads values external to the loop and then begins execution. It is only when the loop completes its execution that data can be passed from inside the loop to objects outside the loop.

4. The iteration counter i can serve as a counter in practical applications. We have used it to generate a staircase waveform in several simulation Exercises.

5. A *Shift Register* can be implemented in the While Loop as well as in the For Loop. It can store data that has occurred one iteration ago, two iterations ago, and so forth. It can be used to find the average of several values.

6. The *Case structure* is a selection structure that corresponds to the If/Else structure in a high-level language. When a Boolean value is applied to the *Selector terminal (?)* in the border of the Case structure, the Case structure becomes a Boolean type with two Case frames, one *True* and the other *False*. The Boolean input determines which frame will be executed.

    When the input to the Selector terminal is numeric, the Case structure changes to the Numeric type. The case window at the top of the frame indicates the current frame as well as the range of active frames. The numeric input to the selector terminal dictates which frame will be executed.

7. The *Sequence structure* consists of frames beginning with frame 0. When executed it will execute all code in frame 0 and will continue to execute all remaining frames in sequence. Because the order of node execution in LabVIEW depends on the availability of data at the input to the node, there are situations when it is not possible to predict which node in the Block Diagram executes first. Most of the time, the order in which nodes execute is not important. However, as was illustrated in Exercise 2-6, where the order of execution of specific tasks is essential, the Sequence structure is used.

8. The *Formula Node* is a structure used to execute mathematical equations and logical operations. Although many times the same code can be synthesized using the function blocks, the Formula Node is a convenient alternative option for the user.

9. A terminal in the Block Diagram that corresponds to the Front Panel object cannot be duplicated. But there are situations when you need a copy of a terminal. As was demonstrated in the Exercises, the *Local Variable* is the answer.

10. The *Select* function has three inputs, one input being Boolean. The Boolean input (?) determines which of the remaining two inputs will be passed to the output.

11. The timers are useful in providing a time delay or marking time. The *Tick Count* timer may be used to time the operation, the *Wait* timer is useful in generating accurate time delays, and the *Wait Until Next ms Multiple* timer can also be used to provide time delays or time between loop iteration delays.

12. The *Two Button Dialog* function is useful in displaying messages. It has two buttons, *T* and *F*. The default names for the T and F buttons are OK and Cancel, respectively, but the user can change the default names by providing appropriate string inputs. When the Two Button Dialog function is executed, the message that the user created is displayed on the screen. It outputs a *TRUE* when you click on the T button, and *FALSE* when you click on the F button. In Exercise 2-5 this function was used to alert the user to the fact that the roots are imaginary.

Chapter 3
Arrays

One-Dimensional Arrays

An array is a collection or an orderly arrangement of objects. The objects may be numbers, square LEDs, switches, and many others. In LabVIEW the members of an array may not be arrays, charts, or graphs. A one-dimensional array of integers may look like this:

$$4 \quad 3 \quad 1 \quad 0 \quad 12 \quad 15 \quad 9 \quad 6$$

There are eight integer values in this array. Each member is identified by its value and by its numerical position within the array. Let's give this array a name, **Bunch**. Now each member of this array can be uniquely identified. For example:

    Bunch[0] = 4
    Bunch[3] = 0
    Bunch[5] = 15

and so on. Notice that all members of the array have the name Bunch. What sets them apart is the *index* value. The index is the number inside the brackets. Arrays are zero based, which means that the maximum index value is one less than the size of the array. For an eight member array such as the array Bunch above, the index range is from 0 to 7.

> Note: *Members of an array must be a homogeneous collection of objects. This means that if an array is composed of integer values, as is array Bunch, every member must be an integer. For example, an array made up of integer numbers, floating point numbers, and square LEDs would be illegal.*

Two-Dimensional Arrays

A two-dimensional array has rows and columns, and thus resembles a rectangular structure. For example, a 4 x 6 array of integers may look like this:

    7   3   12   0   10   9
    0   7   20   6   4   1
    3   6   4   2   19   0
    6   9   5 12   16   7

Suppose that the name of this array is **Box**. All members of this array must have the name Box. Two indexes, one for the row and one for the column, must be attached to Box in order to identify uniquely each member of the array. Hence, Box[i,j] can uniquely identify any member of the array Box; the **i** index is used for the row and the **j** index is used to represent the column.

Here are some members of the array Box:

$$Box[0,3] = 0$$
$$Box[1,4] = 4$$
$$Box[2,5] = 0$$
$$Box[3,2] = 5$$

Notice, once again, that both indexes are zero based; the first row is row 0 and the first column is column 0, so Box[0,0] = 7. That means that the first index i can take on values from 0 to 3 in the above example, and the second index j can take on values from 0 to 5.

The preceding discussion singled out the integer value arrays. However, exactly the same can be said for an array made up of floating point numbers, or for an array consisting of square LEDs. A 3 x 5 array of square LEDs is shown below:

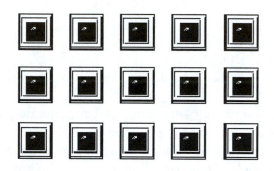

Arrays in LabVIEW

Array Controls and Indicators

To Open an Array Shell in the Front Panel, click on the *Array and Cluster* subpalette in the Controls palette and choose ***Array,***

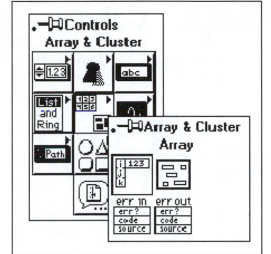

as shown in the illustration to the left.

When you open the array shell in the front panel, it will look like this:

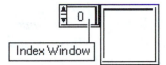

To Create Array Control/Indicator, open an empty array shell in the front panel as shown in the previous illustration, then pop up with the positioning tool inside the array shell and choose the desired object from a subpalette of the Controls palette. This illustration shows two array controls, one Boolean and the other digital.

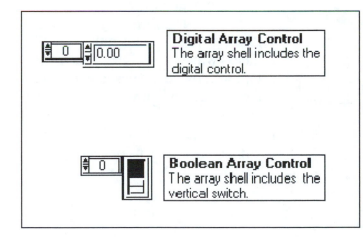

Digital Array Control
The array shell includes the digital control.

Boolean Array Control
The array shell includes the vertical switch.

The *digital array control* was created by popping up inside the empty array shell and then by choosing *digital control* from the *Numeric* subpalette of the Controls palette.

The *Boolean array control* was created in a similar fashion, except the vertical switch was chosen from the *Boolean* subpalette of the Controls palette.

This illustration shows two *array indicators*, one *Digital* and the other *Boolean*. To create the *digital array indicator*, pop up inside the empty array shell and choose the *digital indicator* from the *Numeric* subpalette of the Controls palette

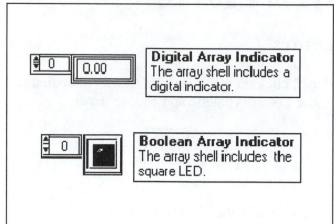

In the case of the *Boolean array indicator,* choose *LED* from the *Boolean* subpalette of the Controls palette.

Objects other than those shown in the above illustrations could also be used inside the empty array shall to create an array control or array indicator. An array, chart, or graph are the only objects that cannot be used inside an empty array shell.

The examples above show *one-dimensional array* controls and indicators. *To create a two-dimensional array control or indicator,* you must add a dimension to the one-dimensional array shell.

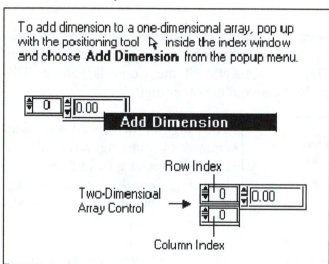

To Add Dimension to the one-dimensional array, pop up with the Positioning Tool inside the index window and choose ***Add Dimension*** from the popup menu.

As shown in this illustration, another index window will be added. The top window indicates the row and the lower window indicates the column. The two indexes uniquely identify the position of an element within a two-dimensional array.

Exercise 3-1: One-Dimensional Array

This exercise illustrates some of the array properties discussed earlier. You will build a VI that creates a one-dimensional array of five random integers between 0 and 20.

1. The Front Panel shown in Fig. 3-1 includes only one object, the *Digital Array Indicator*. You will find the *array* in the *Array & Cluster* subpalette of the Controls palette. Open it in the front panel.

    Pop up inside the array shell and choose **Digital Indicator** from the *Numeric* subpalette of the Controls palette. Move the digital indicator (dotted outline) into the array shell and click.

    Pop up inside the array window on the digital indicator and choose *Representation>I16*. This will change the representation of the digital indicator to a two-byte integer.

    Label the array as **Bunch** with an owned label.

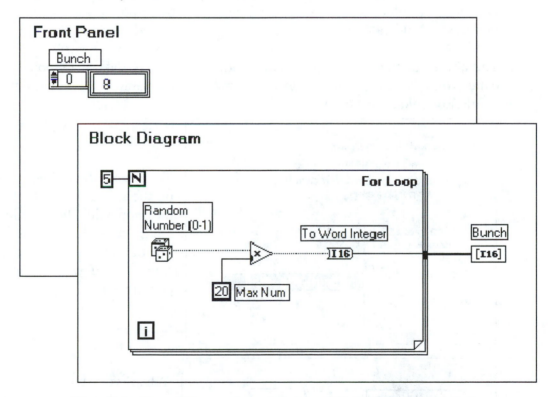

Fig. 3-1 One-Dimensional Array VI for Exercise 3-1

Block Diagram

2. You will find all objects shown in the block diagram in the *Numeric* subpalette of the Functions palette. This includes the *Random Number, Numeric Constant and Multiply* function.

    The **To Word Integer** ⋯I16⟩⋯ conversion function is in the *Conversion* subpalette of the Numeric subpalette. It converts a floating point input to a two-byte integer.

3. *Wire* the block diagram as shown in Fig. 3-1.

4. The numeric constant 20 multiplies the output from the random number generator, thus ensuring that the output from the Multiply function will be in the range of 0 to 20 and will not exceed 20. The *To Word Integer* converts this floating point value to an integer.

    As the For Loop executes, it accumulates array data ***at the boundary*** of the loop. Only after the loop completes execution is the array applied to the *Bunch*, the digital array indicator.

    The ability of the For Loop to index and acquire array data at its boundary is called ***auto indexing***. Auto indexing is enabled by default in the For Loop; in a While Loop you must enable it.

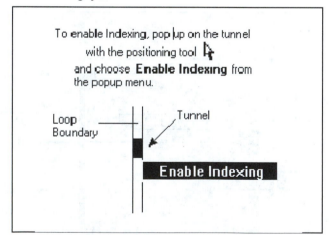

To Enable Indexing, pop up on the tunnel in the border of the loop with the positioning tool, as shown in this illustration, and choose *Enable Indexing* from the popup menu.

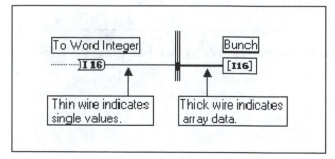

When you wire the digital array indicator Bunch to the tunnel, the wire becomes thick, indicating that the data on that wire is array data, as shown in this illustration.

Should you wire an object from inside the For Loop to an object such as the simple digital indicator where the data passed *is not an array*, you will have a bad wire. To correct this, pop up with the positioning tool on the tunnel and choose ***Disable Indexing***.

> **Note:** *Auto-indexing is enabled in the For Loop by default, but in the While Loop it is disabled by default. When enabled, auto-indexing allows the loop to accumulate and index array data at its boundary.*

Run the VI.

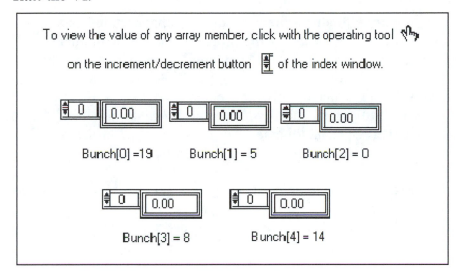

To view the array values, click with the operating tool on the increment/decrement button of the index window, as shown in this illustration,

or

you can resize the digital array indicator in the Front Panel as shown in the illustration below to view the entire array. When you catch the corner of the

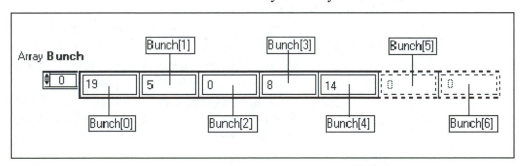

array to resize it, this cursor will appear for resizing the array shell instead of the usual corner cursor for resizing objects. Note the gray elements Bunch[5] and Bunch[6]; these indicate that there are no values there because the array Bunch has only 5 values. Resizing the array for the purpose of viewing element values will obviously not work for large arrays

Save this VI as **1-D Array Example.vi** and close it.

Exercise 3-2: Two-Dimensional Array

In Exercise 3-1 we considered a one-dimensional array. You constructed a one-dimensional array of random numbers. By its very nature, a one-dimensional array is limited to accumulating data in only one dimension. A group of numbers or objects laid out along a straight line represents one dimension.

In practice, however, we will find applications more often for a two-dimensional array. A table of values, for example, is a two-dimensional array.

In this exercise you will build a two-dimensional array that accumulates the values of a sinewave. We will discuss later some of the properties of the two-dimensional array as well as its objects. In this exercise you will learn how to use a sinewave function.

1.   The **Front Panel** contains only one object and that is the two-dimensional **digital array indicator**. The *digital array indicator* has been discussed and illustrated earlier. Open the digital array indicator *shell* in the front panel. You will find it in the *Array & Cluster* subpalette of the Controls palette.

Click inside the shell with the positioning tool and choose *digital indicator* from the *Numeric* subpalette of the Controls palette. Leave the representation as is, because the default is *DBL,* which is fine for this application.

Click inside the *index window* with the positioning tool and choose ***Add Dimension*** from the popup menu. Notice that another index window is added to represent the columns.

Block Diagram

2.   Switch to the block diagram (Ctrl+E) if you are not there already. The Block Diagram includes the following objects:

For Loop	*Structures* subpalette of the Controls palette.
Numeric Constant	*Numeric* subpalette of the Functions palette. There are four numeric constants in this Block Diagram (3, 6, 6, 18)
Pi Constant π	Choose **PI** constant from the *Additional Numeric Constants* subpalette of the *Numeric* subpalette of the Functions palette.
	Note that The Additional Numeric Constants is a *subpalette* of the Numeric *subpalette*. It contains many universal constants such as π, the speed of light, the Planck constant, and others.

Add, Multiply, Divide	These functions are in the *Numeric* subpalette of the Functions palette.
Sine function	This function is in the *Trigonometric* subpalette of the *Numeric* subpalette of the Functions palette.

x (radians) ···········⊏SIN⊐··········· sin(x)   As shown in this illustration, the input to the sine function is in radians.

3.   *Wire* the block diagram objects as shown in Fig. 3-2.

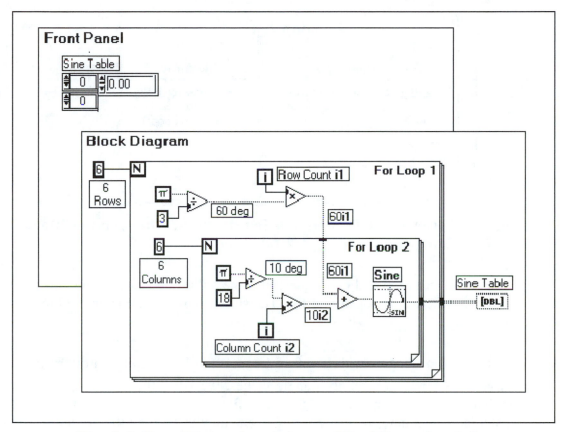

Fig. 3-2 Two-Dimensional Array VI of Exercise 3-2

4.   Let's look at the basic approach to generating a two-dimensional array. Fig. 3-3 shows an abbreviated version of Fig. 3-2 for the purposes of explanation.

To Create a Two-Dimensional Array, you need two loops, which can be For Loops or While Loops, one inside the other. As shown in Fig. 3-3, each loop goes through 6 iterations; in other words, each loop executes 6 times (0 through 5).

For each iteration of Loop 1, Loop 2 executes 6 times. Notice the thickness of the wires from the sine function, from Loop 1, and from Loop 2. The sine function outputs 6 values that are accumulated at the boundary of Loop 2 as a one-dimensional array. Loop 2 accumulates 6 one-dimensional arrays at its boundary as a two-dimensional array. Notice that there are a total of 36 iterations because for each value of i in Loop 1, Loop 2 must execute 6 times. And when both loops finish execution, the two-dimensional array of sine values is passed as a table to the digital array indicator *Sine Table*.

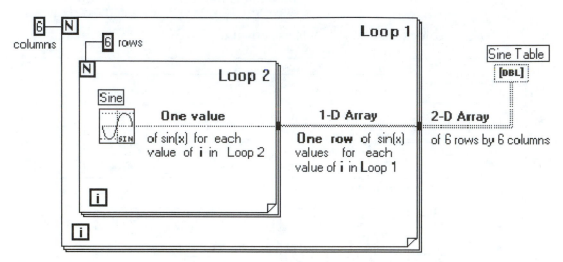

Fig. 3-3   Creating a Two-Dimensional Array, Abbreviated Block Diagram (see Fig. 3-2 for details)

As shown in Fig. 3-2, Loop 1 generates the value $60i_1$ and passes this value to Loop 2, where it is added to $10i_2$. The sum $(10i_2 + 60i_1)$ is applied to the sine function.

The operation begins with $i_1 = 0$ in Loop 1 and i_2 stepping through six values from 0 to 5 and generating 6 angle values: 0°, 10°, 20°, 30°, 40°, and 50° (in radians) that are applied to the sine function. The sine function, in turn, calculates 6 values, one for each angle, and applies them to the border of Loop 2; these then become a one-dimensional array, which is then passed to the border of Loop 1, where the two-dimensional array is being formed.

Next, $i_1 = 1$ and the above process repeats, except this time the 10° steps are added to 60° (the value in radians passed from Loop 1). As i_1 steps once again from 0 to 5, the 10° increments are added to 60° and then applied to the sine function. Thus, the second row of the table is generated and accumulated at the boundary of Loop 1. Notice that one-dimensional arrays (rows) are accumulated at the boundary of Loop 1.

The above process continues as each row of the table is accumulated at the boundary of Loop 1, with the last row generated when $i_1 = 5$. When both loops complete their execution, the two-dimensional array of sin(x) values is passed to the digital array indicator *Sine Table*.

5.   ***Run*** this VI. You may view the individual elements of the Sine Table array by clicking with the operating tool on the increment/decrement arrows of the index windows.

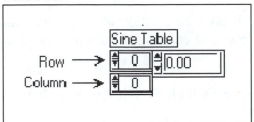

Remember that the upper index is Row and the lower index is Column, as shown in this illustration.

You may also view all values of the *Sine Table* array by resizing the digital array indicator in the front panel. In step 4 of the preceding exercise, you resized a one-dimensional array by dragging the array corner in the horizontal direction. You can resize a two-dimensional array by catching the lower right corner and dragging it downward and to the right until you see all elements of the array. Fig. 3-4 shows the result of a run. The array shows values of sin(x) in 10° increments from 0° to 350°.

Sine Table

0	0.00	0.17	0.34	0.50	0.64	0.77	0.00
0	0.87	0.94	0.98	1.00	0.98	0.94	0.00
	0.87	0.77	0.64	0.50	0.34	0.17	0.00
	0.00	-0.17	-0.34	-0.50	-0.64	-0.77	0.00
	-0.87	-0.94	-0.98	-1.00	-0.98	-0.94	0.00
	-0.87	-0.77	-0.64	-0.50	-0.34	-0.17	0.00
	0.00	0.00	0.00	0.00	0.00	0.00	0.00

Fig. 3-4 Sine Table of the Two-imensional Array.vi

Exercise 3-3: Viewing the Creation of the Two-Dimensional Array

If you would like to see the elements of the last exercise created one by one in slow motion, then you can modify the VI of Exercise 3-2 by adding three digital indicators to the front panel and the time delay to the block diagram as shown in Fig. 3-5.

When you run this VI, you will be able to observe the creation of an element value in the Sine Table array once every second. The row and column of the element being created will also be displayed.

This exercise is optional, and when you are done with it, do not save any changes.

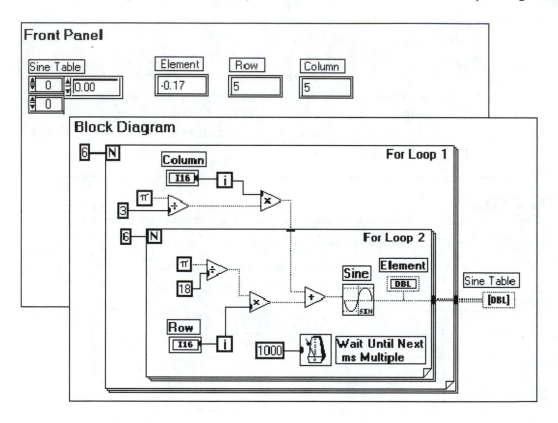

Fig. 3-5 The Front Panel and Block Diagram of Exercise 3-3

Exercise 3-4: Using Array Functions I

The *Array* subpalette of the Functions palette includes many array functions that are useful in performing various operations on arrays. There are functions that find maximum and minimum values in an array, slice away and return a row or a column from a two-dimensional array, find the size of an array, or build an array. In this exercise and in the exercise that follows, we will sample some of these functions to show you how they work. Fig. 3-6 shows the Array subpalette where these functions are to be found.

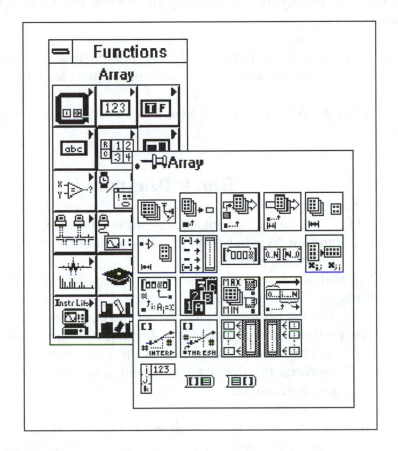

Fig. 3-6 Array Subpalette of the Functions Palette

In this VI we will use two array functions: **Array Max & Min** and **Array Size**.

Front Panel

1. In the front panel you have three *digital indicators* and two *digital array indicators*. The **digital indicators** are in the *Numeric* subpalette of the Controls palette and the **array** is in the *Array & Cluster* subpalette of the Controls palette.

    Open the following objects in the front panel and label them with owned labels as shown in Fig. 3-7:

The **Random Number Array** is a two-dimensional array. Pop up (with the right mouse button) inside the index window with the Positioning Tool and choose **Add Dimension** from the popup menu. To convert this array to *digital array indicator*, pop up inside the array shell with the Positioning Tool and choose *digital indicator* from the *Numeric* subpalette of the Controls palette, then drag and drop the digital indicator inside the array shell. Pop up on the digital indicator inside the array shell and choose *Representation>I8* from the popup palette. Finally, resize this array so that you can view a 5 x 3 (5 rows, 3 columns) array of elements.

The **Size 2** array is a one-dimensional array. For the most part, repeat the above procedure and resize the array so that you can see two elements.

The **Max Value**, **Min Value**, and **Size 1** digital indicators have *Representation* set to I8.

Block Diagram

2.   Open the **For Loop** (you may want to label it with a free label as For Loop 1, as shown in Fig. 3-7). For Loop 1 includes the following objects:

> The **Random Number (0-1)** generator is in the *Numeric* subpalette of the Functions palette.
>
> The **Multiply** function is in the *Numeric* subpalette of the Functions palette.
>
> **Numeric Constant** is in the *Numeric* subpalette of the Functions palette. Use the labeling tool to enter 40 into one constant and 3 into another.
>
> The **To Byte Integer** conversion function is in the *Conversion* subpalette of the *Numeric* subpalette of the Functions palette. This function converts the floating point input from the multiply function to a one byte integer.

> *Wire* all objects inside For Loop 1 as shown in Fig. 3-7. Wire the output of the *To Byte Integer* to the tunnel at the boundary of For Loop 1 (the tunnel will be created automatically when you wire to the For Loop boundary). Don't forget to wire the *Columns* numeric constant to the loop counter terminal *N*.

3.   Open another **For Loop** and label it with a free label as *For Loop 2*. This loop must be positioned so that it includes Loop 1 inside its boundary.

> **The Array Size** function is in the *Array* subpalette of the Functions palette. The Array Size function returns the number of elements in each dimension of input.

> *Complete the wiring* as shown in Fig. 3-7. Wire also the *Rows* numeric constant set to 5 to the loop count terminal of For Loop 2. Notice that the wire from the tunnel of For Loop 1 to the tunnel of For Loop 2 is thicker. That is because the data on that wire is a one-dimensional array.

4.   There are four objects outside For Loop 2.

Array Size is the same object as in step 3 above except this time it is wired to the *Size 2* digital indicator terminal (counterpart of the front panel object).

The **Array Max & Min** function is in the *Array* subpalette of the Functions palette. The *Array Max & Min* function searches the input array for the maximum and the minimum values and outputs these values at their respective terminals.

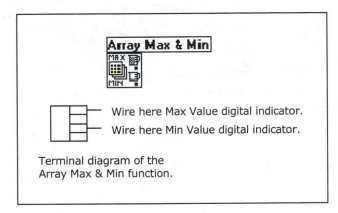

Wire here Max Value digital indicator.
Wire here Min Value digital indicator.

Terminal diagram of the
Array Max & Min function.

Wire the *Max Value and Min Value* terminals as shown in this illustration.

Complete wiring the block diagram as shown in Fig. 3-7. Notice the *thick wires* from For Loop 2 that appear as double lines. These carry the two-dimensional array data.

Run the VI.

The operation begins as both loops read their respective N (loop count) values. The value of N is set to 3 in For Loop 1 and to 5 in For Loop 2. Initially **i=0** in For Loop 2, while For Loop 1 executes 3 times, generating an array of three numbers. This array is accumulated at the boundary of For Loop 1. When For Loop 1 completes execution, the array is passed to the boundary of For Loop 2.

Next, **i** is set to **1** in For Loop 2, and For Loop 1 generates three more numbers (second row). This process continues until the array of five rows is completed. You can view the entire array in the front panel as well as the display of the maximum and minimum values of the array. Notice that the *Size 1* indicator displays 3 because the input to the *Array Size function* in loop 2 is a one-dimensional array of 3 values. On the other hand, the *Size 2* array indicator displays 5, 3 because the input to the *Array Size* function is a two-dimensional array with 5 values in each column and 3 values in each row.

Save this VI as **Array Functions 1.vi** and close it.

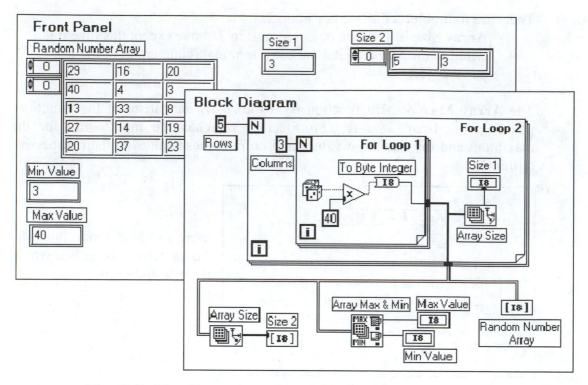

Fig. 3-7 The Front Panel and Block Diagram of Exercise 3-4

Exercise 3-5: Using Array Functions II

In this exercise you will build a VI that illustrates the use of the **Build Array** function.

Front Panel

1.   **Array** is in the *Array & Cluster* subpalette of the Controls palette. As shown in Fig. 3-8, you will need three arrays. The arrays A and B are *digital array controls*. You create a digital array control by popping up inside the array with the positioning tool shell and choosing *digital control* from the *Numeric* subpalette of the Controls palette.

   Similarly, you create the *digital array indictor* by popping up inside the array shell with the positioning tool and then choosing *digital indicator* from the *Numeric* subpalette of the Controls palette.

2.   **Digital Control** C is found in the *Numeric* subpalette of the Controls palette.

   Set the Representation on all objects inside the front panel. Resize the arrays as shown in Fig. 3-8.

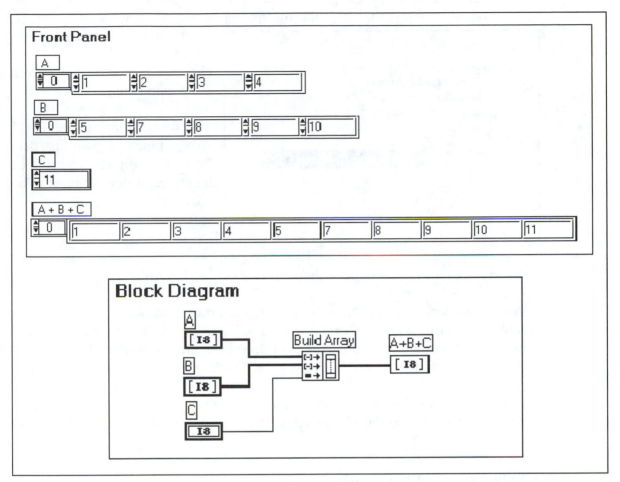

Fig. 3-8 The Front Panel and Block Diagram of Exercise 3-5

Block Diagram

3. The **Build Array** function can be found in the *Array* subpalette of the Functions palette. As shown in the illustration below, the Build Array function must be resized to include three inputs. The inputs to the Build Array function must be configured to accommodate the input data. If the data input is an array, then the inputs on the Build Array function must be changed accordingly.

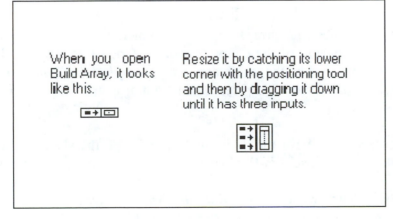

When you open Build Array, it looks like this.

Resize it by catching its lower corner with the positioning tool and then by dragging it down until it has three inputs.

In this exercise two of the inputs are arrays. Changing of the inputs to accommodate arrays is shown in the illustration below. Simply pop up with the positioning tool (right mouse button) on the input and choose **Change to Array** from the popup menu.

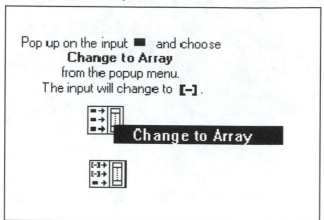

Pop up on the input ■ and choose **Change to Array** from the popup menu. The input will change to [-].

The third input on the Build Array function gets its data from the digital control C; therefore, it doesn't have to be changed.

4. **Complete the wiring** of the block diagram as shown in Fig. 3-8.

5. **Enter** the numerical values into the two digital array controls A and B and the digital indicator C, as shown in the front panel of Fig. 3-8. To enter values click with the positioning tool on the increment/decrement arrows on the left side of the digital control window. You can also use the Labeling tool to type in the values.

    Run the VI. Notice that the two arrays and the digital control's C value have been combined into a single one-dimensional array, as displayed by the digital array indicator A + B + C.

    Save this VI as **Array Functions 2.vi** and close it.

Exercise 3-6:   Using Array Functions III

In this exercise you will build a VI that will use and illustrate the properties of two array functions: **Index Array** and **Transpose 2D Array**.

Front Panel

1. The front panel includes four arrays and one digital indicator. **Array** is in the *Array & Cluster* subpalette of the Controls palette. Open the array shell in the Front Panel. Pop up inside the shell and choose *digital indicator* from the *Numeric* subpalette of the Controls palette. Drag and drop the digital indicator into the array shell.

    Choose Representation>I8 for the digital indicator inside the shell. Click inside the index window and choose *Add Dimension* to make this array two-dimensional.

    Label the array with an owned label as ***Random Number Array***.
    Resize the array to 5 rows by 3 columns as shown in Fig. 3-9.

2.   Follow the procedure in step 1 and create a two-dimensional array *Transposed Random Number Array*. Resize this array to show 3 rows and 5 columns.

3.   Follow the procedure in step 1 and create two one-dimensional arrays. As shown in Fig. 3-9, resize one to show 5 elements and label it with an owned label as *Column 2,* and resize the other to show 3 elements and label it as *Row 3*.

4.   Open a **digital indicator** (*Controls>Numeric*) and label it with an owned label as *Element 4,1*. Also choose *Representation>I8* for this indicator.

Block Diagram

5.   As shown in Fig. 3-9, build For Loop 1 inside For Loop 2. Note that For Loop 1 is exactly the same as that shown in Fig. 3-7.

6.   Open **Index Array** function. You will find it in the *Array* subpalette of the Functions palette. When you first open the Index Array, it will have only one index terminal, the black rectangle in the lower left corner of the icon. Resize the icon to include two index terminals, as shown in this illustration.

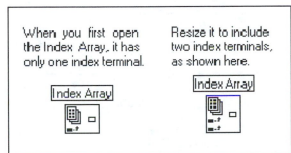

To increase the number of index terminals, catch the lower corner of the icon with the positioning tool and drag in the downward direction. You can also accomplish the same thing by popping up on the index terminal (right mouse button) and choosing **Add Dimension** from the popup menu.

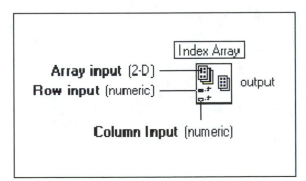

As shown in this illustration, you must wire a two-dimensional array to the input terminal of Index Array and a numeric integer value to either one or both index inputs.

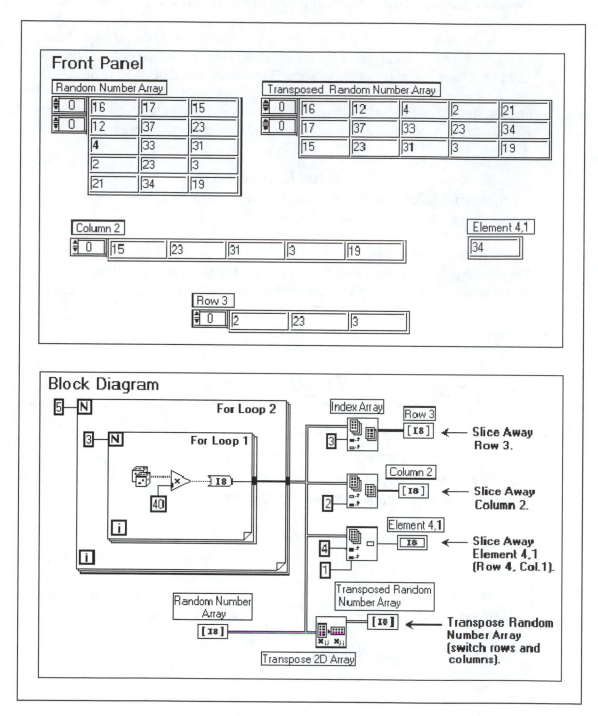

Fig. 3-9 The Front Panel and Block Diagram of Exercise 3-6

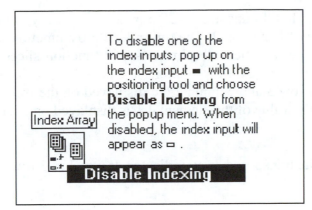

To disable one of the index inputs, pop up on the index input ■ with the positioning tool and choose **Disable Indexing** from the popup menu. When disabled, the index input will appear as ▢ .

When you have to slice away a row or a column from a two-dimensional array, one of the index inputs must be disabled. In this illustration the column index input is disabled, but you can disable the row index input instead.

To Slice Away a Column from a two-dimensional array, you must disable the row index input and apply the numeric value of the column that you wish to slice away to the column index input.

To Slice Away a Row from a two-dimensional array, you must disable the column index input, and apply the numeric value of the row that you wish to slice away to the row index input.

7. Following the guidelines for configuring the Index Array function in step 6, disable the column input and wire a *Numeric constant* whose value is **3** to the row index input, as shown in Fig. 3-9. Wire the array output of For Loop 2 (at the tunnel of For Loop 2) to the array input of the Index Array function, and its output to the *digital array indicator* terminal *Row 3*.

8. Repeat step 7 above, except this time disable the row index input and wire the numeric constant **2** to the column index input. The output of this Index Array function is wired to the digital array indicator *Column 2*. As shown in Fig. 3-9, wire the two-dimensional array from the tunnel of For Loop 2 to the array input of this Index Array function.

9. You need one more Index Array function with both indexes enabled. Wire **4** to the row index input and **1** to the column index input. As was done before, wire the array from the tunnel of For Loop 2 to the array input of this Array Index function. Also wire the output to the *Element 4,1* digital indicator terminal.

10. Open the **Transpose 2D Array** function, which is in the *Array* subpalette of the Functions palette. This function transposes a two-dimensional array by switching rows and columns. Wire the array from the tunnel of Loop 2 to the input of the Transpose 2D Array function and its output to the *Transposed Random Number Array* indicator terminal, as shown in Fig. 3-9.

11. *Wire* the array from the tunnel of Loop 2 to the *Random Number Array* digital array indicator terminal, as shown in Fig. 3-9.

12. *Run* this VI.

As the VI begins execution, For Loops 1 and 2 form a 5 x 3 array, which is accumulated at the boundary of For Loop 2. The top Index Array function in Fig. 3-9 slices away row 3 of the array. Next, the Index Array function slices away column 2, and the last function slices away a single element located in row 4, column 1. Element values of row 3 and column 2 are displayed on the front panel digital array indicators, and the value of the element 4,1 is displayed on the digital indicator.

Note the element values of the transposed array indicator. It now has 3 rows and 5 columns.

Save this VI as **Array Functions 3.vi** and close it.

Exercise 3-7: Using Array Functions IV

In this exercise you will build a VI that explores the properties of the *Array Subset* function. In Exercise 3-5 we used the *Build Array* function to combine arrays A and B plus a digital constant into one array called A + B + C. We can now use the *Array Subset* function to extract the arrays A and B and the constant C from the one-dimensional array A + B + C.

Front Panel

1. As shown in Fig. 3-10, the front panel contains arrays A + B + C, A, and B, as well as the digital indicator C.

    Array A + B + C is a one-dimensional *digital array control*.
    Arrays A and B are one-dimensional *digital array indicators*.
    C is a *digital indicator*.

    Build the front panel as shown in Fig. 3-10. Refer to Exercise 3-5 for creating array objects in the front panel.

Block Diagram

2. The **Array Subset** function is in the *Arrays* subpalette of the Functions palette. As shown in Fig. 3-10, it requires an ***array input*** as well as the **Index** and the **Length** inputs. Its output is the extracted subarray.

    The *Index* specifies the point at which to begin subarray extraction, and the *Length* specifies the number of elements (beginning at the index value) to be extracted.

Open three Array Subset functions. ***Complete wiring*** the block diagram as shown in Fig. 3-10.

3.   As shown in the block diagram of Fig. 3-10, the top Array Subset function extracts array A from array A+B+C because its index is set to **0** and its length to **4**; that means elements 0, 1, 2, and 3 will be extracted. In a similar fashion the middle Array Subset function begins with element **4** and extracts the **5** elements beginning with element 4. Finally the bottom Array Subset function extracts the last element of array A + B + C.

Run this VI after entering the values into the array A + B + C, as shown in the Front Panel.

Save this VI as **Using Array Functions 4.vi** and close it.

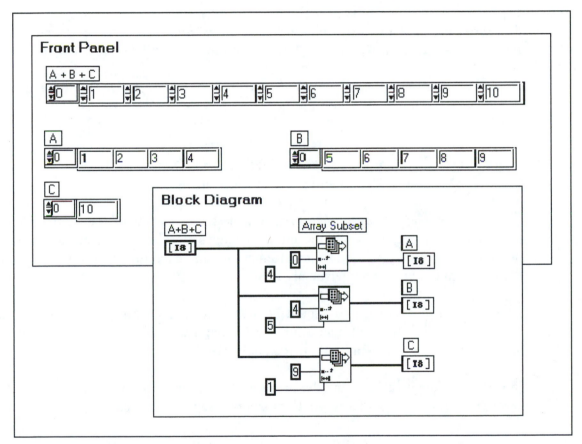

Fig. 3-10 The Front Panel and Block Diagram of Exercise 3-7

Exercise 3-8: Using Array Functions V

In this exercise we will create a VI that illustrates the **Array Subset** function. In some respects this function resembles the *Index Array* function, except that the Index Array function slices away or extracts one row, one column, or one element from an array.

While the Array Subset function can do all that, its more important use is to extract a subset that is two-dimensional from another two-dimensional array. In the preceding exercise we used the Array Subset function on a one-dimensional array, and in this exercise we will use it on a two-dimensional array. Incidentally, the Array Subset function can be used on arrays of any dimension.

Front Panel

1. As shown in Fig. 3-11, the front panel contains arrays *Input Array* and *Subarray*, and four digital controls.
    Input Array is a two-dimensional *digital array control.*
    Subarray is a two-dimensional *digital array indicator.*
    Row, Column, Num. Rows, and Num. Columns are digital controls.

    Build the front panel as shown in Fig. 3-11. Refer to Exercise 3-5 for creating array objects in the front panel.

Block Diagram

2. *Build* the block diagram as shown in Fig. 3-11. Note that the Array Subset function must be resized to include the second dimension.

    To resize the Array Subset function, catch the lower corner of the icon with the positioning tool cursor. When the cursor changes its shape to a corner, drag it down until you see two dimensions. Remember that each dimension must include the *Index* and *Length* input terminals.

3. Before you run this VI, you should be aware of several things. First of all, the Array Subset function extracts a portion (called a subset) of a two-dimensional array. This subset is also a two-dimensional array.

    In order to slice away this smaller array from a larger array, you must tell the VI where to begin and how far to go. For example, suppose you enter

    | Row = 1 | Column = 0 |
    | Num. Rows = 3 | Num. Columns = 2 |

    into the digital controls on the front panel as shown in Fig. 3-11. You are telling the VI to begin with row 1, column 0, and to include three rows (that means rows 1, 2, and 3) and two columns (that means columns 0 and 1). The VI as shown in Fig. 3-11 was executed using these values, and the resulting subset that was extracted from the *Input Array* is displayed inside the *Subarray* digital indicator.

Notice that the values displayed by the Subarray digital indicator is the intersection of rows 1, 2, and 3 and columns 0 and 1 of *Input Array*.

Enter values into Input Array as shown in the Front Panel of Fig. 3-11 or use values of your own choice.

Run the VI. Experiment with different values in the Front Panel digital controls.

Save this VI as **Using Array Functions 5** and close it.

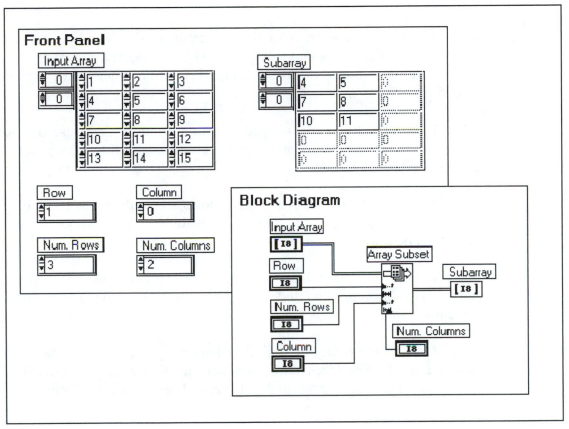

Fig. 3-11   The Front Panel and Block Diagram of Exercise 3-8

Summary

1. An array is a collection of objects such as numbers, square LEDs, Boolean switches, or other objects.

2. In a one-dimensional array objects are placed along a straight line. A two-dimensional array is made up of rows and columns. And in general, a three-dimensional array may be likened to a book, where a two-dimensional array is included on a given page and the page number represents the third dimension. Arrays of dimension higher than 3 are difficult to visualize, although they can still be treated mathematically.

3. In LabVIEW you open an array shell in the front panel. It can be made an array control or an array indicator depending on the type of object that you drop into the shell. By adding a dimension to the array shell, it becomes a two-dimensional array control or array indicator.

4. The For Loop or While Loop is used to create an array. You use one loop to create a one-dimensional array and a nested arrangement of two loops to create a two-dimensional array.

5. Auto-indexing is a feature of the loop that allows accumulation of an array at its boundary. In the For Loop auto indexing is enabled by default, but in the While Loop it must be enabled.

    When auto indexing is enabled, an array will be accumulated at the loop's boundary. By disabling auto indexing, you can pass single data points to an object outside the loop.

6. LabVIEW has a wide variety of array functions that you can use in your VI to manipulate arrays. The Index Array function, for example, can be used to slice away a row, a column, or a single element. The operation of many of these functions has been explored in several exercises.

Chapter 4
Charts and Graphs

Waveform Chart

A Waveform Chart is an indicator for displaying one or more waveforms. It has three update modes. The strip chart mode resembles the old paper strip chart with the scrolling display.

To Open a Waveform Chart, choose ***Waveform Chart*** from the *Graph* subpalette of the Controls palette as shown in this illustration.

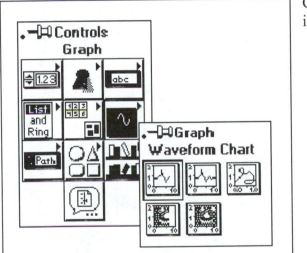

When you open the Waveform Chart in the front panel, it will appear as shown in Fig. 4-1. It has several features that you should be familiar with. Let's first consider the **Palette** shown in the illustration below.

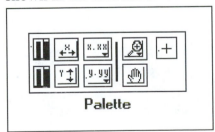

To enable **AutoScale Y**, click with the Positioning tool on the button next to the button. The scale along the vertical scale will adjust automatically to fit the data being displayed.

In the same manner you can also enable the **AutoScale X** by clicking on the button next to the button.

Another way to set the *AutoScale is* by popping up inside the waveform screen and then by choosing *Y Scale>AutoScale Y* or *X Scale>AutoScale X.*

When you click with the Positioning tool on the [image] or the [image] button in the *Palette* window, a popup menu opens. By clicking on the ***Format*** option, as shown in this illustration, the second menu opens, allowing you to choose the desired format. The default is the *Decimal* format.

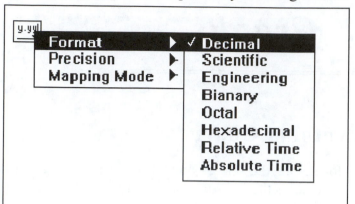

When you click on the ***Precision*** option, the second menu allows you to choose the precision, or number of decimal places, as shown in this illustration.

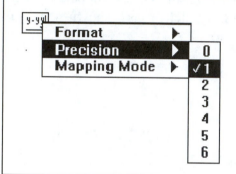

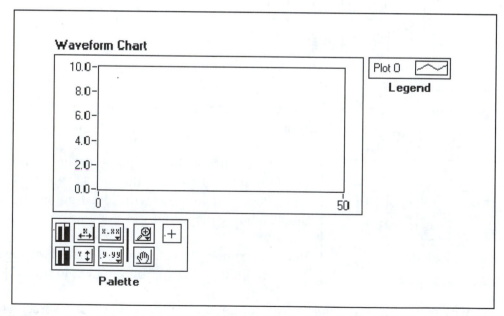

Fig. 4-1 The Waveform Chart

The magnifying glass button inside the palette is the *zooming tool*. As shown in this illustration, by clicking on the zoom button, you open a palette from which you can choose a variety of zooming options.

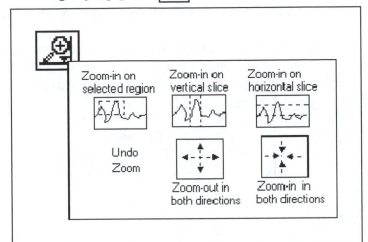

The extended hand button inside the palette is the *waveform positioning tool*. It allows you to move the display horizontally as well as vertically.

The crosshair ⊞ button is the *graph cursor*. You can use this tool to switch from the zoom mode or the waveform positioning mode .

The *Legend* Plot 0 ⌇ that opens by default on the upper right side of the waveform chart shows the characteristics of the waveform that the chart displays. The default name of the waveform is Plot 0; however, you can delete Plot 0 and, using the labeling tool, type the name for the displayed waveform. This legend window can be resized (by catching the lower corner with the positioning tool and dragging it down) to accommodate additional waveforms.

This window ⌇ inside the legend window indicates the *point style* and the *color* of the plot.

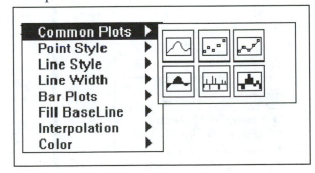

Additional Plot Options, including those of point style and color, are accessed by clicking inside the legend window with the operation tool and choosing the desired option from the popup menu. This illustration shows the *Common Plots* palette.

To access various plot setting accessories, pop up inside the waveform chart window and choose *Show* from the popup menu. A submenu with additional options opens, as shown in this illustration.

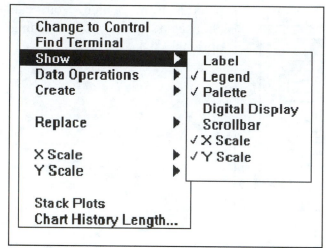

For example, when you click on the *Scrollbar* option, the scrollbar shown below will appear below the waveform chart.

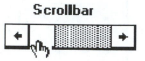

Use the operating tool to scroll through the displayed waveform. The Waveform Chart stores the last 1024 data points. Since the chart window may not display all 1024 points, you can use the scrollbar to show the part of the plot that is not displayed inside the window. The *Chart History Length...* option allows you to change the 1024 point default value.

The *Digital Display* option, when activated, opens the digital indicator next to the legend. It indicates the value of the current point being plotted.

To Format the Y scale, pop up inside the waveform chart and choose *Y Scale>Formatting ...* from the popup menu. The *Y Scale Formatting* window opens, giving you various formatting options as shown in this illustration.

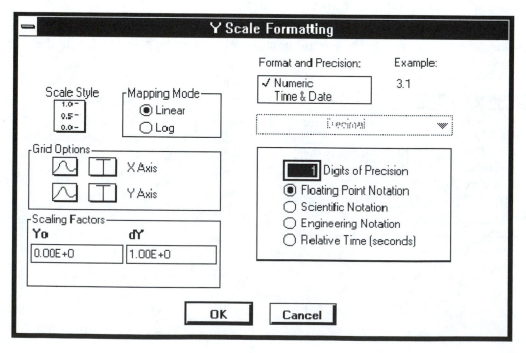

As you can see from this window, you can choose *scale style, linear* or *logarithmic mapping*, and the desired *format* and *precision*. The *Grid Options* give you several choices of grid type and color that you can select individually for the X axis and the Y axis.

Click on this icon ⌐⌐⌐ to display the grid options, and on this icon ⊤ to display the color palette for choosing the grid color. After making all desired selections, click on OK.

You will get almost the same formatting window when you choose **Scale>Formatting...** from the popup menu.

Update Mode
Waveform Chart has three update modes:

Strip Chart
Scope Chart
Sweep Chart

To Access the Update Mode Palette, pop up inside the Waveform Chart and choose *Update Mode* from the *Data Operations* popup menu. The Update Mode palette opens, allowing you to access one of three charts as shown in this illustration. The default is the *Strip Chart*.

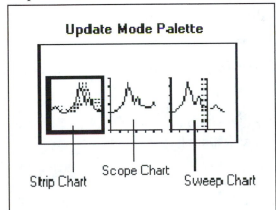

To access the Update Mode palette during VI execution, pop up on the chart and choose Update Mode from the popup menu.

Strip Chart resembles the old paper strip chart that has a scrolling display. As the display reaches the right edge of the display window, it scrolls off and continues to display new data.

Scope Chart has a retracing display that is similar to an oscilloscope. When the display reaches the right edge of the screen, the screen is cleared and the display begins to scroll from the left side of the screen. The screen is cleared every time the display reaches the right edge.

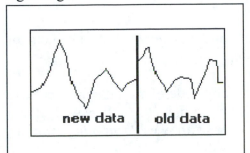

Sweep Chart has a retracing display also, except that the screen is not cleared when the display reaches the right edge. Instead, the vertical line, as shown in this illustration, moves along with the display, adding new data. The plot to the right of the display represents the old data.

Because the strip chart has more software overhead, its display is significantly slower than that of the scope or the sweep charts.

Exercise 4-1: Using the Waveform Chart

In this exercise you will build a VI that illustrates the use of the waveform chart. In this VI the chart will be used to display several cycles of sine and cosine waveforms.

Build this VI. The front panel and the block diagram are shown in Fig 4-2. The following guidelines will help you in the VI construction.

Front Panel

1. **Digital Indicators** are in the *Numeric* subpalette of the Controls palette. Label these with owned labels as shown in Fig. 4-2.

    Vertical Switch is in the *Boolean* subpalette of the Controls palette. Label the switch as **Quit** using an owned label.

    Waveform Chart is in the *Graph* subpalette of the Controls palette. Label the chart with an owned label as **Sine/Cosine Chart**. Resize the *Legend* (that currently shows Plot 0) to include two plots. To resize the legend, catch its lower corner with the *Positioning Tool* and drag it down until you see Plot 0 and Plot 1 inside the legend window. Use the *Labeling Tool* to delete and type **Sine** in place of Plot 0 and **Cosine** in place of Plot 1. Pop up inside the legend window of the Sine waveform and click on the *Color* option from the popup menu. Choose the desired color for the Sine waveform. Repeat this procedure for the Cosine and choose its color. The Default color is white.

    Pop up on the chart, and choose **Show>Digital Display**. Pop up on the chart once more and choose **Show>Scrollbar**.

    When you are done, your waveform chart will resemble that of Fig. 4-2. You may want to resize the waveform chart to make the display larger.

Block Diagram

2. **Multiply, Divide** functions are in the Numeric subpalette of the Functions palette.

    The 2π constant object is in the *Additional Numeric Constants* subpalette of the *Numeric* subpalette of the Functions palette.

    The **Sine** and **Cosine** functions are in the *Trigonometric* subpalette of the *Numeric* subpalette of the Functions palette.

    The **Bundle** function is in the *Cluster* subpalette of the Functions palette. The bundle function is used to combine two or more plots to be displayed. It can be resized by dragging down its lower corner with the *Positioning Tool* to accommodate more than two inputs.

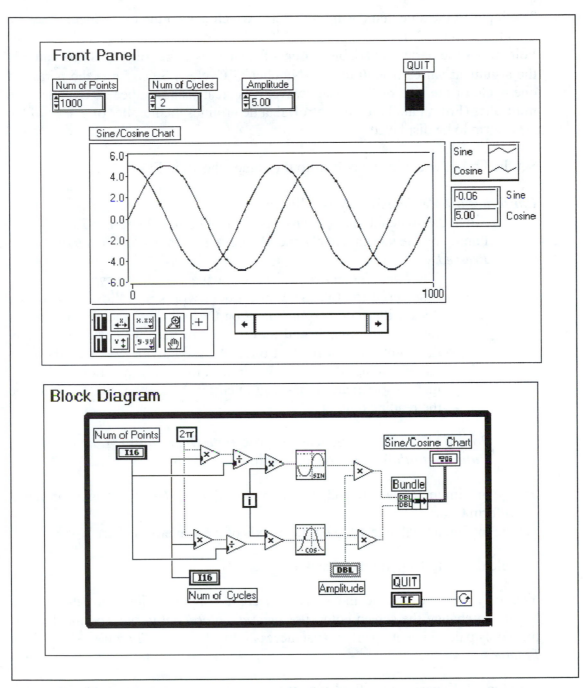

Fig. 4-2 The Front Panel and Block Diagram of Exercise 6-1

3.   **Wire** all objects inside the Block Diagram as shown in Fig 4-2.

4.   The input to the sine and cosine functions, which is also their argument, must be in radians. In this VI, the argument is 2π (Number of Cycles)(i)/(Number of Points). As the iteration variable i counts from 0 to Number of Points, the value of the argument varies from 0 to (2π)(Number of Cycles). Since 2π is the period of one cycle of the sine or the cosine wave, it is clear that Number of Cycles is the multiplier (Front Panel digital control) that determines the number of cycles of the waveform to be displayed.

Set the Front Panel digital controls to the values shown in Fig. 4-2.

Configure the Vertical Switch as follows:
>   Move the switch to the *up* or *true* position with the *Operating Tool*.
>   Pop up on the switch and choose ***Mechanical Action>Latch When Pressed.***
>>   The *Latch When Pressed* mechanical action of the switch, which is often used to stop the While Loop, works as follows: when you click with the *Operating Tool* on the switch as the While Loop is running, the switch will move down to its *false* position and stay there until the VI reads it once. As soon as the VI takes the *false* value reading, the switch will return to its *true* position. It doesn't matter how many times you click on the switch; the VI will take the reading only once.

>   Pop up on the switch again and choose ***Data Operations>Make Current Value Default*** from the popup menu.

Set the maximum value to 6 and the minimum value to −6 on the Y-axis of the Waveform Chart.
Set the maximum value to 1000 on the X-axis of the waveform chart.

Disable the AutoScale for the X and Y axes.

You may want to change the waveform chart window color from the default black. To change the screen color, use the *Set Color Tool* in the Tools palette. If the Tools palette is not open, you can access it by choosing *Windows>Show Tools Palette*.

5.   ***Run*** the VI. Experiment with different settings of the front panel digital controls.
>   While the VI is running, pop up on the waveform chart and choose the ***Update Mode*** option. Select ***Scope Chart*** or ***Sweep Chart*** from the palette that opens. Observe the difference in execution speeds as compared to the *Strip Chart*. As was mentioned earlier, the strip chart is slower because of its software overhead.

Save this VI as **Sine/Cosine Chart.vi** and close it.

Waveform Graph

The Waveform Graph's appearance in the Front Panel is similar to that of the Waveform Chart. However, there are significant differences between the two.

To Open Waveform Graph, choose *Waveform Graph* from the *Graph* subpalette of the Controls palette. As shown in Fig. 4-3, the Waveform Graph is very similar in appearance to the Waveform Chart shown in Fig. 4-1.

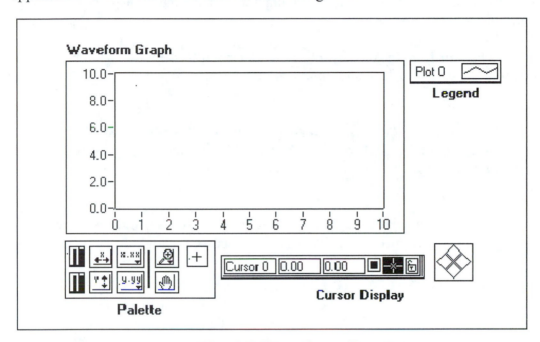

Fig. 4-3 Waveform Graph

By popping up on the Waveform Graph and choosing ***Show***, you will observe that most of the options are the same as those of the Waveform Chart. Notice that the ***Cursor Display*** is an option for the Waveform Graph and not for the Waveform Chart.

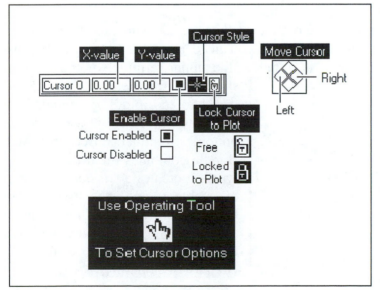

As shown in this illustration, the X-value, Y-value digital indicators provide precise values of the cursor coordinates.

The ***Enable/Disable*** button allows you to enable or disable the cursor. This control toggles.

When you click on the **Lock Cursor to Plot** button, a menu opens with several options. The *Lock to Plot* option forces the cursor movement along the waveform. The *Free* option restricts the cursor movement along the X-axis and not along the waveform.

The **Cursor Style** menu allows you to choose, among other things, the cursor's shape, its point style, and its color.

To Move the cursor, click with the positioning tool inside the *Move Cursor* diamond.

To Remove the Cursor from the display, click on the Cursor Style button and choose **none** from the *Cursor Style* palette and **none** from the *Cursor Point* palette.

> *The most important difference between the Waveform Chart and the Waveform Graph is in the type of applied data. The Waveform Chart accepts data on a point by point basis. The Waveform Graph, however, accepts data only in an **array** form.*

The waveform graph can display a single plot or multiple plots. The single plot and multiple plot graphs require special consideration, as discussed below.

Single Plot Waveform Graph

Probably the simplest way of applying the array data to the Waveform Graph terminal

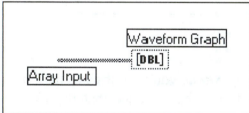

inside the X=1. This means that the plot will begin at the origin with unity spacing between the X values. Notice that the Waveform Graph terminal assumes the one-dimensional array symbol [DBL] .

You can, however, specify the initial value of X as well as the ΔX value, as shown in the illustration below. To combine this information you need a bundle structure. In this illustration the initial value of 0 for X is wired to the upper terminal of the Bundle

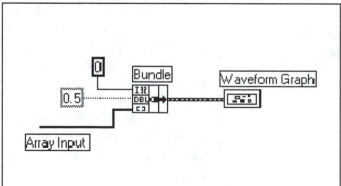

structure, the delta X of 0.5 is wired to the middle terminal, and the array (Y-values) is wired to the bottom terminal. Because the Bundle combines the different data types, it creates a *Cluster* (a cluster of values) and the Waveform Graph icon assumes the cluster symbol [꞉꞉꞉] (the wire between the bundle and the Waveform Graph will be purple).

Multiple Plot Waveform Graph

When two or more waveforms are to be displayed, you need a **Build Array** structure that combines the input arrays into a multidimensional array. In this illustration each of the

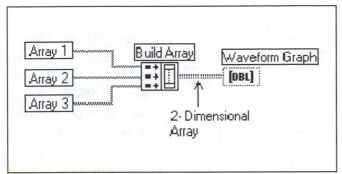

one-dimensional input arrays (Arrays 1, 2, and 3) represents a plot or a waveform to be displayed on the Waveform Graph. As shown, they are applied to the *Build Array* structure, which combines them into a two-dimensional array that is applied to the Waveform Graph terminal.

Notice the difference in the wire thickness of the one-dimensional and two-dimensional arrays. Also note the difference in the icons representing the one-dimensional and two-dimensional array terminals. The one-dimensional array terminal was shown in the earlier illustration as $[DBL]$, and the two-dimensional array terminal $[DBL]$ is shown in the above illustration. This may be a subtle point, but what makes them different is the thickness of the brackets ([ ] versus [ ]). If you wire the arrays as shown in the above illustration, the initial value of X will be taken as 0 and the spacing ΔX will be taken as 1.

If you want to use custom settings for the initial X value and the spacing, you have to use the **Bundle** function again. As shown in this illustration, each input array has its own

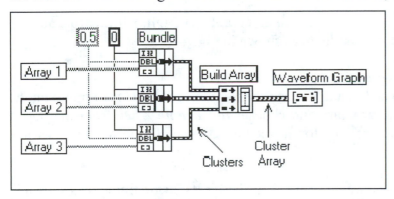

Bundle structure. The same values of $X_o = 0$ and $\Delta X = 0.5$ are applied to the three Bundle functions, and the respective arrays are applied to the bottom terminal of each Bundle function.

The output from each Bundle is a *cluster* of 0, 0.5, and the array data. The three clusters are applied to the *Build Array* structure, which creates a *Cluster Array*. The Cluster Array that includes the X-axis settings and data for three individual plots is finally applied to the Waveform Graph terminal. Note once again the wire thickness of the data lines as well as the color, which you will be able to see once you build a VI. Also notice the cluster representation of the Waveform Graph icon.

Instead of using *Build Array* after the *Bundle* structure, as was done in the preceding illustration, just the opposite may be done, which eliminates two *Bundle* structures and renders a less cluttered configuration. As shown in this illustration, the three input arrays are converted to a two-dimensional array by the *Build Array* structure.

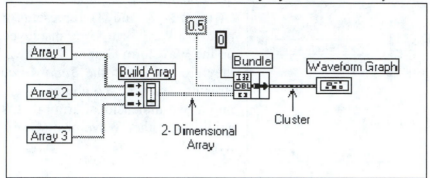

Then the *Bundle* structure combines the two-dimensional array with the X-axis settings into a cluster, which is applied to the Waveform Graph terminal.

Exercise 4-2:   Using the Waveform Graph as a Control

One expects that a Waveform Graph will be used for display purposes only. However, the plot that is displayed is also stored by the Waveform Graph. If that is the case, then it should be possible to output the plot from the Waveform Graph, using it as a control object rather an indicator object. You may wonder, why display it in the first place? What's the point of that? A possible use of this technique will be discussed at the end of this exercise.

Build this VI as shown in Fig. 4-4.

1.   Inside the *front panel* open two **Waveform Graphs**, a **Digital Indicator,** and the **Vertical switch**, and configure them as shown in Fig. 4-4.

2.   ***Switch*** to the ***block diagram***.
        Open the **While Loop** and inside it, open the **Sequence structure**. Pop up on the border of the Sequence structure and choose ***Add Frame After*** from the popup menu. Click on the left arrow button inside the sequence frame window that just appeared to select frame 0.

        Open the **For Loop** inside frame 0 and wire all objects inside frame 0 as shown in Fig. 4-4.
        Open frame 1 and inside it open and wire all objects as shown in Fig. 4-4.
        In frame 2 create a local variable for the *Waveform Graph Indicator* and make it a *Read Local*.

        Complete wiring all objects inside the Block Diagram.

3.   In Frame 0 the For Loop acquires a 1000 point sinewave array at the border of the For Loop. It is displayed, when the loop completes execution, on the Waveform Graph Indicator.

Frame 1 produces a 2-second delay, and in frame 3 the waveform graph indicator, which is now a control, outputs the array generated in frame 0 to the *Waveform Graph Indicator Output* waveform graph for display.

This VI appears to be useless from a practical standpoint; however, it has interesting and important use in data acquisition applications. In data acquisition applications where the software is unable to keep up with the rate at which data is being acquired due to its overhead, the concept of this VI can be inserted as an intermediate step, making the data acquisition process quasi-real time. A sample of data, for example, can be stored in the waveform graph and processed; then the second sample can be stored and processed, and so on. This is a topic of an advanced nature.

Run this VI, **save it** as **Waveform Graph Control.vi,** and close it.

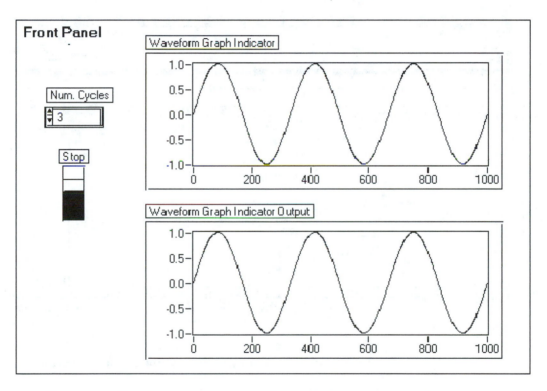

Fig. 4-4 The Front Panel and Block Diagram of Exercise 4-2

Block Diagram

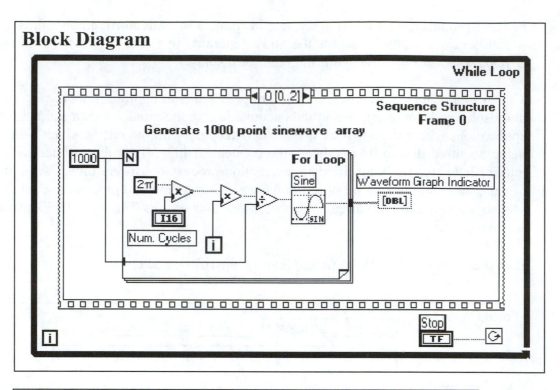

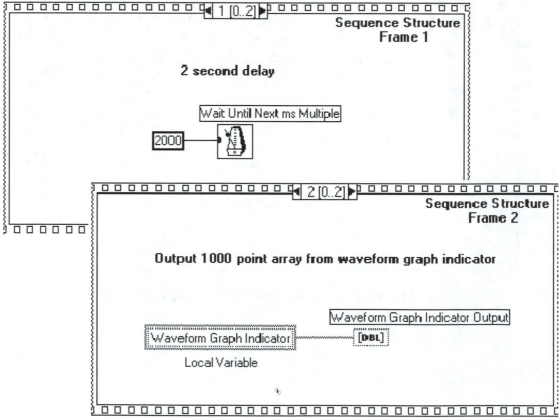

Fig. 4-4 The Front Panel and Block Diagram of Exercise 4-2 (continued)

Exercise 4-3: Using the Waveform Graph (Waves)

In this exercise we will build a VI that uses the waveform graph to plot some of the popular waveforms in electronics. These waveforms include the square wave, the triangle wave, and the sawtooth wave. This will be a relatively simple VI to build because you will use some VIs from the Analysis library that generates these waveforms.

The Front Panel and the Block Diagram of the VI that you are to build next is shown in Fig. 4-5.

Front Panel

1.   **Waveform Graph** is in the *Graphs* subpalette of the Controls palette. Open one Waveform Graph and configure it as follows:

Scale: Enter the Y min and max values as shown in Fig. 4-5.

AutoScale: Disable the Y-axis only. The X-axis autoscale will be enabled by default.

Label the graph as shown in Fig. 4-5:

   Owned Label: **Waveform Graph**

   Free Label: **Samples**

   Menu Ring is in the *List & Ring* subpalette of the Controls palette. Open one Menu Ring and *label it using an owned label* as **Wave Menu**. Using the labeling tool, type into the Menu Ring the following items in the following order:

   Square Wave
   Triangle Wave
   Sawtooth Wave

   Remember that you have to enter these items one at a time. After adding an item, pop up on the Menu Ring and choose **Add Item After** from the popup menu.

   Digital Control is in the *Numeric* subpalette of the Controls palette. Open three digital controls and label them with owned labels, as shown in Fig. 4-5.

   Representation: **I16** for *Num Samples* digital control.

   Increment:

   5 for *Amplitude* digital control.

   50 for *Num Samples* digital control.

   To set the increment, pop up on the digital control, choose **Data Range...** from the popup menu, and type the desired value in the Increment box.

   Vertical Switch is in the *Boolean* subpalette of the Controls palette. Open one Vertical Switch and label it using an owned label as **STOP**.

   Mechanical Action: **Latch When Pressed**.

   Choose **Data Operations>Make Current Value Default** from the popup menu.

Block Diagram

2.　Open the **While Loop** and the **Case structure** inside the While Loop as shown in Fig. 4-5. Resize both structures as necessary.

Wire the *STOP* terminal to the condition terminal of the While Loop.

Wire the *Wave Menu* terminal to the *Selector* [?] terminal of the Case structure.

Numeric Case 0

Open **Square Wave.vi**. It is in the *Signal Generation* subpalette of the *Analysis* subpalette of the Functions palette (Functions>Analysis>Signal Generation).

Wire all terminals to the Square Wave.vi as shown in Fig. 4-5. To identify the

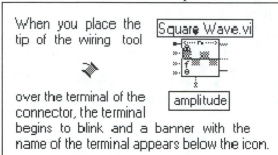

terminal to be wired on the connector, place the tip of the wiring tool over the terminal, which will begin to blink, and the banner with the terminal name will appear below the icon.

In this illustration the tip of the wiring tool was placed on the terminal labeled as A, and the banner with the name Amplitude appeared below the icon.

Wire the Samples, Amplitude and Duty Cycle (%) terminals to the corresponding terminals on the Square Wave.vi connector and the *Square Wave* output terminal to the *Waveform Graph* terminal, as shown in Fig. 4-5.

Numeric Case 1

Open the *Numeric Case 1* frame by clicking inside the numeric case window at the top of the structure.

Open the **Triangle Wave.vi**, which is found in the *Signal Generation* subpalette of the *Analysis* subpalette of the Functions palette.

Wire all terminals inside the *Numeric Case 1*, as shown in Fig. 4-5. Note that the *Samples, Amplitude,* and *Waveform Graph* terminals are all local variables.

Numeric Case 2

Open Numeric Case 2 by popping up inside the case window at the top of the structure and then by choosing **Add Case After** from the popup menu.

Open the **Sawtooth Wave.vi**, which is found in the *Signal Generation* subpalette of the *Analysis* subpalette of the Functions palette.

Wire all terminals inside the *Numeric Case 2*, as shown in Fig. 4-5. Note that the *Samples, Amplitude, and Waveform Graph* terminals are all local variables.

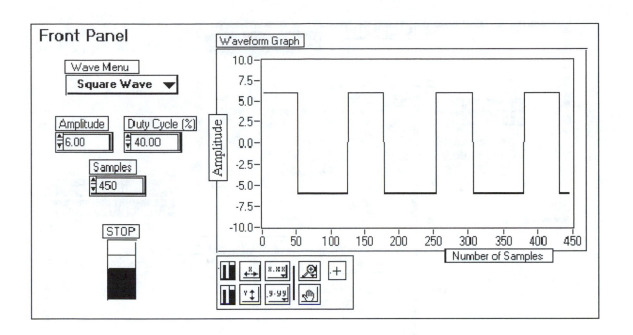

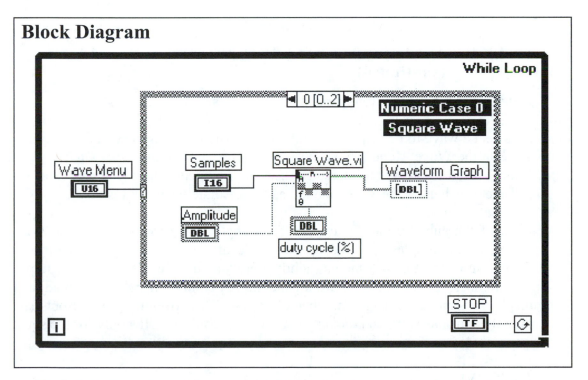

Fig. 4-5 The Front Panel and Block Diagram of Exercise 4-3

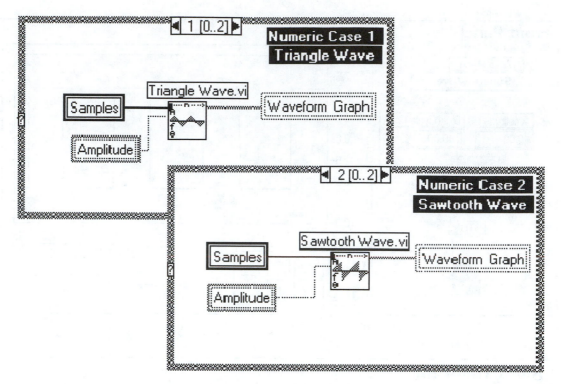

Fig. 4-5 The Front Panel and Block Diagram of Exercise 4-3 (continued)

3. *Enter* the values of the front panel digital controls as shown in Fig. 4-5.

4. To select a wave to be displayed on the waveform graph, click with the *Operating Tool* on the Wave Menu (menu ring) and choose an item. Once again, the While Loop is used to create an interactive user environment where you can enter new parameter values as the VI is running, and immediately see the results.

    Each of the subVIs, the Square Wave, Triangle Wave, or the Sawtooth Wave, requires 128 samples to make up one cycle of a waveform. The square wave display shown in Fig. 4-5 shows slightly more than three cycles over 450 samples. The number of samples is the front panel digital control, which you can adjust to display a smaller or larger number of cycles. The Amplitude is a front panel digital control as well, and the duty cycle digital control works only for the square wave.

    Run the VI.

    Save this VI as **Waves.vi** and close it.

X-Y Graph

The X-Y Graph differs from the Waveform Chart or the Waveform Graph that was just covered because it is used for plotting curves on the Cartesian coordinates. When you open the X-Y Graph in the front panel it will have the appearance of the waveform graph shown in Fig. 4-3. The associated terminal for the X-Y graph in the block diagram has the appearance of the following cluster symbol .

The type of inputs required by the X-Y graph makes it very different compared to the

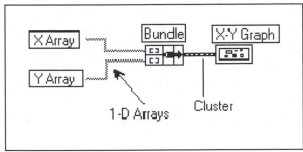

Waveform Graph and the Waveform Chart. As shown in this illustration, both inputs to the X-Y Graph must be arrays, one array for the X-axis and the other for the Y-axis. The arrays must be combined in the Bundle structure before being applied to the X-Y Graph. This configuration is used for one plot.

When two or more plots are to be displayed on the X-Y Graph, you need three pairs of

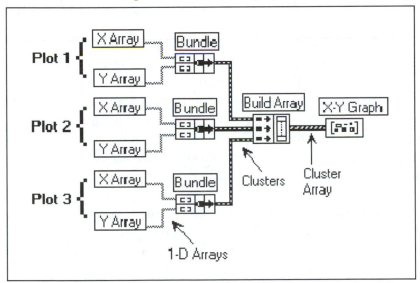

arrays, as shown in this illustration. Each pair is first combined in the Bundle structure and then applied to the Build Array structure. The output from the Build Array structure, as shown in the illustration, is a cluster array.

Summary

1. A Waveform Chart is used to display one or more waveforms. It has three update modes.

   The *Strip Chart* update mode resembles the old paper strip chart with the scrolling display. When the display reaches the right edge of the screen, it scrolls off the screen and continues to display new data.

   The *Scope Chart* update mode has a retracing display similar to that of an oscilloscope. When the display reaches the right edge of the screen, the screen is cleared and the display begins to scroll from the left side of the screen. The display is erased every time it reaches the right side of the screen.

   The *Sweep Chart* update is similar to the Scope Chart because it has a retracing display. However, it differs from the Scope Chart in the execution of the retracing display. When it reaches the right edge of the screen, the screen is not cleared. Instead, a vertical line moves along with the display, adding new data. The data on the left side of the line is new data and the data on the right side of the line is the old data. As the line moves, it erases some of the old data to make room for the new data.

2. Waveform Graph is also used to display waveforms. The data to be displayed on the waveform graph must be in the form of an array. The x-axis scale will be the default setting or it can be specified by the user. The default sets the initial value to 0 and the interval between points to 1. The user may change the default settings by using the Bundle structure. The Bundle icon is resized to accommodate three inputs, one for the initial value of x, one for delta x, and the bottom input is reserved for the wave array.

   When two or more waveforms are to be displayed on the same Waveform Graph, the arrays for each waveform are first combined in the Build Array structure, and the output of the Build Array, a two-dimensional array, is then applied to the waveform graph. The scale for the X-axis will assume the default setting as mentioned above. The user may change the default setting by using the Bundle structure. Each array will require one Bundle structure with three inputs, and the Bundle outputs are combined in the Build Array structure.

3. The X-Y Graph differs considerably from the Waveform Chart or the Waveform Graph because it is used to plot mathematical functions or curves using the Cartesian coordinates. The X-Y Graph requires two arrays as its inputs: one array for the Y-axis and the second array for the X-axis. Since a point in the Cartesian system is described by the (x,y) coordinate pair, it is assumed that for each value in the X-array there is a corresponding value in the Y-array.

To display a single plot on the X-Y Graph, the X-array and the Y-array are first combined in the Bundle structure, and then the output of the Bundle is wired to the X-Y Graph terminal in the Block Diagram. The output of the Bundle structure in this case is a cluster of two arrays (x-array and y-array). To signify this, the X-Y Graph icon assumes a cluster symbol.

To display two or more plots on the same X-Y Graph, the x-array/y-array pair associated with a plot is combined in the Bundle structure. The next x-array/y-array pair is also combined in the Bundle, as are all remaining pairs. The outputs of the bundles are applied to the Build Array. Finally, the Build Array output is applied to the X-Y graph terminal.

At each level of this combining process the *data type* is different. The inputs to the Bundle structure are one-dimensional arrays, whereas the outputs from the Bundle structures are clusters of one-dimensional arrays. The output of the Build Array structure is an array of clusters.

4.   The Waveform Chart, Waveform Graph, and the X-Y Graphs have a number of interesting accessories. These are special tools that help you to configure and operate the chart or the graph. Most of these are common to the chart and the graph.

The *Palette* is one such tool. The controls on the palette can be used to choose precision or format, to set the auto scale of the X and Y axes, or to enable zooming.

The *Legend* is another tool. It can be used to identify displayed plots as well as to choose plot characteristics such as point and line style, line width, type of interpolation between points, plot color, and so on.

The *Scrollbar* tool is used to scroll the display on the Waveform Chart. Waveform Chart stores the last 1024 data points. Since the chart window may not display all 1024 points, you may use the scrollbar to show that part of the plot not shown inside the window. The *Chart History...* option in the popup menu allows you to change the 1024 point default value.

The *Cursor Display* tool is available only in the Waveform Graph and the X-Y Graph from the popup menu (*Show>Cursor Display*). You can place a cursor on the plot and move it along the plot. The coordinates of the cursor are displayed with any degree of precision on the X-value, Y-value digital indicators.

Chapter 5
Strings

What Is a String?

A string is a sequence or a group of ASCII characters. When a character is encoded with the ASCII code, it is represented by a sequence of 0's and 1's (the eighth bit is usually the parity bit used for checking transmission errors). The uppercase A, for instance, is represented by the ASCII code 1100001 in binary or 61 hex (usually written 61H). And so it is for most of the other characters on the keyboard: each character is assigned an ASCII code. The ASCII code is one of the most popular codes (another code that is also used quite frequently is the eight-bit EBCDIC code) used in transporting information from one point to another. You will find other characters in the ASCII table, such as SOH, STX, ETX, and ACK, that are used in network protocols.

In LabVIEW a string is a collection of ASCII characters. LabVIEW uses strings for:
- Text messages.
- Instrument control: data and control string messages are transported over the GPIB interface between the instrument and the computer.
- Storing numeric data to disk.

Storing information as strings of ASCII characters makes it easily accessible by other programs.

String Controls and Indicators

String controls and indicators are front panel objects. String controls are used to pass data to the block diagram, and string indicators display data generated by the block diagram.

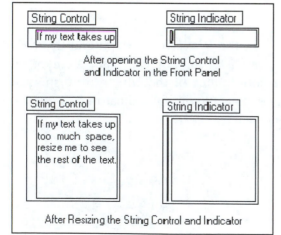

To Open the String Control or String Indicator, choose *String Control* or *String Indicator* from the *String & Table* subpalette of the Controls palette. This illustration shows the appearance of the String Control and the String Indicator as you open them in the Front Panel. Notice that the string indicator has a bar on the left side.

If the text inside the String Control takes too much space and is off the display area, you may resize it. This illustration also shows the String Control and the String Indicator after resizing.

To Activate Scrollbar, pop up on the String Control or String Indicator and choose *Show>Scrollbar*. You can type a lot of text into a small size display area and then use the Scrollbar to scroll through the text.

In the exercises that follow, you will build VIs that use various string functions.

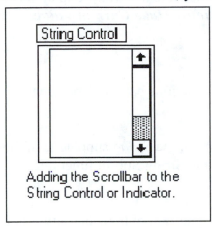

Adding the Scrollbar to the String Control or Indicator.

Exercise 5-1: Using String Functions I

In this exercise you will build a VI that illustrates the use of the string functions **String Subset** and **String Length**, as well as the *String Control* and the *String Indicator*. *Build* the VI whose front panel and Block Diagram are shown in Fig. 5-1.

Front Panel

1.   **String Control/Indicator** objects are in the *String & Table* subpalette of the Controls palette. Open one string control and two string indicators and configure them as follows:

Label: the string control as **Input Text**, one string indicator as **Output Text**, and the other string indicator as **String Subset** using owned labels.

Text: type the text as shown in Fig. 5-1 into the Input Text string control using the *Labeling Tool*. Resize the string control so that the text fits in two lines, as shown in Fig. 5-1.

Resize the *Output Text* string indicator to be the same size as the *Input Text* string control.

Digital Control/Indicator objects are in the *Numeric* subpalette of the Controls palette.

Open two digital controls and one digital indicator and configure them as follows:

Label: the two digital controls as **Length** and **Offset** and the digital indicator as **Num. ASCII Char.** using owned labels.

Representation: **I16** for the two digital controls and the digital indicator.

Vertical Switch is in the *Boolean* subpalette of the Controls palette. Configure the switch as follows:

Label: **STOP** using an owned label.

Mechanical Action: **Latch When Pressed**

Setting: Using the *Operating Tool,* set the switch to *true* (up). Pop up on the switch and choose **Data Operations>Make Current Value Default**.

Block Diagram

2. **While Loop** is in the *Structures* subpalette of the Functions palette. Open one While Loop and resize it as necessary.

The **String Length** and **String Subset** functions are in the *String* subpalette of the Functions palette. Open the String Length and String Subset functions in the Block Diagram.

The **String Length** function has one input and one output. Its output is a count (integer) of the number of ASCII characters in the input string.

String Subset is a function that slices away a portion of the input string. The integer applied to the **offset** input (shown in this illustration)

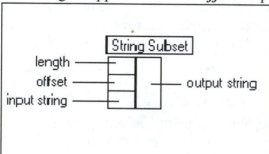

determines where in the input string slicing begins, and the **length** input, also an integer value, specifies the number of characters to slice away. The **output string** contains the portion of the input string that was extracted.

3. *Wire* all objects as shown in the block diagram of Fig. 5-1.

4. When executed, this VI displays in the Output Text string indicator the text that you type in the Input Text string control. It also counts and displays on Num. ASCII Char. The integer values that you enter in the Length and Offset digital controls specify the portion of the Input Text to be extracted. For example, when you set Length = 9 and Offset = 14, "Chapter 7" will be extracted from the input string. The use of the While Loop provides an interactive environment where you make front panel adjustments and see the immediate results while the VI is running.

Run this VI and experiment with different input texts and Offset and Length values.

Save this VI as **Strings 1** and close it.

When you close this VI and open it again later, the text that you entered into the Input Text string control and numeric setting in digital controls will be gone. If it is important to you that this information be retained the next time you open this VI, pop up on the object and choose **Data Operations>Make Current Value Default** before closing the VI.

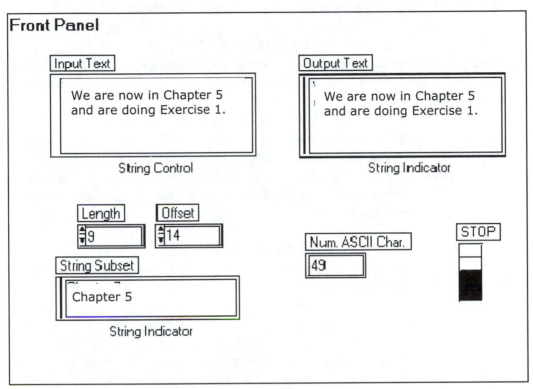

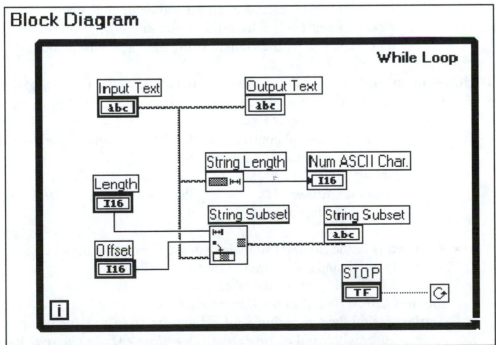

Fig. 5-1 The Front Panel and Block Diagram of Exercise 5-1

Exercise 5-2: Using String Functions II

In this exercise you will build a VI that illustrates the use of the **Concatenate Strings** and **To Fractional** string functions.

Build the VI whose Front Panel and the Block Diagram are shown in Fig. 5-2.

Front Panel

1. **String Control/Indicator** objects are in the *String & Table* subpalette of the Controls palette. Open one string control and one string indicator and configure them as follows:

    Label: the string control as **Dialog** and label the string indicator as **Output String** using owned labels. The *String Control* and *String Indicator* labels are free labels intended as object descriptive information.

    Text: type the text as shown in the front panel of Fig. 5-2 into the Dialog string control using the *Labeling Tool*. Resize the Output String indicator as shown in Fig. 5-2.

    This part is optional, but if you wish, you can separate the numerical value from the rest of the text in the Output String indicator. Notice that the 35.13 value is displayed in the middle of the line that follows the text, as shown in the Front Panel of Fig. 5-2. To accomplish this, place the *Labeling Tool* cursor after the last word "to" in the Dialog string control and press the *Enter* key on the keyboard, then enter approximately 28 spaces using the space bar on the keyboard. To make the text inside the Output String indicator bold, select the string indicator and choose the *Dialog Font* or enter *Ctrl+3* from the keyboard. When you run the VI, the text displayed by the Output String indicator will be bold.

    Digital Control/Indicator is in the *Numeric* subpalette of the Controls palette. Open one digital control and one digital indicator and configure them as follows:

    Label: the digital control as **Pick a Number** and digital indicator as **Num. ASCII Char. in Output String** using the *Labeling Tool* and owned labels.

    Representation: **I16** for digital indicator and **DBL** (default) for digital control.

    Vertical Switch is in the *Boolean* subpalette of the Controls palette. Open one Vertical Switch and configure the switch as follows:

    Label: *STOP* using an owned label.

    Mechanical Action: *Latch When Pressed*

    Setting: Using the *Operating Tool*, set the switch to *true* (up). Pop up on the switch and **choose Data Operations>Make Current Value Default**.

Block Diagram

2. **While Loop** is in the *Structures* subpalette of the Functions palette. Open one While Loop and resize it as necessary.

Concatenate Strings is in the *String* subpalette of the Functions palette. As

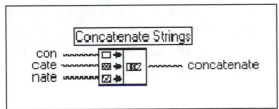

shown in this illustration, the Concatenate Strings function concatenates input strings into a single output string.

Open one Concatenate Strings function.

To Fractional is in the *Additional String to Number* subpalette of the *String*

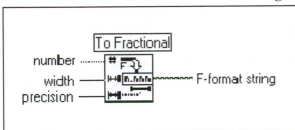

subpalette of the Functions palette. Open one *To Fractional* string function. As shown in this illustration, the inputs to this string function are *number*, *width*, and *precision*.

The ***number*** input is an integer or a floating point numeric value that is to be converted to a string.

Width is an integer representing the width of the output string. It defaults to **0** if you leave it unwired, which will make the output string as small as possible.

Precision is an integer that determines the number of decimal places allowed in the output string for the *number* input.

F-format string is the output that represents the input number converted to a fractional format floating point string.

String Length is in the *String* subpalette of the Functions palette. Open one *String Length* string function.

Square Root is in the *Numeric* subpalette of the Functions palette. Open one Square Root function.

3. **Wire** all objects as shown in the Block Diagram of Fig. 5-2. Notice that a numeric constant, whose value is **2**, is wired to the Precision input of the *To Fractional* function, thus allowing the number input two decimal places in the output string. The Numeric Constant can be found in the *Numeric* subpalette of the Functions palette.

4. As this VI begins execution, the Square Root function calculates the square root of the number that you enter into the *Pick a Number* digital control in the Front Panel. This number is converted to a string and is concatenated with the string from the *Dialog* string control. The resulting string is displayed by the *Output String* string indicator. The number of ASCII characters in the output string is determined by the *String Length* string function and displayed on the *Num. ASCII Char. in Output String* digital indicator.

   The use of the While Loop provides for you once again an interactive environment that allows you to make front panel adjustments and see the immediate results.

   Run this VI. Experiment with different values.

   Save this VI as **Strings 2.vi** and close it.

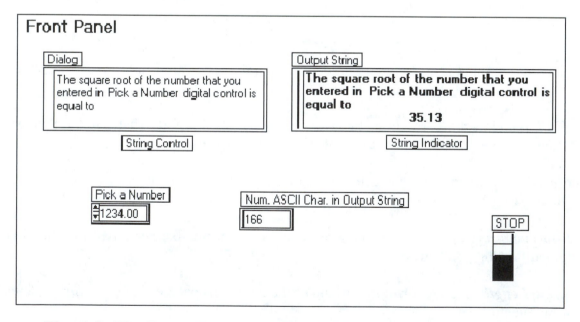

Fig. 5-2 The Front Panel and Block Diagram of Exercise 5-2

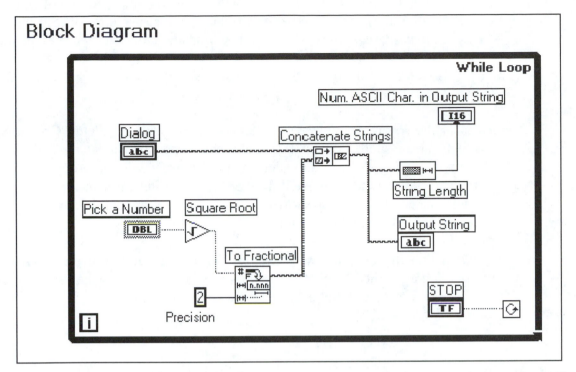

Fig. 5-2 The Front Panel and Block Diagram of Exercise 5-2 (continued)

Exercise 5-3: Using String Functions III

In this exercise you will build a VI that offers another illustration of the **Concatenate Strings** and **To Fractional** string functions.

Build the VI whose Front Panel and Block Diagram are shown in Fig. 5-3.

Front Panel

1. **String Control/Indicator** objects are in the *String & Table* subpalette of the Controls palette. Open two string controls and one string indicator and configure them as follows:

   Label: One string control as ***Dialog 1*** and the other string control as ***Dialog 2***, and label the string indicator as ***Output String*** using owned labels. The *String Control* and *String Indicator* labels are free labels intended for your information only.

   Text: type the text as shown in the front panel of Fig. 5-3 into the Dialog 1 and Dialog 2 string controls using the *Labeling Tool*. Resize the Output String indicator as shown in Fig. 5-3.

   This part is optional, but if you wish, you can make the Output String display more or less in the center of the string indicator and be in bold letters. To do that use the space bar on the keyboard to make nine spaces before the word *Random* and a couple of spaces after the word *Number* in the Dialog 1 string control. Similarly, make two spaces before and two spaces after the word *is* in the Dialog 2 string control. Then choose *Dialog Font* from the *Text Settings* menu in the Control Bar or enter *Ctrl+3* from the keyboard.

Array is in the *Array & Cluster* subpalette of the Controls palette. Open one empty array shell. Pop up inside the shell and choose *String Indicator* from the *String & Table* subpalette of the Controls palette. Drop the string indicator inside the array shell. Resize the string indicator inside the array shell to a smaller size, as shown in the front panel of Fig. 5-3. Also resize the array shell so that you can see at least ten elements.

Label this String Array as **Random Number Array** using the *Labeling Tool* and an owned label. The *String Array* label shown in the Front Panel of Fig. 5-3 is a free label intended for your information only.

Block Diagram

2.   **While Loop** is in the *Structures* subpalette of the Functions palette. Open one While Loop and resize it as necessary.

Concatenate Strings is in the *String* subpalette of the Functions palette. Open one Concatenate Strings function.

To Fractional is in the *Additional String to Number* subpalette of the *String* subpalette of the Functions palette. Open two *To Fractional* string functions. As shown in the front panel of Fig. 5-3, in both of these functions, a value of **0** is wired to the precision input, causing the output string equivalent of the *number* input to have no decimal places.

Wait Until Next ms Multiple is in the *Time & Dialog* subpalette of the Functions palette. In this VI it creates a time delay of 1 second between iterations, as determined by the value of the numeric constant wired to the input of this function. To change the delay time, enter another value into this numeric constant.

Random Number, Increment, Multiply, and **Numeric Constant** functions are in the *Numeric* subpalette of the Functions palette. Open five *Numeric Constants* and one of each of the remaining functions.

Carriage Return ⏎ is in the *String* subpalette of the Functions palette. It causes the first line to be skipped in the output of the Concatenate Strings function. The Dialog 1 text will thus be placed on the second line.

3.   *Wire* all objects as shown in the Block Diagram of Fig. 5-3.

4. As this VI begins execution, the output of the *Concatenate Strings* function is displayed on the *Output String,* a string indicator. The five inputs to the Concatenate Strings function are responsible for forming its output string. *Carriage Return*, the first input, causes a line to be skipped. The *Dialog 1* string control outputs "Random Number" text in the second line. The spacing of this text was discussed in step 1 of this exercise. The third input is the value of i, the iteration terminal, which is first converted to a string by the *To Fractional* function. The fourth input is the text *is* from the Dialog 2 string control. The fifth input is a random number from 0 to 9, which is converted to a string by the *To Fractional* function.

The For Loop executes 10 iterations with a 1 second time delay between iterations. After each iteration the Output String indicator displays *Random Number* followed by the value of *i*, followed by *is,* followed by the value of the random number, as shown in the front panel of Fig. 5-3.

Run the VI. Experiment with different parameter values.

Save this VI as **Strings 3.vi** and close it.

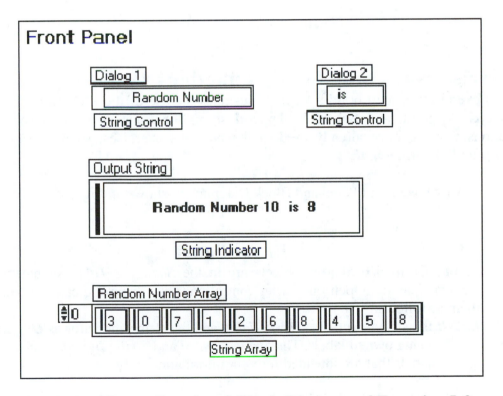

Fig. 5-3 The Front Panel and Block Diagram of Exercise 5-3

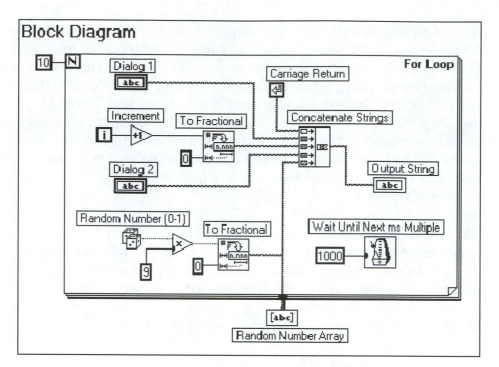

Fig. 5-3 The Front Panel and Block Diagram of Exercise 5-3 (continued)

Exercise 5-4: Using String Functions IV

In this exercise you will build a VI that illustrates the use of the **Pick Line & Append**, **Index & Append**, and **Select & Append** string functions as well as some of the functions that we have already used in the previous exercises, such as *Concatenate Strings* and *To Fractional*.

Build the VI whose Front Panel and Block Diagram are shown in Fig. 5-5.

Front Panel

1. **String Control/Indicator** objects are in the *String & Table* subpalette of the Controls palette. Open one string control and one string indicator and configure them as follows:

    Label: the string control as ***Test Dialog*** and the string indicator as ***Output String*** using owned labels. The *String Control* and *String Indicator* labels are free labels that are intended for your information only.

    Vertical Pointer Slide is in the *Numeric* subpalette of the Controls palette. Open one *Vertical Pointer Slide* and configure it as follows (see Fig. 5-4):

    Pop up on the slide (vertical portion) and choose ***Text Labels*** from the popup menu.

Erase *min* inside the digital control window, click with the labeling tool inside the digital control window, and type **Motor Speed**. The digital control window should be resized so that you can see the entire text being typed. After you type *Motor Speed* and press enter, it will appear as the option at the bottom of the slide.

Pop up inside the digital control window and choose **Add Item After** from the popup menu. The pointer will move up to the next position along the slide. Click with the labeling tool inside the digital control window and type **Load Voltage**. After clicking on the **enter** button, the Load Voltage will appear as the second option from the bottom.

This step is optional. Resize the Vertical Pointer Slide to a larger size. Pop up on the vertical part of the slide and choose **Fill Options**.

Select the **Fill to Minimum** option from the palette.
Using the *Operating Tool* click on the increment/decrement arrow buttons on the left side of the digital control until you see *max*.
Erase *max* and type **Impedance** using the *Labeling Tool* and click on the *Enter* button.

Pop up inside the digital control window next to the slide and choose **Add Item Before** from the popup menu. Click with the labeling tool inside the digital control window and type **Load Voltage**. Click on the *Enter* button.

This step is optional. Pop up on the vertical portion of the slide and choose **Show>Text Display** from the popup menu. This will remove the digital control.

Vertical Toggle Switch is in the *Boolean* subpalette of the Controls palette. Open one Vertical Toggle Switch and configure it as follows:
Resize: As desired.
Mechanical Action: **Switch When Pressed**
Labels: **AC/DC Set** using the labeling tool and an owned label. **DC** and **AC** on the side of the switch are free labels.

Vertical Switch is in the *Boolean* subpalette of the Controls palette. Open one Vertical Switch and configure it as follows:
Label: **STOP** using an owned label.
Mechanical Action: **Latch When Pressed**.

Setting: Using the *Operating Tool,* set the switch to *true* (up). Pop up on the switch and choose **Data Operations>Make Current Value Default**

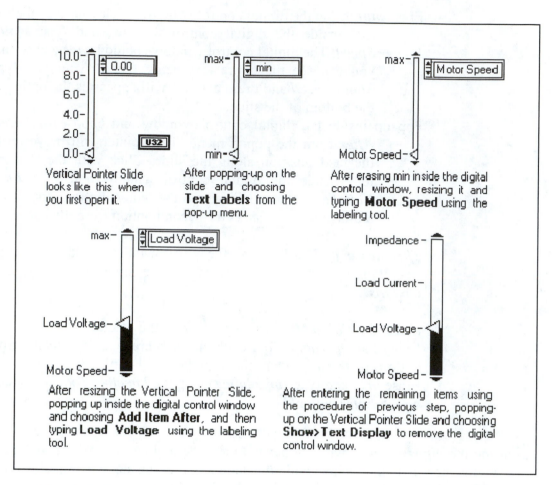

Fig. 5-4 The Steps in Configuring the Vertical Pointer Slide

Block Diagram

2.   **While Loop** and the **Case Structure** are in the *Structures* subpalette of the Functions palette. Open one While Loop and one Case structure and resize them as necessary.

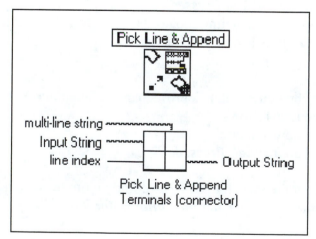

The **Pick Line & Append** string function is in the *String* subpalette of the Functions palette. Open one Pick Line & Append string function. The icon for this function and its connector with four terminals is shown in this illustration.

The value of the *line index* determines which line of the *multi-line string* is appended to the *input string*. The input string and the appended line will appear in the output string.

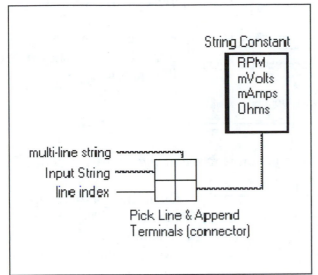

Next, open a **String Constant**, which you will find in the *String* subpalette of the Functions palette, enlarge it, click with the *Labeling Tool* inside the string constant, and begin typing.

As shown in this illustration, type **RPM** on the first line, and press the *Enter* key on the keyboard; then type **mVolts** on the second line, and so on. Each line must be terminated by pressing the *Enter* key.

Wire the string constant to the *Output String* terminal on the connector, as shown here.

Select & Append string function is in the *String* subpalette of the Functions palette. Open one *Select & Append* string function. The icon for this function and

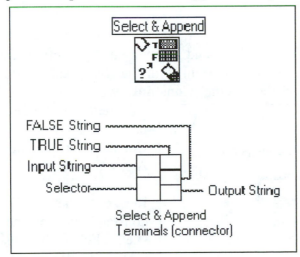

its connector with five terminals is shown in this illustration.

The **Selector** input must be a Boolean object. If the Selector input is *true*, then the string wired to the *true* String terminal will be appended to the *Input String*. If the Selector input is *false*, then the string wired to the *false* String terminal will be appended to the Input String.

Thus, the *Output String* includes the Input String and either the *false* or the *true* string appended to it.

Open two **String Constants**. Type **AC** inside one of them and **DC** inside the other.

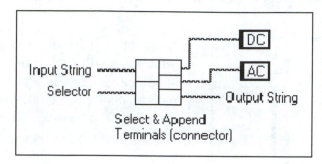

Wire the *DC* string constant to the *true* String terminal and the *AC* string constant to the *false* String terminal, as shown in this illustration.

The **Index & Append** string function is in the *String* subpalette of the Functions palette.

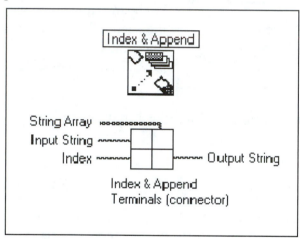

Open one *Index & Append* string function. The icon for this function and its connector with five terminals is shown in this illustration.

The value of the integer *Index* input determines which element is picked from the *String Array* input and appended to the *Input String*. The *Output String* contains the combined string.

Next, create a String Array for the Index and Append string function. Open the **Array Constant** empty shell. Array Constant is in the *Array* subpalette of the Functions palette. Then open the **String Constant**, which is in the *String*

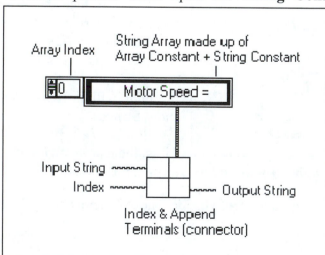

subpalette of the Functions palette. Drop the String Constant inside the empty Array Constant shell.

Resize (enlarge) the string constant inside the array shell, click with the labeling tool inside the array constant, and begin typing.

As shown in this illustration, first insert about 9 or 10 spaces using the space bar on the keyboard and then type ***Motor Speed*** =. Click with the *Operating Tool* on the Array Index increment or decrement arrow button to advance the String Array to element 1 and type ***Load Voltage*** = . In element 2 type ***Load Current,*** and finally in element 3 type ***Impedance*** = .

Wire the String Array that you just created to the *String Array* input terminal on the Index & Append connector, as shown in the illustration.

Concatenate Strings is in the *String* subpalette of the Functions palette. Open two Concatenate Strings functions. Resize them so that one of them has *three inputs* and the other *four inputs*.

To Fractional is in the *Additional String to Number* subpalette of the *String* sub-palette of the Functions palette. Open one *To Fractional* string function.

The **Carriage Return** character is in the *String* subpalette of the Functions palette. Open four Carriage return characters. They will be used as inputs to Concatenate Strings.

The **Random Number** generator, **Multiply** function, and **Numeric Constant** are in the *Numeric* subpalette of the Functions palette. Open one **Random Number** generator, one *multiply function*, and three *numeric constants*. Two numeric constants have a value of ***1000***, and the one wired to the *Precision* input of the *To Fractional* function has a value of ***1***.

Wait Until Next ms Multiple is in the *Time & Dialog* subpalette of the Functions palette. In this VI it creates a time delay of 1 second between iterations, as determined by the value of the numeric constant wired to the input of this function. To change the delay time, enter another value into this numeric constant.

3. **Wire** all objects as shown in the block diagram of Fig. 5-5. Notice that *Numeric Case 0* contains the *Output String* (front panel string indicator) terminal and *Numeric Cases 1, 2,* and *3* contain the *Local Variable* associated with the *Output String*. For more information on creating a local variable, refer to the Numeric Case Structure in LabVIEW in Chapter 2.

    Remember that the Case structure is Boolean by default, but once you wire the *Test Function* terminal to the Case selector [?] terminal, the Case structure changes to Numeric Case. To advance to Case 1, click on the arrow button in the case window that is at the top of the case structure. To advance to Numeric Case 2 or 3, pop up inside the case window and choose *Add Case After* from the popup menu.

4. When you run this VI, the While Loop allows you to interact with the front panel controls and observe immediate results.

Use the *Operating Tool* to change the front panel settings. You can even change the text while the VI is running by clicking with the operating tool inside the string control, typing new text or deleting old text, and then clicking on the *Enter* button.

As soon as this VI begins execution, the Output String indicator will display ***This Test is now in progress...*** followed by the Test Function name and its value, which changes once every second, followed by the units and the AC or DC qualifying parameter.

The *Vertical Pointer Slide* is configured with four settings. When the slide is set to *Motor Speed*, its terminal in the block diagram has a value of **0**. Its *Load Voltage*, *Load Current*, and *Impedance* settings produce values of 1, 2, and 3, respectively, for its terminal in the Block Diagram. Notice that the *Test Function* terminal in the block diagram is wired to the *Index* input of the *Index & Append* and *Pick Line & Append* string functions, and also to the *Selector* terminal of the Numeric Case structure. Consequently, if you set the Test Function to *Load Voltage*, the Index & Append function picks the *Load Voltage* = , the second line in the String Array constant, and appends it to the output of the Concatenate Strings function. So far this will create the text ***System Test is now in progress...*** followed by the ***Load Voltage* =** text.

What happens next is the formation of a random number between 0 and 1000. This random number is converted to a string by the *To Fractional* function and applied to the *String Input* terminal of the *Pick Line & Append* function, whose index input is still 1 (assuming that Test Function is set to Load Voltage). The index of 1 causes the Pick Line & Append function to pick the second line, *mVolts,* from the string constant and append it to the Input String. As a result, the Output String of the Pick Line & Append function will contain ***System Test is now in progress...*** on one line and ***Load Voltage* =** followed by a numeric value and ***mVolts*** on the line below. All this is displayed by the *Output String* string indicator on the front panel.

The string created thus far is applied next to the *Select & Append* function, whose Selector terminal gets a *true* or a *false* input from the *AC/DC Set* terminal (front panel toggle switch). If set to *true* (up position), *DC* will be appended to the incoming string.

The resulting string is finally applied to the *Output String* local variable in Numeric Case 1. The Numeric Case 1 will be executed because the Test Function being set to Load Voltage this time has a value of 1.

We use Numeric Case structure in this VI with four cases, and each case has the same object, the Output String terminal. It may seem puzzling at first why one would want to do that, since there is a minor tactical problem. Motor Speed, for instance, must have RPM as its units, but the *Pick & Append* function will also append *AC* or *DC*, depending on the setting of the *AC/DC Set*, the front panel Boolean control. The same is true of *Impedance,* whose Ohm units will also be appended *AC* or *DC*. This does not make sense.

The Numeric Case solves this dilemma. The strings associated with the Motor Speed and Impedance selections will be applied to Numeric Cases 0 and 3, to be displayed on the Output String indicator before AC or DC is appended. The Load Voltage and Load Current strings will be appended AC or DC and then applied to Numeric Cases 1 or 2, to be displayed on the Output String indicator. The numeric value of the Test Function terminal makes the decision about which one of four cases is to be executed.

Run this VI assuming that the Test Dialog string control contains the appropriate text. Experiment by changing the control settings or changing text.

Save this VI as **Strings 4.vi** and close it.

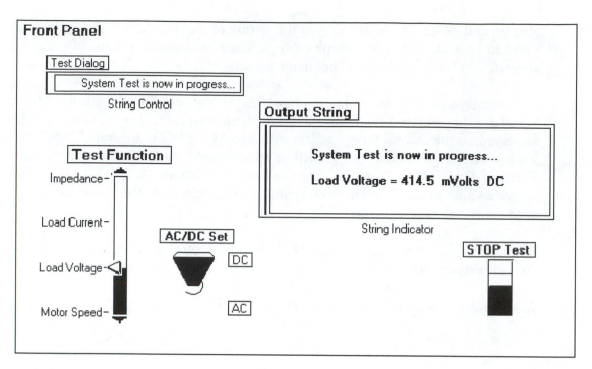

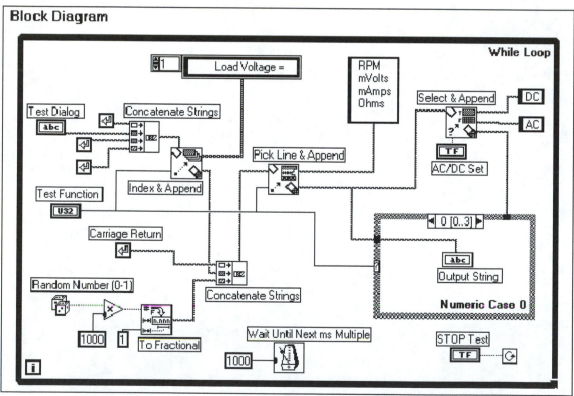

Fig. 5-5 The Front Panel and Block Diagram of Exercise 5-4

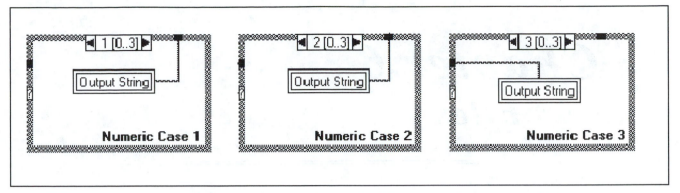

Fig. 5-5 The Front Panel and Block Diagram of Exercise 5-4 (continued)

Summary

1.   A string is a collection of ASCII characters.

2.   Strings are used by LabVIEW for text messages and instrument control and for storing data to disk.

3.   A string indicator is used to display a string generated by the block diagram.

4.   A string control is used to pass string data to the block diagram.

5.   A string constant is an object used in the block diagram. It stores a string that seldom has to be changed.

6.   The Function palette includes a String subpalette. Here one can find a variety of functions that operate on and manipulate strings.

7.   A string array is an array of string indicators, string controls, or string constants.

Chapter 6
Files

The text as well as data that your VI generates can be saved to a file on the disk. In LabVIEW special procedures must be followed to write text or data to a file on the disk or to retrieve information from the disk.

Writing to a New File

A file that does not exist and cannot be found in any directory is a *New File*. The procedure for saving data to a New File is illustrated in Fig. 6-1. The first object that you create contains complete **path** information. The path tells LabVIEW the directory and the name of the file where the data is to be saved.

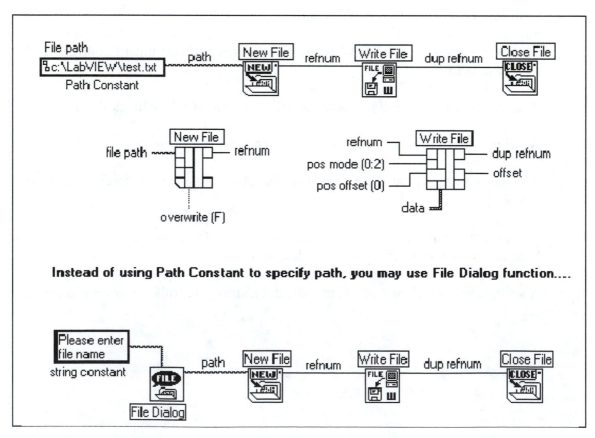

Fig. 6-1 Block Diagram for Writing to a New File

File Path is created in one of two ways: using the *Path Constant* , which can be found in the *File Constants* subpalette of the *File I/O* subpalette of the Functions palette, or using the *File Dialog* function shown in Fig. 6-1.

The *File Dialog* function is easy to use. Type a message such as *Please Enter the Name of Function to Create* inside a string constant and wire the string constant to the *Prompt* input on the File Dialog connector. The function will work even if you didn't wire the prompt.

When the File Dialog function executes, it gives you access to directories and offers you the opportunity to type the name of the new function. Once the path has been created by the File Dialog function, it will be available as output at its *path* terminal.

The File Dialog file/directory selection has the following restrictions that can be wired to the *Select Mode* terminal:
  0 Select existing files
  1 Select a new file
  2 Select an existing or new file
  3 Select an existing directory
  4 Select a new directory
  5 Select an existing or new directory
If the Select Mode terminal is unwired, it defaults to **2**, thus allowing you to select either an existing file or a new file.

As shown in Fig. 6-1, the path just created is applied to the **New File** function. The New

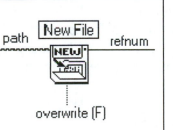

File function creates the file specified in the input path and opens this file for writing or for reading. In this section we are concerned with writing data to the file.

The **Overwrite** input, whose default value is *false*, is shown in this illustration to the left. It offers the user the ability to overwrite *(true)* or not to overwrite *(false)* the file specified by the path. If, for example, the file specified by the path already exists and you set the Overwrite input to *true*, then the New File function will erase the contents of the file, thus making it a new file. The New File function also outputs a **refnum** that provides the reference number to other functions about the opened file.

The **Write File** function is the next block in line in Fig. 8-1. It operates on the file specified by the input refnum. Basically, it takes the data in string format that is applied to the *data* terminal and writes that data to the file specified by the input refnum.

Pos mode and **pos offset** together specify where the write operation begins:

pos mode = 0 and the value of *pos offset* specify how many characters to skip from the beginning of the file before writing new data to the file. For example, if you set pos offset = 4 and pos mode = 0, the write operation will begin after the first four characters in the file.

pos mode = 1 and the value of *pos offset* specify how many characters to skip after the end of the file before writing new data to the file.

pos mode = 2 and the value of *pos offset* specify where to begin the write operation relative to the current position of the file mark. It is possible to set the file mark at points in the file other than at the beginning or the end of the file.

If pos offset is not wired, pos mode defaults to 0. If pos offset is wired, pos mode defaults to 2 and the offset is measured relative to the current file mark.

What this means for the simple file I/O operations that we will consider in this chapter is that if you want to:

> **Overwrite** *or replace the old file data with new data, then don't wire anything to pos mode and pos offset inputs. You will get the same effect as if you wired **0** to both inputs.*
>
> **Append** *the new data to the end of the file, then wire a **1** to the pos mode input and **0** or nothing to the pos offset input.*

The **offset** output terminal on the Write File connector (shown in Fig. 6-1) specifies the number of bytes or ASCII characters to the end of the file.

Dup refnum is the duplication refnum assigned to the current open file for other file functions to use.

The **Close File** function is the last block in the line illustrated in Fig. 6-1. It closes the file specified by the dup refnum. When you open a file, you must remember to close it at the end. This is your responsibility because LabVIEW will not do it by default. High-level languages such as C also have this type of requirement.

Writing to an Existing File

If you compare Figures 6-1 and 6-2, they appear almost the same. The only difference is that the *Open File* function shown in Fig. 6-2 replaces the New File function of Fig. 6-1.

The **Open File** function requires the complete path of the file to be opened as its input and generates a *dup refnum* for other functions to use. It has the following restrictions on opening a file that you can wire to the *open mode* terminal:

- 0 Read and write
- 1 Read only
- 2 Write only
- 3 Write only (truncate first)

If this terminal is not wired, it defaults to **0**, allowing the file to be opened for both read and write operations.

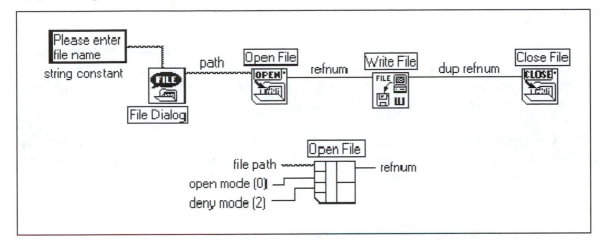

Fig. 6-2 Block Diagram for Writing to an Existing File

The Open File function also has the following *deny mode* restrictions:

0 Deny Read or Write
1 Permit Read, Deny Write
2 Permit Read and Write

If not wired, the *deny mode* defaults to **2**.

Reading from a File

Any file that has been saved on the disk can also be opened and its contents retrieved or read. As shown in Fig. 6-3, you still have to specify the path of the file that you wish to open, open the file, and, when you are done reading, close the file.

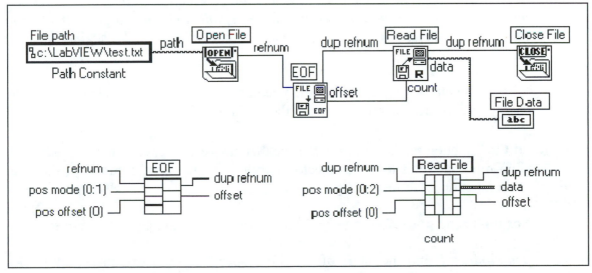

Fig. 6-3 Block Diagram for Reading from a New File

File Path specifies the location of the file that you want to open. The *Path Constant* or *File Dialog* function can be used to accomplish this, as discussed in the preceding section.

The ***Open File*** function opens the file specified by the path string applied to its input and generates a refnum to be used by other file functions that will also operate on the opened file. You might have noticed in examining Fig. 6-2 that, whether you write to or read from a file, the same Open File function is used.

The **EOF** function sets and returns the *EOF, the end of file mark*, for the file that has been specified by the input refnum. Refer to the terminal configuration on the connector of the EOF function, shown in Fig. 6-3.

Pos mode and **pos offset** together specify where the EOF is to be placed:

> *pos mode = 0* sets the EOF mark at the beginning of the file plus the value of *pos offset*. For example, if pos offset = 15, the EOF mark will be set at the sixteenth ASCII character, counting from the beginning of the file.
>
> *pos mode = 1* sets the EOF mark to the end of the file plus the value of *pos offset*. Suppose that the length of the file is 1000 ASCII characters and pos offset = 5. These settings (also pos mode = 1) will place the EOF mark at the 1006th character position in the file, counting from the beginning of the file.
>
> *pos mode = 2* sets the EOF mark at the current location of the file mark plus the value of the *pos offset*.
>
> *If pos offset is left unwired, it defaults to **0** and pos mode defaults to **1**. If you wire a value to the pos offset terminal, pos mode defaults to **0** and the placement of the EOF mark is determined by the pos offset value relative to the beginning of the file.*
>
> The ***Offset*** output specifies the number of bytes or ASCII characters to the end of the file, counting from the beginning of the file.

The **Read File** function reads data from the file that has been opened and specified by the dup refnum. Refer to the connector and the terminal configuration of the Read File function as shown in Fig. 6-3.

Pos mode and **pos offset** together specify where the read operation begins:

pos mode = 0 and the value of *pos offset* specify how many characters to skip from the beginning of the file before reading data from the file. For example, if you set pos offset = 4 and pos mode = 0, the read operation will begin after the first four characters in the file.

pos mode = 1 and the value of *pos offset* specify how many characters to skip after the end of the file before reading new data to the file.

pos mode = 2 and the value of *pos offset* specify where to begin the read operation relative to the current position of the file mark. It is possible to set the file mark at points in the file other than at the beginning or the end of the file.

If *pos offset* is not wired, *pos mode* defaults to **0**. If *pos offset* is wired, *pos mode* defaults to **2** and the offset is measured relative to the current file mark.

The *Data* output terminal provides the contents of the file read in a string format.

Offset is the output in bytes that indicates the length of the file from the beginning to the current file mark.

The *Count* input specifies the number of bytes or ASCII characters to read.

Exercise 6-1: Files I (Simple Write and Read)

In this exercise you will build a VI that illustrates how a file can be saved on the disk and read from the disk. The VI is designed so that you can instantly see what has been saved on the disk. All of the file functions discussed above are used by this VI.

Build the VI whose Front Panel and Block Diagram are shown in Fig. 6-4.

Front Panel

1. **String Control/Indicator** objects are in the *String & Table* subpalette of the Controls palette. Open one string control and one string indicator and configure them as follows:
   Label: The string control as **Text to File** and the string indicator as **Text from File**
   Text: Using the *Labeling tool* type **Text line number** in *File String* control. Resize the string indicator as shown in Fig. 6-4.

   Digital Control/Indicator objects are in the *Numeric* subpalette of the Controls palette. Open three digital controls and one digital indicator and configure them as follows:
   Label: the three digital controls as **Number to file, Pos offset,** and **Pos mode;** and the digital indicator as **Count**.
   Representation: **I16** for the three digital controls and the digital indicator.

Block Diagram

2. **Sequence Structure** is in the *Structures* subpalette of the Functions palette. Open one Sequence Structure and resize it to fill all objects, as shown in Fig. 6-4.

Sequence Frame 0

Path Constant in the *File Constants* subpalette of the *File I/O* subpalette of the Function palette. Open one Path Constant and type the complete path to the file where text will be saved. As shown in Fig. 8-4, a suggested path for the file test.txt is A:\test.txt. In your computer this path may be different.

Concatenate Strings and **End of Line** are in the *String* subpalette of the Functions palette and **To Fractional** is in the *Additional String to Number Functions* subpalette of the *String* subpalette. Open one each of the above functions.

The **Write File** and **Close File** functions are in the *File I/O* subpalette of the Functions palette and the **Open File** function is in the *Advanced File Functions* subpalette of the *File I/O* subpalette. Open one each of the above functions.

Sequence Frame 1

Sequence Frame 1 contains objects used by Sequence Frame 0. You will also need the **EOF** function, which can be found in the *Advanced File Functions* subpalette of the *File I/O* of the Functions palette.

3. *Wire* all objects in Sequence Frames 0 and 1, as shown in Fig. 6-4.

4. In this VI, the test.txt file is created and one line of text is written to the file. Then the file is opened, and the contents of the file are read and displayed on a string indicator.

In Sequence Frame 0, the text supplied by the *Text to file* string control is combined with the number from the *Number to file* from the digital control in the *Concatenate Strings* function and the resulting output is applied to the data input of the *Write File* function. The last input to the Concatenate Strings function, the *End of Line* control character, terminates the text line and starts the new line. Since the Concatenate Strings function accepts only strings at its input, the numeric value from the *Number to file* terminal must first be converted to string format before being applied.

The *pos offset* and *pos mode* Front Panel digital controls allow you to experiment with the effect that they have on the *Write File* function.

As always the file must first be opened by the *Open File* function, which requires the complete function path at its input. You can use the *File Dialog* function, *path control* or the *path constant* to generate the path. In this exercise the path constant has been used. And as always you are required to close the file after the write operation is completed. This is done by the Close File function.

Note that in this exercise it is presumed that the file *Test.txt* already exists. If it doesn't, then you can create it either manually or by replacing the *Open File* in sequence frame 0 by the *New File* function. As shown in this illustration, wire the Boolean constant to the *Overwrite* terminal of the New File function. You will find that this constant is FALSE by default. Click with the operating tool to change the state of this constant to TRUE. After you run the VI once, the New File function will create the file specified in the path constant for you.

In Sequence Frame 1 the Open File function opens the file that you have written to in Sequence Frame 0, reads the contents of the file, and displays the file contents on the *Text from file* string indicator. The *EOF* function is used here to produce the *offset* value, which is the length of the file in bytes. This offset value is displayed on the Count digital indicator and it is also wired to the *Count* input terminal of the *Read File* function, thus instructing the Read File function on the number of bytes to read. As usual the file is closed by the *Close File* function after the read operation is complete.

Run this VI by first entering the **Text line number** into the *Text to file* string control and **1** into the *Number to file* digital control. Make sure that the path constant in frames 0 and 1 has the proper path. Also, set *pos mode* and *pos offset* to **0**. Run the VI several times, each time advancing the *Number to file* count by 1. Observe what you probably expected: the file is overwritten each time.

Now set the pos mode = **1**, leaving pos offset at **0** and repeat the above procedure. Notice that this time the new information is appended to the end of the file.

Save this VI as **Files 1.vi** and close it.

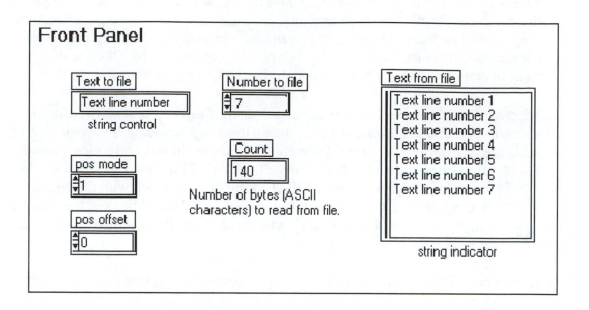

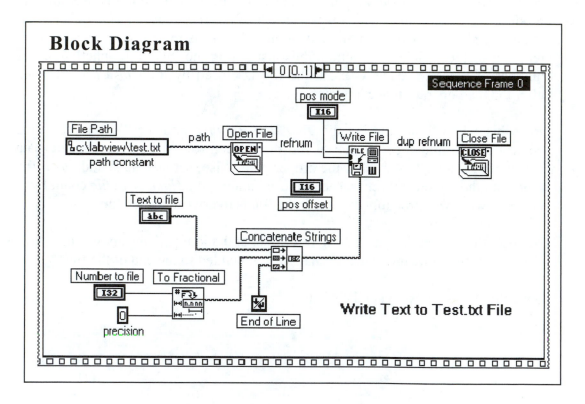

Fig. 6-4 Front Panel and Block Diagram of Exercise 6-1

Block Diagram

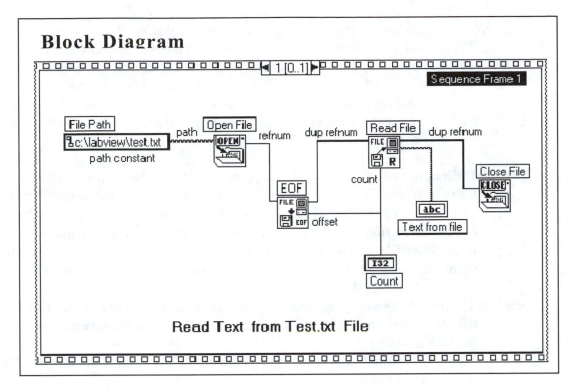

Fig. 6-4 Front Panel and Block Diagram of Exercise 6-1 (continued)

Exercise 6-2: Files II (Collect, Store, and Retrieve Data)

In this VI we will involve more of the library resources in collecting an array of data points, storing them to a file, and retrieving this data. The data collected will be, for illustration purposes, that of sine and cosine waves. The process of data generation will be done in slow motion so that you will be able to view each line of data written to the file on the string indicator and also on the waveform chart.

The VI that you are to build is shown in Fig. 6-5. The Block Diagram consists of four *Sequence* frames. Sequence Frame 0 includes *Numeric Cases* 0, 1, and 2, and Sequence Frame 1 also includes three frames of Numeric Case structure. Notice that the additional Numeric Case frames are shown immediately below the Sequence frame in Fig. 6-5. Following is the detailed, step-by-step procedure for building this VI.

Front Panel

1. **Waveform Chart** is in the *Graph* subpalette of the Controls palette. Open one Waveform Chart, resize it as necessary, and configure it as follows:
   Labels: **Data to file chart** is an owned label and **Angle (degrees)** is a free label.
   Auto Scale X and *Y*: disable
   Y scale: max = 1 and min = −1
   X scale: max = 72 and min = 0
   Palette: disable

Legend: Resize for two plots, and using the *Labeling Tool*, delete Plots 0 and 1 and replace them with **sine** and **cosine**, as shown in Fig. 6-5. Pop up on the inside Legend *sine* window and choose the color you wish for the sine plot. Choose another color for the *cosine* plot by popping up inside the Legend cosine window. Instead of color, you may use a different *point style* for the plots, as shown in the Front Panel of Fig. 6-5.

String Indicator is in the *String & Table* subpalette of the Controls palette. In LabVIEW version 6, it is in the String and Path subpalette of the Controls palette. Open two string indicators and label them using an owned label as **Data line to file**. Label the other string indicator with an owned label as **Data from file**. Resize it and add the *Scrollbar* by popping up inside the string indicator and then choosing **Show>Scrollbar** from the popup menu.

Digital Control/Indicator objects are in the *Numeric* subpalette of the Controls palette. Open one digital control and one digital indicator and configure them as follows:
Label: the digital control as **Num data pts.** and digital indicator as **File length (bytes)** .
Representation: **I16**

File Path Control is in the *Path & Refnum* subpalette of the Controls palette. In LabVIEW version 6, it is in the String and Path subpalette of the Controls palette. Open one File Path Control and, using the *Labeling Tool*, type the complete path of the file. A suggested path is shown in the Front Panel of Fig. 6-5. Label the file path control as **File Path** using an owned label.

Vertical Pointer Slide is in the *Numeric* subpalette of the Controls palette. Open one Vertical Pointer Slide and label it as shown in Fig. 6-5. For more information on configuring the vertical pointer slide, see Fig. 5-4 of Chapter 5. Label the slide using an owned label as **File Status**.

Block Diagram

2. **Sequence Structure** is in the *Structures* subpalette of the Functions palette. Open one Sequence structure and resize it as necessary. You will notice that as you open the sequence frame, there is no frame window at the top to identify this frame as Frame 0. Don't worry about that right now. This unmarked frame is Frame 0. This situation will be corrected when you get to the next frame.

Sequence Frame 0

Open Case Structure inside the Sequence Frame 0 and resize it. You will find Case Structure in the *Structures* subpalette in the Functions palette.

Wire the **File Status** terminal to the *Selector* terminal [**?**] of the Case and observe the Boolean Case change to Numeric Case. You will next build three Numeric Case frames inside the Sequence Frame 0 as follows:

Numeric Case 0

Open one **Open File** function. It is in the *Advanced File Functions* subpalette of the *File I/O* subpalette of the Functions palette. *Wire* all objects as shown in Sequence Frame 0 of Fig. 6-5.

Numeric Case 1

Switch to Numeric Case 1 by clicking on the arrow button in the Case window.

Open one **Open File** function and one **EOF** function, which you will find in the *Advanced File Functions* subpalette of the *File I/O* subpalette of the Functions palette inside Numeric Case 1. *Wire* all objects as shown in Numeric Case 1 of Fig. 6-5.

Numeric Case 2

Switch to Numeric Case 2 by clicking inside, Case window and choose ***Add Case After*** from the pop up menu.

Open one **New File** function, which you will find in the *Advanced File Functions* subpalette of the *File I/O* subpalette of the Functions palette. ***Wire all objects*** as shown in Numeric Case 2 of Fig. 6-5. Notice that the Boolean constant set to *true* is wired to the *overwrite* input terminal of the New File function. You will find the Boolean constant in the *Boolean* subpalette of the Functions palette. Use the operating tool to change its state from *false* to *true*.

Sequence Frame 1

Switch to Sequence Frame 1 by popping up on the frame border and choosing ***Add Frame After***. Notice that the frame window appears and identifies this frame as Frame 1, with Frame 0 behind it.

For Loop is in the *Structures* subpalette of the Functions palette. Open one For Loop inside Sequence Frame 1, resize it, and ***wire*** the ***Num. data pts.*** terminal to the Loop counter terminal *N*, as shown in Sequence Frame 1 of Fig. 6-5.

All of the following objects are to be placed inside the For Loop:
Sine and **Cosine** functions are in the *Trigonometric* subpalette of the *Numeric* subpalette of the Functions palette. Open one of each.

Concatenate Strings is in the *String* subpalette of the Functions palette. Open one Concatenate Strings function and resize it to accommodate eight inputs.

To Fractional is in the *Additional String to Number* subpalette of the *String* subpalette of the Functions palette. Open three *To Fractional* string functions. As shown in Fig. 6-5 a numeric constant **2** is wired to the *precision* input of two of these functions and **0** to the third. Recall that the precision input decides on the number of decimal places that the numerical value will have in the output.

Wait Until Next ms Multiple is in the *Time & Dialog* subpalette of the Functions palette. In this VI it creates a time delay of 0.5 second between iterations, as determined by the value of the numeric constant wired to the input of this function. To change the delay time, enter another value into this numeric constant.

Tab and **End of Line** control characters are in the *String* subpalette of the Functions palette. Open one of each.

The **Bundle** structure is in the *Cluster* subpalette of the Functions palette. Open one Bundle structure. In this VI the Bundle combines the sine and cosine waves for display on the *Data to file chart*.

Multiply, Divide, and **Numeric Constant** are in the *Numeric* subpalette of the Functions palette. Open two Multiply functions, one Divide function, and five Numeric Constants.

Open one **Pi** π constant, which is in the *Additional Numeric Constants* subpalette of the *Numeric* subpalette of the Functions palette.

Wire all objects inside the For Loop as shown in Sequence Frame 1 of Fig. 6-5. This does not include the Numeric Case, which you will build after wiring. Follow the suggested object layout shown in Fig. 6-5, leaving room for the case structure in the lower right corner of the For Loop.

Open **Case Structure** inside the For Loop, resize it, and *wire* the *File Status* local variable to the selector terminal [?] of the Case structure.

Numeric Case 0
Open the **Write File** function, which you will find in the *File I/O* subpalette of the Functions palette.
Wire:

0 and 1 numeric constants to the *pos offset* and *pos mode* input terminals of the Write File function, respectively.

Concatenate Strings output to the *data* input terminal of the Write File function.

Refnum from sequence local input to the *refnum* input terminal of the Write File function. Notice that the tunnel , a black rectangle, is formed   when you wire through the wall of the For Loop and the Numeric Case 0.

Numeric Case 1

Switch to Numeric Case 1 by clicking on the arrow button in the Case window.

Open the **Write File** function, which you will find in the *File I/O* sub-palette of the Functions palette.

Wire:

> from the tunnel (data) to the *data* input terminal of the Write File function.

> from the tunnel (refnum) to the *refnum* input terminal of the Write File function.

Numeric Case 2

Switch to Numeric Case 2 by clicking inside the Case window and choosing ***Add Case After*** from the popup menu.

Open the **Write File** function, which you will find in the *File I/O* sub-palette of the Functions palette.

Wire:

> from the tunnel (data) to the *data* input terminal of the Write File function.

> from the tunnel (refnum) to the *refnum* input terminal of the Write File function.

Sequence Frame 2

Switch to Sequence Frame 2 by popping up anywhere on the frame border and choosing ***Add Frame After*** from the popup menu.

Open one **Close File** function, which is in the *File I/O* subpalette of the Functions palette. *Wire* the *sequence local* (refnum) to the *refnum* input terminal of the Close File function.

Sequence Frame 3

Switch to Sequence Frame 3 by popping up anywhere on the frame border and choosing ***Add Frame After*** from the popup menu.

Create the **File Path** local variable. Open local variable object from the Structures subpalette of Functions palette. Right click on the local variable icon and choose File Path from the Select Item option.

The *Open* **File** and **EOF** functions are in the *Advanced File Functions* subpalette of the *File I/O* subpalette of the Functions palette. Open one of each.

The **Read File** and **Close File** functions are in the *File I/O* subpalette of the Functions palette. Open one of each.

Wire all objects as shown in Sequence Frame 3 of Fig. 6-5.

3.   The sequence structure used in this VI ensures that various operations are executed in the proper order. In Frame 0 the Vertical Pointer Slide specifies the Numeric case number to be executed. The local variable of the Vertical Pointer Slide appears also in Sequence Frame 1 and also controls the Numeric Case in that frame. The three Numeric Cases in Sequence Frames 0 and 1 decide whether an existing file will be opened or a new file will be created. If a New File is created, the new data will be written from the beginning of the file. But if an existing file is opened, then the user is given an option to overwrite or append new data. The Vertical Pointer Slide is used to select these options.

In Sequence Frame 0 the sine and cosine data points are collected in a column format. The value of sine and cosine is calculated in 5-degree increments. The first column includes the value of the angle in degrees and the second and third columns contain the sine and cosine values. Two tabs between columns separate the columns, and the End of Line control character terminates the line of data and starts a new line. With each iteration of the For Loop one line of data is stored in the file, and displayed on the *Data line to file* string indicator and as two points on the *Data to file chart*. The *Num. data pts.* digital control specifies the number of data lines. The *Wait Until the Next ms Multiple* provides the time delay between loop iterations, thus slowing down the operation so that you can follow it. The time delay is set to 0.5 seconds in this VI. You can change that to another value. Upon the completion of the write operation, the file is closed in Sequence Frame 2.

In Sequence Frame 3 the file containing the data that was written in the preceding frames is now opened for read operation. The data extracted from the file is displayed on the *Data from file* string control, and the number of ASCII characters in the file is displayed on the File Length digital indicator. Notice that the file length information is provided by the *offset* output terminal of the EOF function. It is displayed and also applied to the *count* input, thus specifying the number of bytes starting at the beginning of the file that the *Read File* function should read.

In short, the four sequence frames specify the complete file path, open the file or create a new file, write to the file, and then read the content of the file. As you view the execution of this VI, you will see the data lines being stored and displayed on the chart in slow motion. At the end of execution you will see on the string indicator the actual content of the file that was opened.

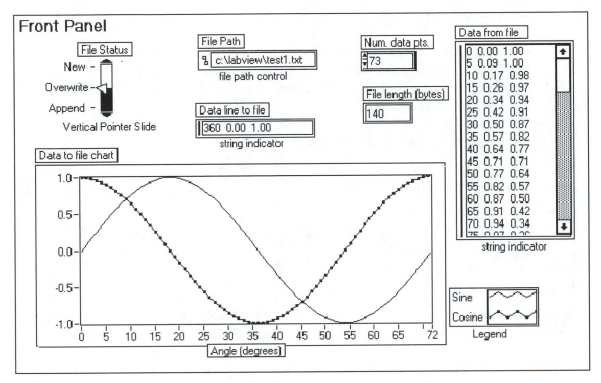

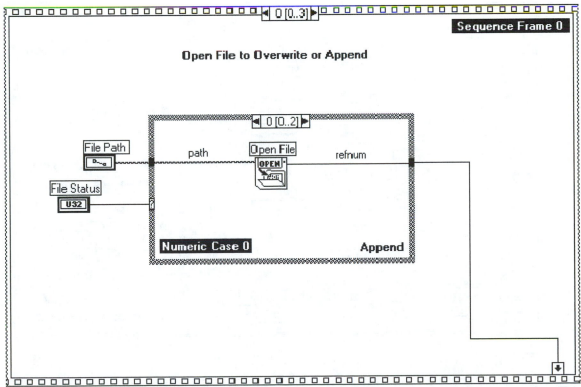

Fig. 6-5 The Front Panel and Block Diagram of Exercise 6-2

In the Front Panel, enter 73 into the *Num. data pts.* digital control, the complete path of the file where data will be saved. With the *Operating Tool* set the File Status (vertical pointer slide) to *Overwrite*.

Run the VI. Experiment with different settings of the Front Panel Controls. Check the content of the file where data was stored. You can use any word processor to open the text file. Microsoft Windows' *Word* or *WordPad*, for example, can be used for this purpose.

Save this VI as **Files 2.vi** and close it.

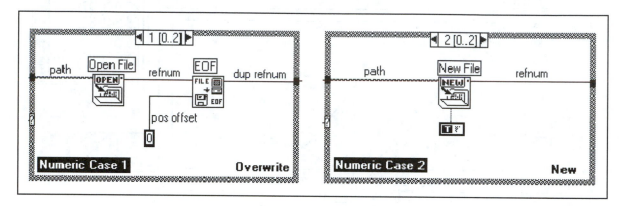

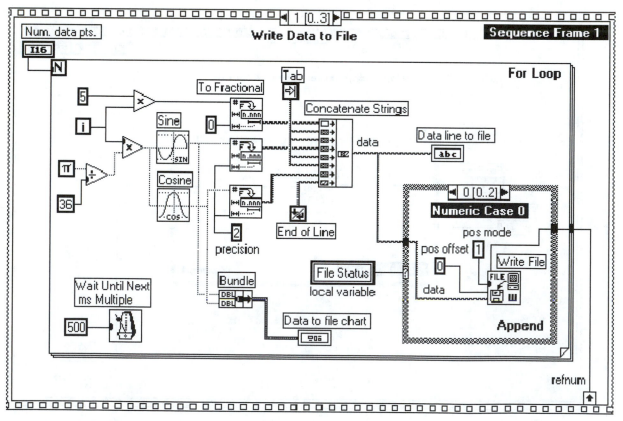

Fig. 6-5 The Front Panel and Block Diagram of Exercise 6-2

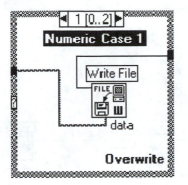

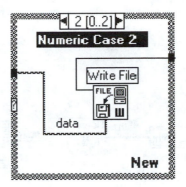

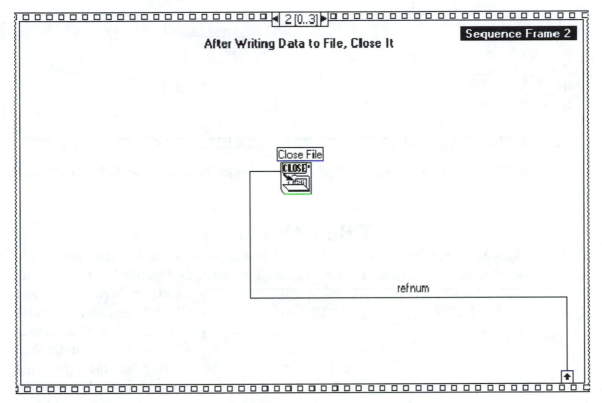

Fig. 6-5 The Front Panel and Block Diagram of Exercise 6-2 (continued)

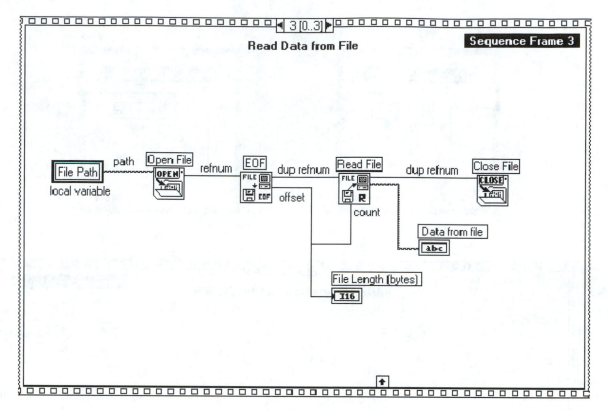

Fig. 6-5 The Front Panel and Block Diagram of Exercise 6-2 (continued)

File I/O VIs

In the previous exercises the procedure of saving data to a file or reading data from a file required the user to configure array file I/O functions such as Open File or New File,

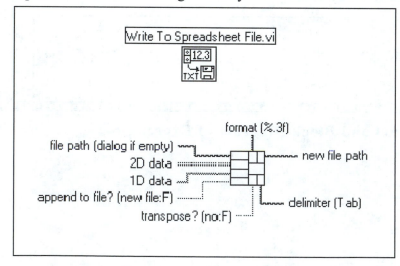

Read or Write File, and Close File. LabVIEW also has utility VIs that include all of those individual steps. Also, the input data is in the form of an array of single precision numbers and does not have to be converted to the string format by the user. This illustration shows the icon and the connector of the *Write To Spreadsheet File.vi*.

The *file path* input is the complete path to the file. If this input is not wired, then the VI will open a dialog box, giving you access to directories and allowing you to choose the file.

The **2D** and **1D** inputs are the array data that are to be saved in the selected file.

The *append to file?* Boolean input can be used to append or overwrite the file. It is *false* by default. If you wire a *true*, the new data will be appended to the file.

The *transpose?* Boolean input is used to transpose the array input data (switch rows and columns) before saving it. It is *false* by default.

The *delimiter* string input is *Tab* by default. The data that you save using this VI is in the spreadsheet format. This means that the VI inserts a *tab* between the columns and an *end of line* control character at the end of the data line. You can override this default setting by wiring another character such as a comma in string format.

The *format* string input specifies the format to be used by this VI on the input data. It is *%.3f* by default. The **%** begins the format specification and is a required control character. Following the period (.) is the precision string specifying the number of decimal places. The conversion character **f** specifies how to convert the input number. The default specifier **f** converts the input number *to a floating point number with fractional format*. You can override this default format specification by wiring another format string. The number before the period, for instance, may be used to specify the field width, and **f** can be replaced by **d**, which converts the input data to *decimal integers*. Refer to the LabVIEW Function Manual for more information.

The *Read From Spreadsheet File.vi* icon and the connector are shown in Fig. 6-6. This VI will open the file specified in the file path input string and output the file data as an array. After the *read* operation is complete, the VI will close the file. If the file path input is left unwired, the VI will open the dialog box, allowing you to choose the file. The inputs and outputs for this VI are described below.

File path is the complete path of the file to be read. If the path is not wired, the VI will open the file dialog box from which you can select the file.

The *number of rows* numeric input specifies the maximum number of rows to be read by the VI. If the Number of Rows < 0, the VI will read the entire file. The default value is −1. A row is defined as a character string ending in a line control character such as carriage return, line feed, carriage return followed by a line feed, or a string that has the maximum line length as specified by the *max characters/row* input.

Start of read offset is a numeric input specifying the number of characters or bytes from the beginning of the file where reading is to begin. The default value of the offset is 0.

Max characters/row is the numeric input specifying the maximum number of characters that the VI is to read. The default value is 0, which means that there is no limit on the number of characters to be read by the VI.

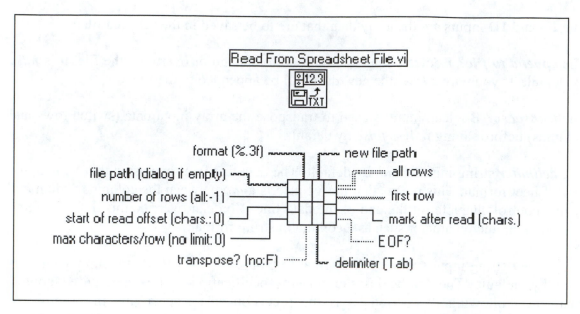

Fig. 6-6 Icon and Connector of the Read From Spreadsheet.vi

The *transpose?* Boolean input specifies whether to transpose the data after converting it from a string format. The default value is *false*.

Mark after read is the numeric output of the number of characters or bytes read from the file.

The *EOF?* Boolean output indicates whether the read operation extended beyond the end of the file.

The *delimiter* string input is used to format columns in spreadsheets. Although the default delimiter is *tab*, you can use other delimiters.

The *format* input string specifies the conversion of characters to numbers. The default format is *%.3f.*

All rows is a two-dimensional array data output. This is the data read from the file.

First row is the first row in a one-dimensional array of the *All Rows* output data.

As shown in the following illustration, there are other Utility File Function VIs. *Write Characters To File.vi* opens a file, writes the input *character string* to the file, and then closes the file.

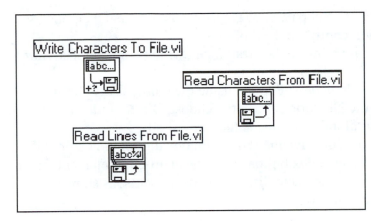

Similarly, **Read Characters From File.vi** opens the file, reads the characters at the specified position in the file, outputs the character string, and closes the file.

Read Lines From File.vi reads the specified line of characters, outputs it as a line string, and then closes the file.

Exercise 6-3: Files III (Using the Utility File I/O VIs)

This exercise illustrates the use of the Utility File I/O VIs. The VI that you will build collects data from three different sources and saves it to a spreadsheet, then reads that data from a file and displays it on the X-Y graph and on the digital array indicator. As the VI runs, you will be able to see immediately the data that was saved to the file.

Build the VI according to the following guidelines. The Front Panel and the Block Diagram are shown in Fig. 6-7.

Front Panel

1.   **X-Y Graph** is in the *Graph* subpalette of the Controls palette. Open one X-Y Graph, resize it as necessary, and configure it as follows:
Labels: **Data from file graph** is an owned label.
Auto Scale X and *Y*: leave enabled (default)
Palette: remove from screen.

Digital Control/Indicator objects are in the *Numeric* subpalette of the Controls palette. Open two digital controls and one digital indicator and configure them as follows:
Label: the two digital controls as **K** and **Num. Cycles**, the   digital indicator as **Num bytes from file**. All are owned labels.
Representation: **I16** for the two digital controls and the digital indicator.

File Path Control is in the *Path & Refnum* subpalette of the Controls palette. Open one File Path Control and, using the *Labeling Tool*, type the complete path of the file. A suggested path is shown in the Front Panel of Fig. 6-7. Label the file path control as **File Path** using an owned label.

Vertical Pointer Slide is in the *Numeric* subpalette of the Controls palette. Open one Vertical Pointer Slide and label it as shown in Fig. 6-7 using an owned label as **Select Plot**.

Array is in the *Array & Cluster* subpalette of the Controls palette. Open one ·empty array shell and configure it as follows:

Dimension: Two. Pop up inside the array index window and choose **Add Dimension** from the popup menu.

Create Digital Array Indicator: Pop up inside the empty array shell, navigate to the Numeric subpalette, choose Digital Indicator, and drop the digital indicator inside the array shell.

Resize: Optional. You can leave the array with one digital indicator and operate the arrow index buttons to view a particular element, or you can resize the array to display many elements, as shown in the Front Panel of Fig. 6-7.

Block Diagram

2. **Sequence Structure** is in the *Structures* subpalette of the Functions palette. Open one Sequence structure and resize it as necessary. You will notice that as you open the sequence frame, there is no frame window at the top to identify this frame as Frame 0. Don't worry about that right now. This unmarked frame is Frame 0. This situation will be corrected when you get to the next frame.

Sequence Frame 0 *(Save Data From File)*

For Loop is in the *Structures* subpalette of the Functions palette. Open one For Loop, resize it, and **wire objects inside the For Loop as follows**:

Formula Node is in the *Structures* subpalette of the Functions palette. Open one Formula Node and type inside the Formula Node : **x = (2*K/N)*(i–N/2);**

The **Numeric subpalette** of the Functions palette has all of the remaining objects that you need inside the For Loop as follows:

Multiply function: You need three.

Divide Function: You need one.

Random Number function

Sine and **Sinc** functions are in the *Trigonometric* sub-palette of the Numeric subpalette.

The **To Word Integer** function is in the *Conversion* sub-palette of the Numeric subpalette of the Functions palette.

2π is in the *Additional Numeric Constants* subpalette of the Numeric subpalette of the Functions palette.

Numeric Constant: you need three: **50, 100,** and **100**. Notice that one of the 100 numeric constants is wired from outside the For Loop to the loop counter *N* and *N* input of the Formula Node.

|i| is the loop iteration terminal. It was inside the loopwhen you opened it.

Wire all objects inside the For Loop as shown in Sequence Frame 0 of Fig. 6-7.

Build Array is in the *Array* subpalette of the Functions palette. Open one Build Array structure and resize it to accommodate six inputs as shown in Sequence Frame 0 of Fig. 6-7.

Transpose 2D Array is in the *Array* subpalette of the Functions palette. Open one Transpose 2D Array function.

Write To Spreadsheet File.vi is in the *File I/O* subpalette of the Functions palette. Open one Write To Spreadsheet File VI.

Wire all objects inside Sequence Frame 0 as shown in Fig. 6-7.

Sequence Frame 1 *(Read Data from File)*

Switch to Sequence Frame 1 by popping up on the frame border and choosing *Add Frame After.* Notice that the frame window appears and identifies this frame as Frame 1, with Frame 0 behind it.

Case Structure is in the *Structures* subpalette of the Functions palette. Open one Case Structure inside Sequence Frame 1 as shown in Fig. 6-7, resize it, and **wire** the ***Select Plot*** terminal to the selector terminal **[?]** of the Case Structure.

Numeric Case 0

You will need the following objects inside Numeric Case 0:

Index Array is in the *Array* subpalette of the Functions palette. Open two Index Array functions, and resize them by dragging the lower corner down to accommodate two index inputs. Disable the upper index input by popping up on the index and choosing *Disable Indexing.* The black rectangle will change to white as shown here

Bundle is in the *Cluster* subpalette of the Functions palette π is in the *Additional Numeric Constants* subpalette of the *Numeric* subpalette of the Functions palette.

Numeric Constant is in the *Numeric* subpalette of the Functions palette. You will need three numeric constants: **3, 4,** and **180**.

The **Multiply** and **Divide** functions are in the *Numeric* subpalette of the Functions palette. You will need one of each.

Wire all objects inside Numeric Case 0, as shown in Fig. 6-7.

Numeric Case 1

Switch to Numeric Case 1.

Wire all objects inside Numeric Case 1 as shown in Fig. 6-7. All objects in this Numeric Case were covered in Numeric Case 0 above. *Data from file graph* is a **local variable**. Open the Local Varibale object from the Structures subpalette of the Functions palette. Right-click on the icon and choose "Data from file graph" from the Select Item option.

Numeric Case 2
Switch to Numeric Case 2 by popping up inside the Case window at the top of the case and choosing *Add Case After*.
Wire all objects inside Numeric Case 2 as shown in Fig. 6-7. All objects in this Numeric Case were covered in Numeric Case 0 above. *Data from file graph* is a **local variable**. Right-click on the icon again and choose the "Change to Read" option from the pop-up menu.

Read From Spreadsheet File.vi is in the *File I/O* subpalette of the Functions palette.

Wire all objects inside Sequence Frame 1 and outside the Numeric Case structure, as shown in Fig. 6-7. Note that *File Path* is a local variable.

3. In this VI the function of Sequence Frame 0 is to collect data from three different sources and save this data in the spreadsheet format to the text file as specified by the path control. 100 data points (0 to 99) of the Sinc function are collected as an array at the border of the For Loop. Also, 100 points for the sine wave and 100 points for the random number generator have been collected at the border of the For Loop. The x, I, and angle arrays are formed at the border of the For Loop as well. These arrays are completed when the For Loop executes 100 times.

The six arrays are combined in the Bundle structure and transposed by the Transpose 2D Array function before being applied to the Write To Spreadsheet File.vi utility. Transposing of the arrays is necessary because they are formed by the For Loop horizontally. Transposing switches rows and columns so that the x array is in the first column, the i array in the second column, the Sinc array in the third column, and so on.

The purpose of Sequence Frame 1 is to extract the data array of interest from the spreadsheet and display it on the X-Y graph. The Read From Spreadsheet File.vi utility opens the file specified by the path control, reads the data, and outputs it as a two-dimensional array. This two-dimensional array is applied to the Numeric Case structure.

The Select Plot Vertical Slide decides which Numeric Case will be executed. If you set the Vertical Pointer Slide with the operating tool to the Sinewave position, its corresponding terminal in Sequence Frame 1 has a value of 0 that is applied to the selector terminal [?] of the Numeric Case, thus forcing Numeric Case 0 to be executed.

Inside Numeric Case 0 the two Index Array functions slice away columns 3 and 4 from the incoming two-dimensional array. Column 3 is the 100 points of the 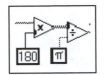 sinewave and column 4 is the corresponding 100 points of the angle, or the argument of the sinewave. In this detail the angle in radians is converted to degrees to be used by the X-Y graph for the X-axis. The X-array and the Y-array are combined in the Bundle structure and then applied to the X-Y graph.

In a similar fashion, choosing the Sinc function on the Vertical Pointer Slide will force Numeric Case 1 to be executed, and the Random Number setting will cause Numeric Case 2 to be executed.

The setting of the **K** value on the Front Panel determines the range of the X-axis. Notice that the setting of 20 in the Front Panel of Fig. 6-7 results in X-axis range −20 to +20. K is used only for the Sinc function display.

The *Num. Cycles* digital control setting specifies the number of cycles of the sine wave to be collected and displayed. This control applies only to the sine wave data.

The six columns of data are also displayed on the *Data from file array* Digital Array Indicator. The total number of bytes or characters read from the file is displayed on the *Num. bytes from file* digital indicator.

Run this VI. Be sure to enter first the file path in the path control and the values of K and Num. Cycles.

Save this VI as **Files 3.vi** and close it.

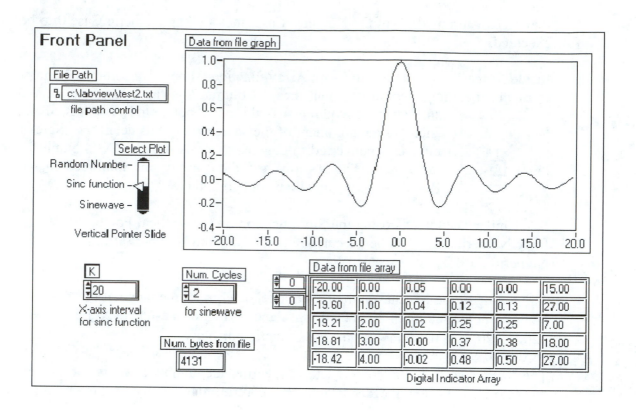

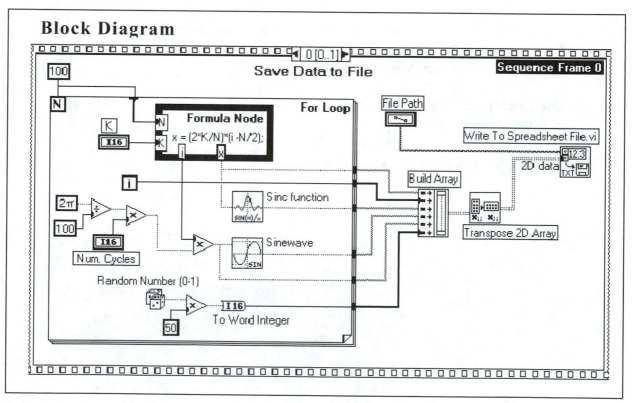

Fig. 6-7 Front Panel and Block Diagram of Exercise 6-3

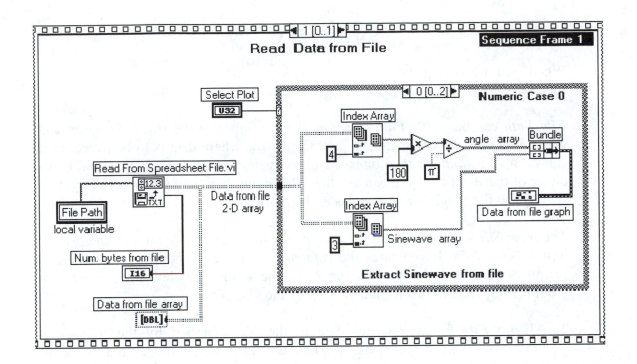

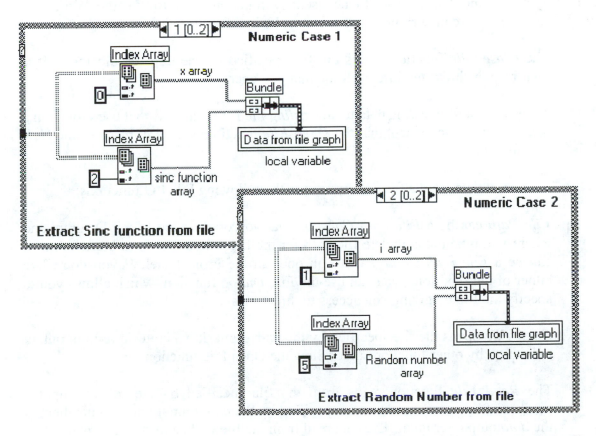

Fig. 6-7 Front Panel and Block Diagram of Exercise 6-3 (continued)

Summary

1. When you are saving data to a *new file*, you must wire the following File I/O functions:

    File Path can be a *File Path Constant* that you open in the Block Diagram and where you type the complete file path for the file where data is to be saved, or it can be a *File Path Control* that you open in the Front Panel. If you don't like either of these options, you can use the *File Dialog* function, which allows you to specify the file by giving you access to directories.

    The **New File** function creates the file specified by the path input and opens it for writing new data. It generates the *refnum*, the reference number that uniquely identifies the file, to be used by functions that follow the New File. If the path input is not wired, it uses its default setting, the File Dialog.

    The **Write File** function writes data to the file specified by the refnum. It generates the *dup refnum*, the duplication reference number that uniquely identifies the opened file, to be used by functions that follow the Write File function. The data written to the file must be in string format.

    The **Close File** function closes the file specified by the dup refnum input. It is your responsibility to close each file that you open.

2. The procedure for saving data to an *existing file* is the same as that used for saving data to a new file, except that the **Open File** function is used in place of the New File function.

3. To extract data from a file you must wire the following File I/O functions:

    File Path can be a *File Path Constant* that you open in the Block Diagram and where you type the complete file path for the file where data is to be saved; or it can be a *File Path Control* that you open in the Front Panel. If you don't like either of these options, you can use the *File Dialog* function, which allows you to specify the file by giving you access to directories.

    Open File opens the file specified by the path input. It generates a refnum output to be used by other functions that follow the Open File function.

    The **Read File** function reads from the file specified by the refnum input a number of bytes or characters as specified by the *count* input and outputs them at the *data* output terminal. The data read from the file will be in string format.

    The **EOF** function returns the End of File mark for the file specified by the refnum input. *Pos mode* and *pos offset* together specify where the EOF mark is placed. The EOF reverts to its default setting, placing the EOF mark at the end of the file if pos mode and pos offset are left unwired.

This will force the EOF function to produce at its *offset* output the file length in bytes (a count of bytes from the beginning of the file to the EOF mark at the end of the file), making it convenient to wire the *offset output* of the EOF function to the *count input* of the Read File function and thus specifying that the entire file is to be read.

The ***Close File*** function closes the file specified by the dup refnum input. It is your responsibility to close each file that you open.

4.   Utility File VIs relieve you of the responsibility of wiring the opening, closing, and read/write functions. They do all that and don't require that you present the data for writing in string format. They also save your data to a spreadsheet file that other spreadsheet applications can read.

Write To Spreadsheet File.vi accepts the input data as a one-dimensional or two-dimensional array and saves it to the file specified by the *file path* input in the form of a spreadsheet. It places *tab* characters between the columns and the *end of line* control character constant at the end of each line. You may choose to *transpose* the data or to *append* the data to the end of the file by wiring the appropriate Boolean inputs.

Read From Spreadsheet File.vi reads data from the file specified by the *file path* input and outputs the data as a two-dimensional array of single precision numbers, including all rows or the *first row* of the two-dimensional array. If the file path is not wired, the utility will open the dialog box for you to choose the file.

Write Characters To File.vi writes the character string to the file specified by the file path input.

Read Characters From File.vi reads the characters at the specified position in the file and outputs the character string.

Read Lines From File.vi reads the specified line of characters from the file and outputs it as a character string.

All Utility File I/O VIs first open the file, perform the read or write operation, and then close the file.

Chapter 7
Fourier Analysis and Geometry

Exercise 7-1: Polynomial Plot

This VI provides another illustration in the use of the X-Y graph. As mentioned, the X-Y graph uses Cartesian coordinates to create the plot. In order for the X-Y graph to work properly, you must generate the X-array of points and Y-array of points and apply them to the X-Y graph terminal as inputs. The two arrays are combined by the Bundle structure, whose output, a cluster, is applied to the X-Y terminal. As you build this VI, note the color and the thickness of the wires at various points. They represent the type of data that these wires carry.

Build the VI whose Front Panel and Block Diagram are shown in Fig. 7-1.

Front Panel

1.   **X-Y Graph** is in the *Graph* subpalette of the Controls palette. Open one X- Y Graph and configure it as follows:

   Scale: Enter the minimum and maximum X-axis values (–5 and 5). Don't worry about the Y-axis.

   AutoScale: Disable for the X-axis and enable for the Y-axis (enabled by default).

   Label: **Polynomial Plot** with owned label; X, Y with free labels.

   Digital Control is in the *Numeric* subpalette of the Controls palette. Open ten digital controls and configure them as follows:

   Representation: **DBL** for all digital controls (DBL is the default setting).

   Increment: **0.5** for the X-Range digital indicator and 1 for all remaining digital indicators (1 is the default setting)

   Label: Label all digital controls with owned labels, as shown in Fig. 7-1.

   Vertical Switch is in the *Boolean* subpalette of the Controls palette. Open one Vertical Switch and configure it as follows:

   Label: **STOP** using owned label.

   Mechanical Action: **Latch When Pressed**.

   Data Operations (from popup menu): Choose *Make Current Value Default.*

   All other labels inside the Front Panel are free labels and are optional.

Block Diagram

2. **While Loop, For Loop,** and **Formula Node** are all in the *Structures* subpalette of the Functions palette. Open the While Loop, the For Loop inside the While Loop, and the Formula Node inside the For Loop. Resize all structures as necessary.

3. *Wire* the numeric constant whose value is **1000** to the loop counter *N*. You will find the numeric constant in the *Numeric* subpalette of the Functions palette.

   Wire the **STOP** terminal to the condition terminal of the While Loop.

   Add the following to the Formula Node border:
   > *Inputs*: i, c, N, C0, C1, C2, C3, C4, E1, E2, E3, E4.
   > *Outputs*: x, y.
   >> To add an input or output to the formula node, pop up on the border of the Formula Node and choose *Add Input* or *Add Output* as the case may be.

   Type the x and y expressions, as shown in Fig. 7-1, inside the Formula Node using the Labeling Tool.

   Wire all terminals to the Formula Node inputs as shown in Fig. 7-1. Also wire the x, y outputs to the border of the For Loop.

   Bundle function is in the *Cluster* subpalette of the Functions palette. Open one Bundle function and wire to its inputs the x and y arrays from the tunnels of the For Loop. Wire the output of the Bundle structure to the *Polynomial Plot* terminal.

4. *Enter* all values into the Front Panel digital controls as shown in Fig. 7-1.

   Note how the X-array is generated. We use the expression $X = (2c/N)[i - N/2]$ that we have used before to generate an array of points ranging from $-c$ to $+c$, where the value of c is the setting of the X-range Front Panel digital control. Notice that this range is independent of the value of N, the number of samples or data points. For example the X-Range setting in Fig. 7-1 is 4.5, and the number of data points is 1000 (the numeric constant wired to the iteration counter N of the For Loop). This means that there will be an array of 1000 points between -4.5 and $+4.5$ (for each value of x, one value of y will be calculated inside the Formula Node, thus forming a 1000-point y array). Should you change the value of 1000 to, say, 5000 in the Block Diagram, the X range will remain unchanged, extending from -4.5 and $+4.5$, with more points that are closer together. The advantage of this approach is that we deal only with the values of x in decimal format, as we have always done, without worrying about the sampling rates and the arrays of samples that make up our plots.

Run this VI. Experiment with different parameter values.
Save this VI as **Polynomial Plot.vi** and close it.

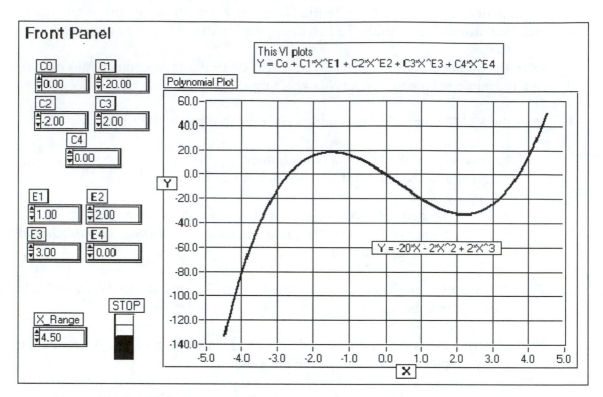

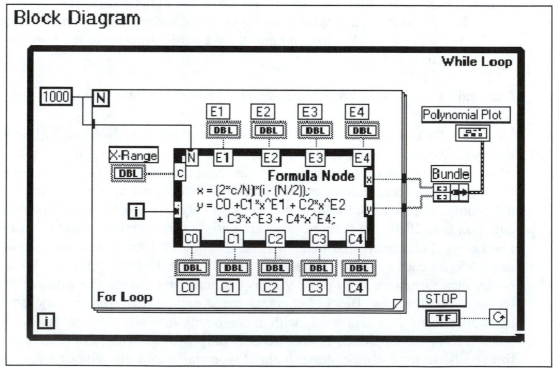

Fig. 7-1 Front Panel and Block Diagram of Exercise 7-1

Exercise 7-2: Conic Sections

In this exercise we will once more illustrate the operation of the X-Y graph. You will build a VI that will plot a straight line and some conic sections. As mentioned earlier the X-Y graph differs considerably from other graphs because it requires array inputs for the X and the Y axes. The X-Y graph uses Cartesian coordinates to plot single-valued as well as multi-valued functions.

You are to build the VI whose Front Panel and Block Diagram are shown in Fig. 7-2.

Front Panel

1.   **X-Y Graph** is in the *Graph* subpalette of the Controls palette. Open the X-Y Graph and configure it as follows:

Scale: Enter the minima and maxima (−10 to +10) X-axis and Y-axis values as shown in Fig. 7-2.

Resize: Resize the X-Y graph to the shape of a **square**. It is important that the X-axis and the Y-axis are the same length; otherwise a circle will appear as an ellipse.

AutoScale: Disable auto-scale for both the X and Y axes.

Label: *X-Y Graph* is an owned label.

Palette: Disable (optional).

Plot Settings: Resize the Legend to accommodate two plots and label them using the labeling tool as **y1** and **y2**, as shown in Fig. 7-2. Pop up on the *Legend* and choose one of the following settings from the popup menu options:

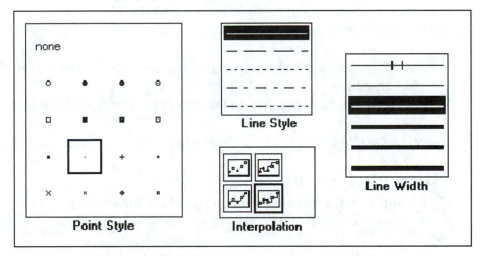

Digital Control/Indicator is in the *Numeric* subpalette of the Controls palette. Open nine digital controls and three digital indicators, and configure them as follows:

Representation: **I16** for *Num Data Points* digital control and **DBL** for the remaining digital controls and the digital indicators.

Increment: Configure all digital controls as follows:

Increment = 1 for Xo, Yo, Cx, and Cy.

Increment = 0.5 for m, b, X-Range, and p.
Increment = 500 for Num Data Points.
*To set the increment, pop up on the digital control, choose **Data Range...** from the popup menu, and type the desired value in the **Increment box**.*

Menu Ring is in the *List & Ring* subpalette of the Controls palette.
Label the *Menu Ring* with an owned label as **Plot Menu**. Using the Labeling tool, type into the Menu Ring the following items in the given order:

> *Line*
> *Parabola (Y-axis)*
> *Parabola (X-axis)*
> *Ellipse/Circle*

Remember that you have to enter these items one at a time. After adding an item, pop up on the *Menu Ring* and choose **Add Item After** from the popup menu.

Vertical Switch is in the *Boolean* subpalette of the Controls palette.
Label the switch as **Stop** using an owned label.
Mechanical Action: **Latch When Pressed**
Choose **Data Operations>Make Current Value Default** from the pop-up menu.

Block Diagram

2. **While Loop** and the **Case structure** are in the *Structures* subpalette of the Functions palette. Open the *While Loop* and the *Case structure* inside the While Loop. Resize both as necessary.

Wire the **Plot Menu** terminal to the *Selector terminal* [?] of the Case structure.
Wire the **Stop** terminal to the *Condition* terminal of the While Loop.

Next, you are to build the four frames of the Numeric Case.

Numeric Case 0

For Loop is in the *Structures* subpalette of the Functions palette. Open the For Loop inside Numeric Case 0 and resize it as necessary.

Wire the *Num Data Points* terminal to the loop iteration counter *N* through the border of the Numeric Case as shown in Fig. 7-2.

The **Build Array** structure is in the *Array* subpalette of the Functions palette.

The **Bundle** structure is in the *Cluster* subpalette of the Functions palette.
Open one *Build Array* and one *Bundle* structure inside Numeric Case 0 and outside the For Loop, as shown in Fig. 7-2.

Formula Node is in the *Structures* subpalette of the Functions palette.

> *Open* a Formula Node inside the For Loop and resize it as necessary.
>
> *Add Inputs and Outputs* to the Formula Node as shown in Fig. 7-2.
>
> > *To add an input or an output, pop up on the border of the Formula Node and choose* **Add Input** *or* **Add Output** *from the pop-up menu.*
> >
> > *Inputs:* m, i, c, b, N, Xo, Yo
> >
> > *Outputs*: x, y
>
> *Wire* all terminals to the Formula Node inputs, as shown in Fig. 7-2. As shown, the Xo and Yo terminals are wired from outside the For Loop.
>
> *Wire* the x and y outputs to the border of the For Loop.
>
> *Type* with the *Labeling Tool* the formulae inside the Formula Node as shown in Fig. 7-2. *Don't forget the semicolon at the end of each formula.*

Wire the x and y outputs from the border (tunnels) of the For Loop to the *Bundle* structure.

Wire the output of the *Bundle* structure to the input of the *Build Array* structure and the output of Build Array to the *X-Y Graph* terminal, as shown in Fig. 7-2.

Numeric Case 1

Open Numeric Case 1 by clicking inside the Case window at the top of the structure.

For Loop is in the *Structures* subpalette of the Functions palette. Open the For Loop inside Numeric Case 1 and resize it as necessary.

Wire the *Num Data Points* terminal to the loop iteration counter *N* from the tunnel in the border of the Numeric Case, as shown in Fig. 7-2.

Build Array structure is in the *Array* subpalette of the Functions palette.

Bundle structure is in the *Cluster* subpalette of the Functions palette.

> *Open* one *Build Array* and one *Bundle* structure inside Numeric Case 1 and outside the For Loop, as shown in Fig. 7-2.

Formula Node is in the *Structures* subpalette of the Functions palette.

> *Open* Formula Node inside the For Loop and resize it as necessary.
>
> *Add Inputs and Outputs* to the Formula Node as shown in Fig. 7-2.
>
> > *To add an input or an output, pop up on the border of the Formula Node and choose* **Add Input** *or* **Add Output** *from the pop-up menu.*
> >
> > *Inputs*: i, c, p, N, Xo, Yo
> >
> > *Outputs*: x, y, Xf, Yf
>
> *Wire* all terminals to the Formula Node inputs, as shown in Fig. 7-2. As shown, the Xo and Yo terminals are *Local Variables* that are wired from outside the For Loop. Before wiring Xo , popup on Xo and choose *Change to Read Local;* otherwise, you will get a bad wire. Recall that the local variables are *Write Locals* by default when you open them. Do the same for Yo.

Wire the **x** and **y** outputs to the border of the For Loop.

Wire the Xf, Yf outputs of the Formula Node to the Xf, Yf terminals, which are *local variables* outside the For Loop. Pop up on the tunnel used by the Xf wire in the border of the For Loop and choose ***Disable Indexing*** from the popup menu. Do the same for the Yf wire tunnel.

Type with the *Labeling Tool* the formulae inside the Formula Node as shown in Fig. 7-2. *Don't forget the semicolon at the end of each formula.*

Wire the **x** and **y** outputs from the border (tunnels) of the For Loop to the *Bundle* structure.

Wire the output of the *Bundle* structure to the input of the *Build Array* structure and the output of Build Array to the *X-Y Graph* terminal, as shown in Fig. 7-2. Note that the X-Y Graph terminal is a local variable.

Numeric Case 2

Open **Numeric Case 2** by popping up inside the case window at the top of the structure and choosing ***Add Case After***.

For Loop is in the *Structures* subpalette of the Functions palette. Open the For Loop inside Numeric Case 2 and resize it as necessary.

Wire the *Num Data Points* terminal to the loop iteration counter *N* from the tunnel in the border of Numeric Case, as shown in Fig. 7-2.

The **Build Array** structure is in the *Array* subpalette of the Functions palette.

The **Bundle** structure is in the *Cluster* subpalette of the Functions palette. Open one *Build Array* and two *Bundle* structures inside Numeric Case 2 and outside the For Loop, as shown in Fig. 7-2.

Formula Node is in the *Structures* subpalette of the Functions palette.

Open Formula Node inside the For Loop and resize it as necessary.

Add Inputs and Outputs to the Formula Node as shown in Fig. 7-2.

*To add an input or an output, pop up on the border of the Formula Node and choose **Add Input** or **Add Output** from the pop-up menu.*

Inputs: i, c, p, N, Xo, Yo.

Outputs: x, y1, y2, Xf, Yf.

Wire all terminals to the Formula Node inputs, as shown in Fig. 7-2. As shown, the Xo,Yo, p, and X-Range terminals are *Local Variables* that are wired from outside the For Loop.

Before wiring Xo, pop up on Xo and choose ***Change to Read Local;*** otherwise, you will get a bad wire. Recall that the local variables are *Write Locals* by default when you open them. Do the same for Yo, p, and X-Range.

Wire the x, y1, and y2 outputs to the border of the For Loop.

Wire the Xf, Yf outputs of the Formula Node to the Xf, Yf terminals, which are *local variables* outside the For Loop. Pop up on the tunnel in the border popup menu. Also disable Indexing for the Yf wire tunnel.

Type with the *Labeling Tool* the formulae inside the Formula Node as shown in Fig. 7-2. *Don't forget the semicolon at the end of each formula.*

Wire the x, y1, and y2 outputs from the border (tunnels) of the For Loop to the *Bundle,* as shown in Fig. 7-2.

Wire the two outputs of the *Bundle* structures to the input of the *Build Array* structure and the output of Build Array to the *X-Y Graph* terminal, as shown in Fig. 7-2. Note that the X-Y Graph terminal is a local variable.

Numeric Case 3

Open **Numeric Case 3** by popping up inside the Case window at the top of the structure and choosing *Add Case After.*

For Loop is in the *Structures* subpalette of the Functions palette. Open the For Loop inside Numeric Case 2 and resize it as necessary.

Wire the *Num Data Points* terminal to the loop iteration counter N from the tunnel in the border of Numeric Case, as shown in Fig. 7-2.

The **Build Array** structure is in the *Array* subpalette of the Functions palette.

The **Bundle** structure is in the *Cluster* subpalette of the Functions palette.

Open one *Build Array* and two *Bundle* structures inside Numeric Case 2 and outside the For Loop as shown in Fig. 7-2.

Formula Node is in the *Structures* subpalette of the Functions palette.

Open Formula Node inside the For Loop and resize it as necessary.

Add Inputs and Outputs to the Formula Node as shown in Fig. 7-2.

> *To add an input or an output, pop up on the border of the Formula Node and choose* **Add Input** *or* **Add Output** *from the popup menu.*
> **Inputs**: i, c, Cx, Cy, N, Xo, Yo.
> **Outputs**: x, y1, y2, f

Wire all terminals to the Formula Node inputs, as shown in Fig. 7-2. As shown, the Xo, Yo, and X-Range terminals are *Local Variables* that are wired from outside the For Loop. Before wiring Xo, pop up on Xo and choose **Change to Read Local**; otherwise, you will get a bad wire. Recall that the local variables are *Write Locals* by default when you open them. Do the same for Yo, f, and X-Range.

Wire the x, y1, and y2 outputs to the border of the For Loop.

Wire the f output of the Formula Node to the *Distance between foci* terminal outside the For Loop. Pop up on the tunnel in the border of the For Loop used by the f output wire and choose **Disable Indexing** from the popup menu.

Type with the *Labeling Tool* the formulae inside the Formula Node as shown in Fig. 7-2. *Don't forget the semicolon at the end of each formula.*

Wire the x, y1, and y2 outputs from the border (tunnels) of the For Loop to the *Bundle* structures, as shown in Fig. 7-2.

Wire the two outputs of the *Bundle* structures to the input of the *Build Array* structure, and the output of Build Array to the *X-Y Graph* terminal, as shown in Fig. 7-2. Note that the X-Y Graph terminal is a local variable.

3.   The combination of the *Menu Ring* and the *Numeric Case* is used once again in this exercise in order to use a single X-Y graph for displaying a variety of plots.

Each Numeric Case frame has a Formula Node inside a For Loop. You type inside the Formula Node the equations representing the curves or plots that are to be displayed on the X-Y graph. The For Loop is used for the sole purpose of creating the X-array and Y-array associated with the plot to be displayed. Notice that the y-output and the x-output are wired from the Formula Node to the border of the For Loop, where the array is created. The number of points in the array is set by the user in the *Num Data Points,* a Front Panel digital control. For each plot, the X-Y graph requires an X-array and a Y-array as its inputs.

The X-array is created in a special way. The formula $x = (2c/N)[i - n/2]$ produces an array of N points ranging from $-c$ to $+c$. Both N and c are Front Panel digital controls; *Num Data Points* is wired to N and *X-Range* is wired to c. In this arrangement the x-values will always fall between $-c$ to $+c$ and the end points $-c$ to $+c$ are independent of N. The value of N determines only the number of data points between $-c$ to $+c$.

Functions that we plot in this VI can be either single valued or double valued. As shown in this illustration, a *single valued* function such as a straight line or a

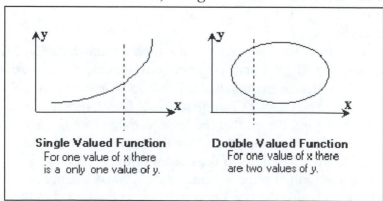

parabola symmetric about the Y-axis has one value of y for each value of x.

A *double valued* function, on the other hand, such as a circle or an ellipse, has two values of y for each value of x.

When we plot a double valued function such as an ellipse or a parabola symmetric about the x-axis, we generate the y1 segment to represent the upper part of the curve and the y2

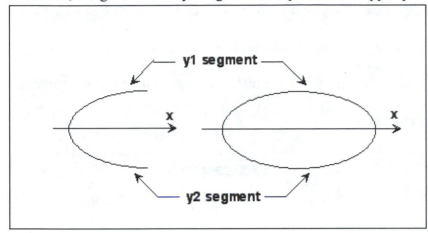

segment to represent the lower part of the curve. When the two are plotted, they are joined to form the whole curve. See the illustration.

Both Numeric Cases 2 and 3 of Fig. 7-2 must plot double valued functions. This illustration shows a portion of Numeric Case 3, where the arrays y1 and y2 are combined

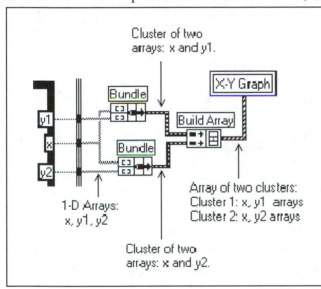

with the x-array in the *Bundle* structure. The *Cluster* outputs of the *Bundle* structures are then combined in the *Build Array* structure before being applied to the X-Y graph.

Note the *data types* at each point. Data from the tunnels of the For Loop are *one-dimensional arrays*. The Bundle structure outputs *clusters of arrays*. The upper Bundle outputs a cluster of y1 and x arrays, and the lower Bundle outputs a cluster of y2 and x arrays. The Build Array structure outputs an *array of two clusters*: the first cluster contains the x and y1 arrays, and the second cluster includes the x and y2 arrays.

It is possible to apply the output of the Bundle structure, whose **output data type is cluster of arrays**, directly to the X-Y graph. It is also possible to apply the output of the Build Array structure, whose **output data type is array of clusters**, to the X-Y graph. But it is not permissible to apply the two different data types to the same X-Y graph. Notice that Numeric Case 0 uses the X-Y Graph terminal, but Numeric Cases 1, 2, and 3 use the local variables associated with that terminal.

This means that if you wire the Bundle output in Numeric Cases 0 and 1 to the X-Y graph and the output of the Build Array structure to the X-Y graph in Numeric Cases 2 and 3, you will get a bad wire because two different data types are applied to the same X-Y graph. To remedy this problem, a single-input Build Array structure is inserted between the X-Y graph and the Bundle structure in Numeric Cases 0 and 1 to adjust the data type so that the data type applied to the X-Y graph is an array of clusters in all Numeric Cases.

Enter the numeric values into digital controls in the Front Panel, as shown in Fig. 7-2.

Run the VI. Experiment with different values and different plots while the VI is running.

Save this VI as **Conics.vi** and close it.

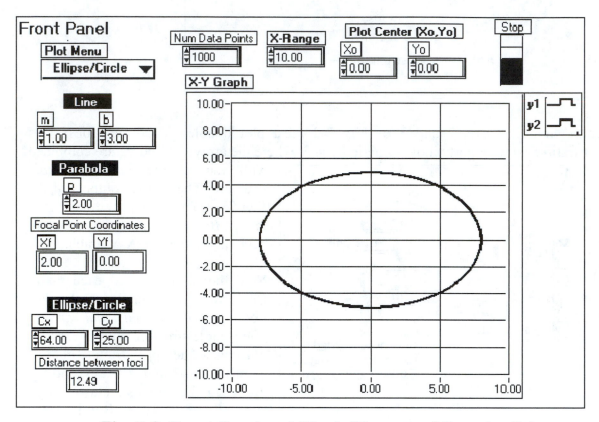

Fig. 7-2 Front Panel and Block Diagram of Exercise 7-2

Block Diagram

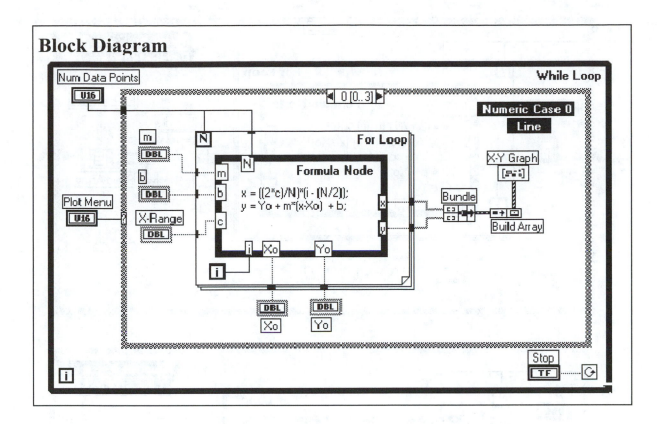

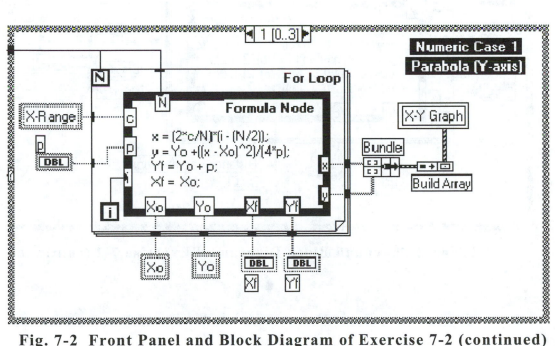

Fig. 7-2 Front Panel and Block Diagram of Exercise 7-2 (continued)

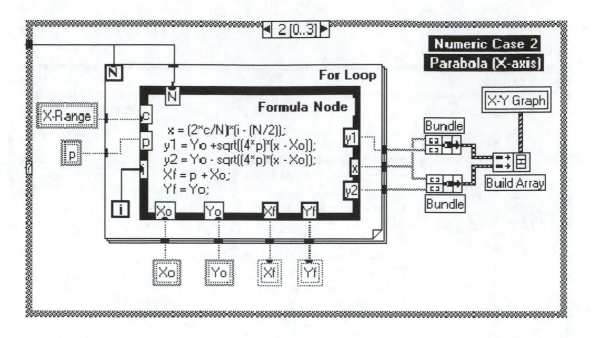

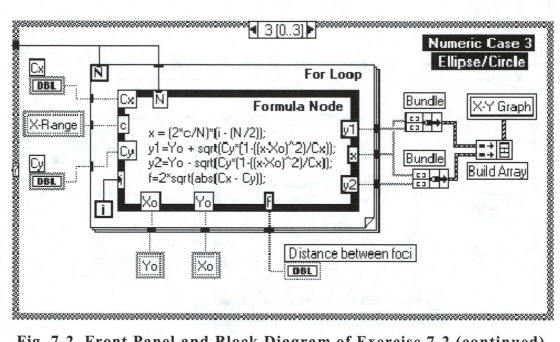

Fig. 7-2 Front Panel and Block Diagram of Exercise 7-2 (continued)

Exercise 7-3: Fourier Spectrum

In this exercise we will illustrate the use of the waveform graph in yet another practical VI. Any periodic waveform can be expressed, according to Fourier, as an infinite sum of sines and cosines. The closeness of the Fourier approximation to the actual waveform depends on the number of terms being added. The more terms you add, the better is the approximation. In this VI you will be able to add as many terms as you like and see the resulting Fourier approximation.

Our main VI, called Fourier.vi, has two subVIs: *Amplitude.vi* and *Cos(x).vi*. As you know, any VI can be used as a subVI as long as it has the *connector* and the icon, although the icon has a cosmetic rather than a functional value.

The first item on the agenda is to build two VIs with an icon and a connector so that later they can be used as subVIs. The first of these subVIs to be built is Amplitude.vi.

Amplitude.vi

The Front Panel and Block Diagram for Amplitude.vi is shown in Fig. 7-3. Build this VI, then create the connector and design the icon. The suggested icon is shown below.

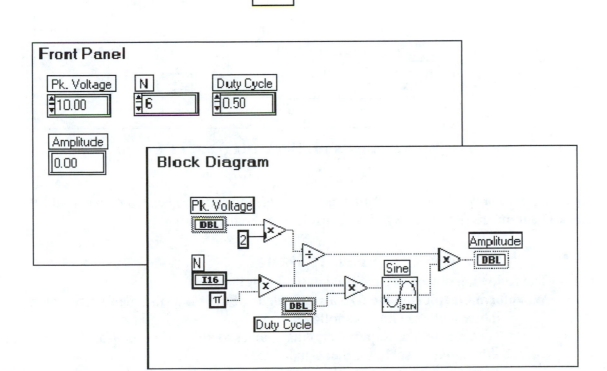

Fig. 7-3 Front Panel and Block Diagram of Amplitude.vi

Cos(x).vi

This the second VI that will be used later as a subVI. The Front Panel and Block Diagram for this VI shown in Fig. 7-4. Build this VI, create the connector, and design an icon. A suggested icon for this VI is shown below.

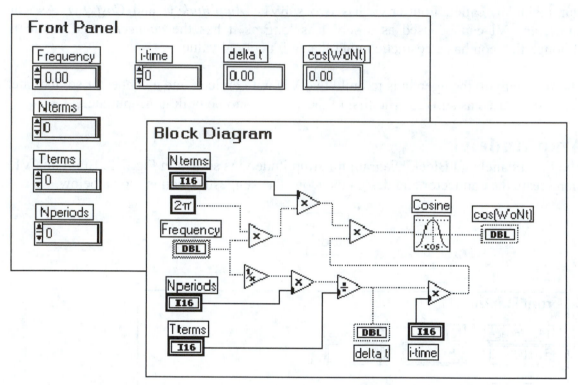

Fig. 7-4 Front Panel and Block Diagram of Cos.vi

Now let's return to the main VI of this exercise, Fourier.vi. The Front Panel and the Block Diagram are shown in Fig. 7-5. Consider the Front Panel first.

Front Panel

1.   The following objects are required in the Front Panel:

Waveform Graph is in the *Graphs* subpalette of the Controls palette. Configure the Waveform Graph as follows:

Scale: Enter the X and Y min/max values as shown in Fig. 7-5.

AutoScale: Disable for the X and Y axes.

Remove from the screen the waveform graph's Palette and Legend.

Label: Fourier Wave using owned labels and Time (seconds) using free labels as shown in Fig. 7-5.

Color: Use the *Set Color* Tool from the Tools palette to change the plot background color and the *Color* palette from the Legend to change plot color.

Six **Digital Controls**. They are found in the *Numeric* subpalette of the Controls palette.

The *Representation* for 5 controls is *I16* and *DBL* for the Duty Cycle digital control.

Block Diagram

2. The following objects are required in the Block Diagram:

One **Sequence structure** and four **For Loops**. They are found in the *Structures* subpalette of the Functions palette. Two *For Loops* are required in Frame 0 and two *For Loops* in Frame 1.

The **Multiply, Add,** and **Increment** functions are found in the *Numeric* subpalette of the Functions palette.

SubVIs: Amplitude.vi and Cos(x).vi. Click on the **Select a VI...** button in the Function palette and navigate to the directory where the subVIs were saved earlier.

Index Aray, Array Subset, Transpose 2-D Array, and **Array Constant** are found in the *Array* subpalette of the Functions palette.

When you open the *Array Constant*, drop the Numeric Constant 0 inside the empty array shell to set its constant value to 0.

When you open the *Index Array*, resize it to include two index inputs, disable the row (upper) index, and wire **i** to the column index (lower), as shown in Fig. 7-5.

Build Array is in the *Cluster* subpalette of the Functions subpalette. Resize it to include three inputs. You will need it in Frame 2 of the Sequence structure.

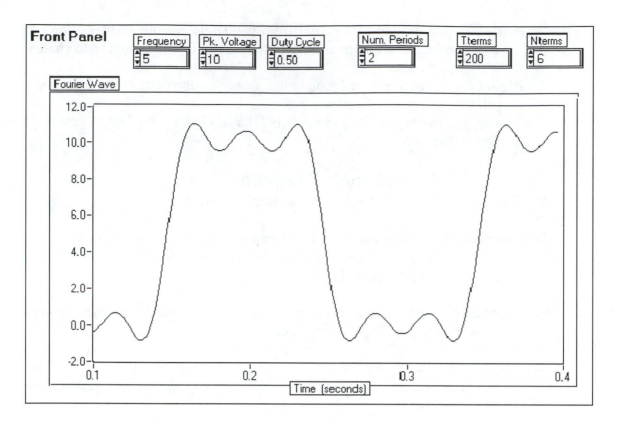

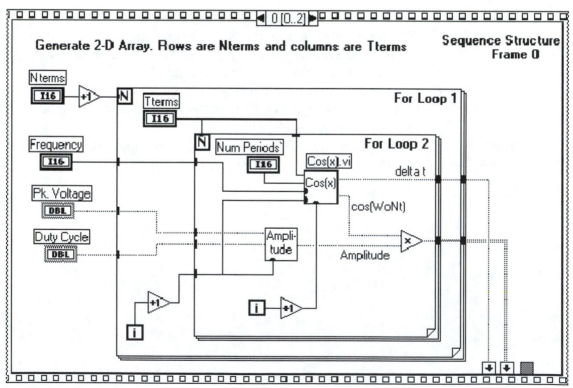

Fig. 7-5 Front Panel and Block Diagram of Exercise 7-3

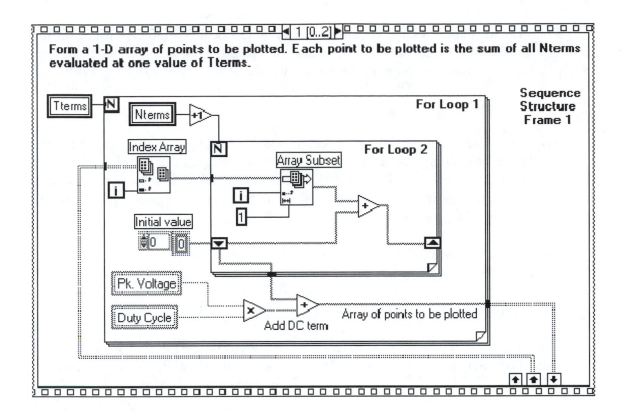

Form a 1-D array of points to be plotted. Each point to be plotted is the sum of all Nterms evaluated at one value of Tterms.

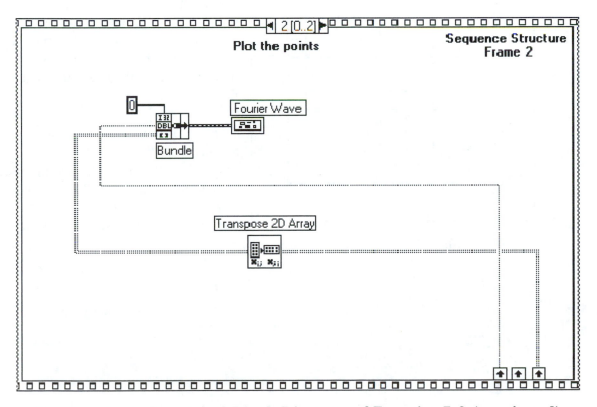

Fig. 7-5 Front Panel and Block Diagram of Exercise 7-3 (continued)

Four **Local Variables** are required in Frame 1. Local Variable is found in the *Structures* subpalette of the Functions palette. As was mentioned before, you have to pop up on the local variable and choose *Select Item* from the pop-up menu. Then choose the local variable name from the menu that opens.

Sequence Local, as mentioned before, provides a means of passing data between frames. This sequence local ⬆ indicates that data is entering the frame and this one ⬇ indicates that data is leaving the frame. The direction arrow with its color representing the data type will appear as soon as you wire something to it.

To create a sequence local, pop up on the border of the sequence structure and choose *Add Sequence Local.*

3. *Wire* all objects as shown in Fig. 7-5.

4. The objective of Frame 0 is to generate the two-dimensional array of Fourier data. For each value of Nterms (Front Panel control) in For Loop 1, Loop 2 executes Tterm (Front Panel control) number of times, thus generating a row of values, as shown by the darkened row in this illustration. When Loop 2 completes execution, all rows in this two-dimensional array will be filled.

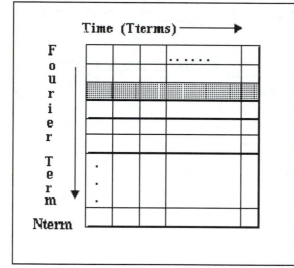

The size of this array is (Nterms)(Tterms).

The two-dimensional array is acquired at the boundary of For Loop 2. When the loop completes execution, the two-dimensional array is passed over the sequence local to Frame 1 to be processed.

The objective of Frame 1 is to create an array of points to be plotted in the time domain. The way to do this is to slice away a column (such as the darkened

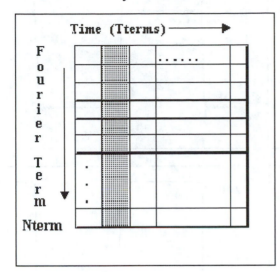

column in this illustration) and add all the terms in that column. The sum represents one point to be plotted. The Index Array accomplishes the slicing away task in For Loop 1 of Frame 1. The one-dimensional array is then passed to For Loop 2, where all elements in that array are added with the help of the Array Subset structure and the shift register. As the point to be plotted exits, the DC value is added in For Loop 2. This point and all remaining points generated as For Loop 2 executes Tterm number of times are acquired at the boundary of For Loop 2. This one-dimensional array is then passed to Frame 2. The Transpose two-dimensional Array structure switches the rows and columns of the array, a formatting step necessary for displaying the array on the waveform graph. The Bundle function combines the array of points to be plotted, the initial value of X, and the time values generated in Cos(x).vi at which the Y points to be plotted are evaluated.

Enter parameter values as shown in the Front Panel of Fig. 7-5.

Run the VI. Experiment with different parameter values in the Front Panel.

Fig. 7-6 shows the Fourier Wave plots with different settings of Nterms. Notice how the Fourier approximation approaches the square wave as the sum of the Fourier terms is increased.

Save this VI as **Fourier.vi** and close it.

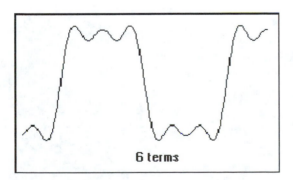
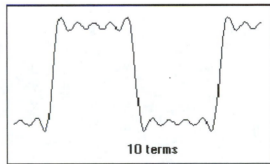

6 terms   10 terms

Fig. 7-6 Waveshape of the Square wave as a Function of Number of Terms in the Fourier Series

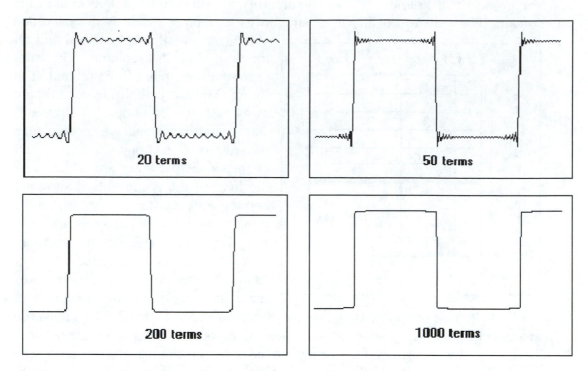

Fig. 7-6 Waveshape of the Square wave as a Function of Number of Terms in the Fourier Series (continued)

Chapter 8
Simulation in Electronic Communication

Amplitude Modulation

In this section we will discuss amplitude modulation (AM), including the following types of AM:

- Double Sideband Full Carrier (DSBFC)
- Double Sideband Suppressed Carrier (DSBSC)
- Single Sideband (SSB)

At the end of this segment you will experiment with LabVIEW software that demonstrates various AM properties.

Introduction

What is modulation? Modulation is a process in which one of the characteristics of a sine wave is being varied by the information signal. The sine wave whose characteristic is being varied is called the **carrier**. A typical equation for the carrier may be expressed as

$$E_c = E_{cm}\sin(\omega_c t + \theta) \tag{8-1}$$

Thus, if the amplitude of the carrier, E_{cm}, is being varied by the information signal, we get AM (Amplitude Modulation). If the carrier's frequency, ω_c, is being varied by the information signal, the result is FM (Frequency Modulation). Similarly we get PM (Phase Modulation) when the phase, θ, of the carrier is being varied by the information signal.

Why do we use modulation? The primary reason for using modulation is to reduce the length of the transmitter antenna. According to theory, the length of an antenna should be one half of a wavelength ($\lambda/2$), and the wavelength is defined as the distance that the wave travels in time duration of one cycle. Suppose that we didn't use modulation and applied 1 kHz audio signal directly to the antenna. The wavelength can be calculated from

$$\lambda = v_c/f = 3 \times 10^8/1 \times 10^3 = 3 \times 10^5 \text{ m} = 300 \text{ km} \tag{8-2}$$

where v_c is the speed of light: 186,000 miles/hr. or 3×10^8 m/s. The antenna converts the applied electric power to electromagnetic waves, which propagate through space at the speed of light. Thus, it can be seen from the above equation that the wavelength of 300 km corresponds to the frequency of 1 kHz. This means that the antenna length, $\lambda/2$, is 150 km (93.75 miles). This is much too large for a practical antenna size. By using a high-frequency carrier, say, 1 MHz

$$\lambda = v_c/f = 3 \times 10^8/1 \times 10^5 = 300 \text{ m} = 900 \text{ ft}$$

and $\lambda/2$ is 450 feet. The Marconi, antenna which uses a ground as a reflecting medium reduces the antenna size to only $\lambda/4$ or 225 ft. With top loading this size could be reduced further to 200 ft or less. Even if it is 150 ft or so, the antenna would require a structure to support it, which is possible to build.

AM Theory (Tone Modulated Carrier)

Let's consider the simplest type of AM signal: *tone modulated carrier.* As shown in Fig. 8-1a, the carrier is a high-frequency sine wave whose amplitude is 500 V_{pk} and whose frequency is 100 kHz . Notice the time of one cycle is 10 μ s and the reciprocal of 10 μ s is 100 kHz.

The modulating signal in Fig. 8-1b is single frequency (10 kHz) whose amplitude is 250 Vpk. When only one frequency modulates the carrier, we call that frequency a *tone.* Checking its frequency, notice that its period is 100 μ s and the reciprocal of 100 μ s is 10 kHz.

The AM wave that is a tone modulated carrier is shown in Fig. 8-1c.

$$E_{max} = E_{cm} + E_{sm} \tag{8-3}$$
$$E_{min} = E_{cm} - E_{sm} \tag{8-4}$$

Using values from Fig. 8-1c in Eq. 8-3 and 8-4, confirm that E_{max} = 750 V_{pk} and E_{min} = 250 V_{pk}.

The modulation index m is defined by

$$m = E_{sm}/E_{cm} = (E_{max} - E_{min})/(E_{max} + E_{min}) \tag{8-5}$$

Using values from Fig. 8-1c in Eq. 8-5, confirm that m = 0.5.

Also, look closely at the time scale in Fig 8-1c which is in μ s. The time duration between any two peaks, as shown in this detail,

is 10 μ s (period) and its reciprocal is 100 kHz. It is, of course, the frequency of the carrier.

Also notice that the time between the two consecutive Emax peaks is $100\,\mu$ s. This is the period of the modulating signal, and the reciprocal of $100\,\mu$ s is 10 kHz, the frequency of the modulating signal.

The AM wave shown in Fig. 8-1 is in the Time Domain because the horizontal scale represents time.

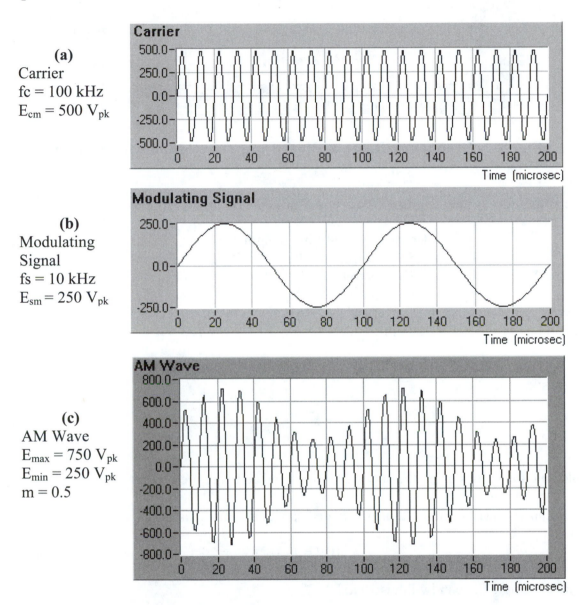

(a)
Carrier
fc = 100 kHz
E_{cm} = 500 V_{pk}

(b)
Modulating
Signal
fs = 10 kHz
E_{sm} = 250 V_{pk}

(c)
AM Wave
E_{max} = 750 V_{pk}
E_{min} = 250 V_{pk}
m = 0.5

Fig. 8-1   Time Domain Wave (DSBFC) of the Tone Modulated Carrier

Let's look now at the ***Frequency Domain*** (notice, the horizontal scale is frequency in kHz) representation of the above tone modulated AM wave, as shown in Fig. 8-2. In this illustration, there are three spectral lines: the carrier is at 100 kHz, the upper sideband (USB) is at 110 kHz, and the lower sideband (LSB) is at 90 kHz.

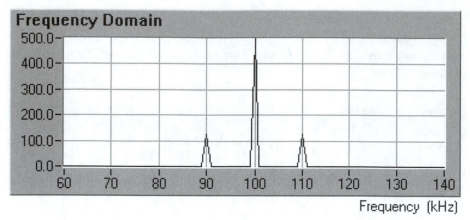

Fig. 8-2 Frequency Domain Plot of the Tone Modulated Carrier

The amplitudes or the heights of the spectral lines are defined as follows:

$$\text{Carrier Spectral Line Length} = E_{cm} \qquad (8\text{-}6)$$
$$\text{USB Spectral Line Length} = E_{USB} = \tfrac{1}{2}\, E_{sm} \qquad (8\text{-}7)$$
$$\text{LSB Spectral Line Length} = E_{LSB} = \tfrac{1}{2}\, E_{sm} \qquad (8\text{-}8)$$

Check Fig.8-2 and use the above equations to confirm that the spectral line length is 500 V, and that of the lower and upper sidebands $E_{LSB} = E_{USB} = 125$ V.

For the tone modulated carrier, the frequencies of the USB and the LSB are defined as follows:

$$f_{USB} = f_C + f_S \qquad (8\text{-}9)$$
$$f_{LSB} = f_C - f_S \qquad (8\text{-}10)$$

Calculate the values using the above equations and compare them to those in Fig. 8-2.

The bandwidth (BW) of the tone modulated carrier is defined as

$$BW = f_{USB} - f_{LSB} = 2f_S \qquad (8\text{-}11)$$

In this example the bandwidth is 20 kHz.

AM Carrier Modulated by Multiple Frequency Information

As shown in Fig. 8-3, the modulating signal now contains many frequencies between f_{min} and f_{max}. This is the signal that will now be used to amplitude modulate the carrier resulting in the DSBFC type of AM wave. The Time Domain representation of the AM wave in this case is difficult to show graphically, but its Frequency Domain plot can be easily done:

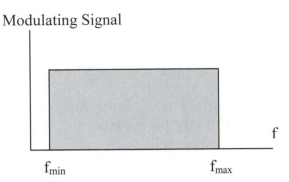

Fig. 8-3 Modulating Signal Includes Frequencies from f_{min} to f_{max}

The resulting Frequency Domain representation of the AM (DSBFC) wave is shown in Fig. 8-4. Remember that the AM wave now consists of the carrier modulated by a signal containing many frequencies. The carrier shown by a single spectral line is at a frequency f_c, the USB extends from f_c+f_{min} to f_C+f_{max}, and the LSB extends from f_c-f_{max} to f_C-f_{min}.

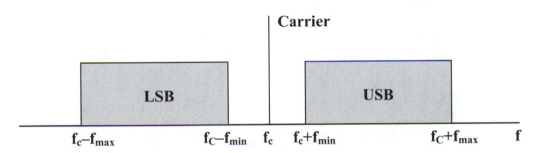

Fig. 8-4   Frequency Domain Plot of Carrier Modulated by Multi-Frequency Signal

If, for example, the modulating signal extends from $f_{min} = 2$ kHz to $f_{max} = 20$ kHz and the carrier frequency is 100 kHz, then our Frequency Domain representation of this AM wave might look like that shown in Fig. 8-5.

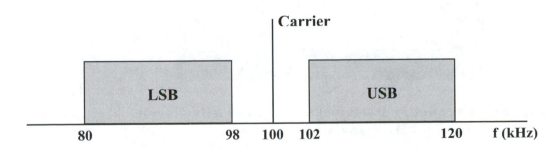

Fig. 8-5 Frequency Domain Plot of a 100 kHz Carrier Modulated by a 2–20 kHz Signal

The bandwidth of the AM wave where the carrier is modulated by a multifrequency signal can be expressed as follows:

$$BW = 2f_{max}$$

(8-12)

And in this example the bandwidth is 40 kHz (120 – 80 or 2 × 20).

In the computer simulation (LabVIEW) shown in Fig. 8-6, a 100 kHz carrier is amplitude modulated by two signals, one at 10 kHz and the other at 20 kHz. First, notice the complex Time Domain wave; this is something that would be difficult to sketch.

But the Frequency Domain plot shows exactly what we would expect: the carrier frequency spectral line at 100 kHz, the USB includes the 110 and 120 kHz spectral lines, and the LSB consists of the 90 and 80 kHz spectral lines with the BW of this AM wave being 40 kHz (120 – 80).

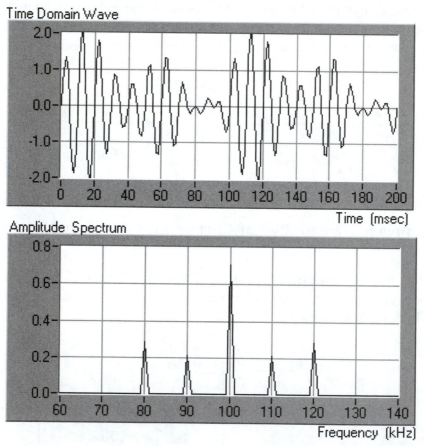

Fig. 8-6 Frequency Domain Plot of a 100 kHz Carrier Modulated by 10 and 20 kHz Signals

Power Distribution in AM Wave (Tone Modulated Carrier)

Power in the upper and lower sidebands is the same and may be expressed as

$$P_{USB} = P_{LSB} = (E_{sm})^2/(8R_L) = \tfrac{1}{4}\,P_c m^2 \qquad (8\text{-}13)$$

where m is the modulation index expressed as a decimal, P_c is the carrier power, E_{sm} is the peak value of the modulating signal and R_L is the load resistance. The carrier power is expressed by

$$P_c = (E_{cm})^2/(2R_L) \qquad (8\text{-}14)$$

where E_{cm} is the peak value of the carrier voltage and R_L is the load resistance. The total power in the AM wave is

$$P_T = P_c + P_{USB} + P_{LSB} \qquad (8\text{-}15)$$

After substituting Eq. 8-13 and 14 into Eq. 8-15 and factoring out P_c, we get Eq. 8-16 (another equation for the total AM wave power) that is equivalent to Eq. 8-15.

$$P_T = P_c(1 + \tfrac{1}{2}\,m^2) \qquad (8\text{-}16)$$

Suppose that the peak value of the carrier voltage is 1000 Vpk, $R_L = 100\,\Omega$, and $E_{sm} = 600\ V_{pk}$. Then

$$P_c = (E_{cm})^2/2R_L = (1000)^2/(200) = 5\ kW$$
$$m = E_{sm}/E_{cm} = 600/1000 = 0.6$$
$$P_{USB} = P_{LSB} = (E_{sm})^2/(8R_L) = (600)^2/(800) = 450\ W$$

and the total power

$$P_T = P_c + P_{USB} + P_{LSB} = 5000 + 450 + 450 = 5900\ W$$

Let's check the total power by using Eq. 8-16:

$$P_T = P_c(1 + \tfrac{1}{2}\,m^2) = 5000(1 + \tfrac{1}{2}(0.6)^2) = 5900\ W$$

As expected, the results are the same.

This means that the total sideband power (upper plus the lower) is 900 W, and the total power in the AM wave is the sum of the carrier plus the sideband powers or 5900 W. Remembering that both sidebands contain the same information and no information is in the carrier, the above example shows that the AM wave contains only 450 W of useful power out of the total of 5900 W. Since the amount of transmitted power determines the distance that the AM wave will travel, the basic AM configuration double sideband full carrier (DSBFC) is very inefficient. Commercial AM radio signals are also DSBFC.

To show this inefficiency further, consider the DSBFC AM wave where m = 1. The ratio of the power in the carrier to the total power may be expressed as

$$P_c/P_T = P_c/(P_c (1 + \frac{1}{2})) = 2/3 \qquad (8\text{-}17)$$

This means that 2/3 or about 67% of the total AM wave power is in the carrier and only 33% in the sidebands. Since both sidebands contain the same information, only half of 33, or 16.5%, is the useful power. Its hard to imagine that of 100 W of transmitted power, 84 W is lost and only 16 W is useful. That is why single sideband suppressed carrier (SSBSC) has the advantage of being very efficient. But before covering SSBSC (also SSB), we must first consider the DSBSC.

Double Sideband Suppressed Carrier (DSBSC) AM Wave

The practical circuit that produces the DSBSC waveform is called the Balanced Modulator. As shown in Fig. 8-7, the Balanced Modulator multiplies the two inputs: carrier $e_c(t)$ and the information or the modulating signal $e_s(t)$. Its output is the DSBSC

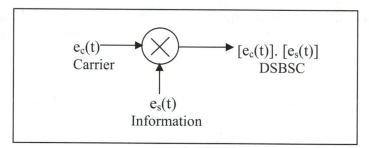

Fig. 8-7 A Balanced Modulator Circuit Produces DSBSC Wave

signal that includes both sidebands but not the carrier. This can be shown mathematically with trigonometric identities

$$\cos(A - B) = \cos(A)\cos(B) + \sin(A)\sin(B) \qquad (8\text{-}18)$$
$$\cos(A + B) = \cos(A)\cos(B) - \sin(A)\sin(B) \qquad (8\text{-}19)$$

After subtracting Eq. 8-19 from Eq. 8-18 and solving for the product of the sines, we get

$$\sin(A)\sin(B) = (0.5)[\cos(A - B) - \cos(A + B)] \qquad (8\text{-}20)$$

After replacing A by $\omega_c t$ and B by $\omega_s t$, where the former is associated with the carrier and the latter, the modulating signal, it is clear from the expression in Eq. 8-20 that the product of two sines includes only two terms, the sum of two frequencies (USB) and the difference of two frequencies (LSB) and no carrier, hence DSBSC. As shown in Fig. 8-8, both sidebands are present but the carrier is gone.

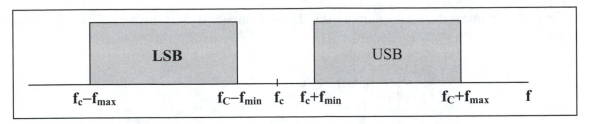

Fig. 8-8 Frequency Domain Plot for DSBSC. Carrier is Suppressed by the Balanced Modulator.

As an illustration of the DSBSC, a computer simulation is shown in Fig. 8-9 where a 200 kHz carrier is amplitude modulated in a Balanced Modulator by a 10 kHz modulating signal. The resulting Time Domain wave and the Frequency Domain Spectrum are shown in Fig. 8-9. First notice that in the Frequency Domain plot, the carrier at 200 kHz is missing, as it should be, and the sidebands occur at 190 kHz (200 – 10) and 210 kHz (200 + 10) as expected.

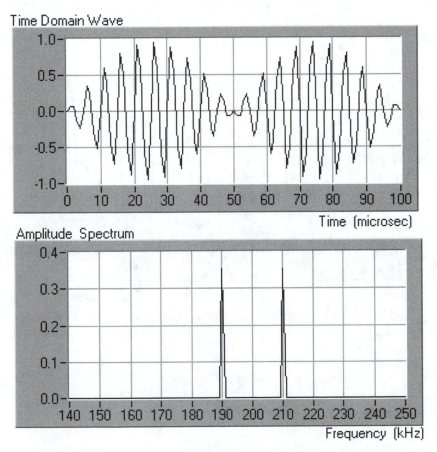

Fig. 8-9 DSBSC Time Domain and Frequency Domain Plots for a 200 kHz Carrier Modulated by a 10 kHz Signal

Of particular interest is this part of the Time Domain wave. It always has the same shape:

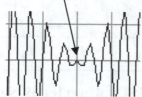

regardless of the carrier or the modulating signal amplitudes. Let's take another look at the Time Domain wave in Fig. 8-9 and see if we can determine why it looks the way it does.

The frequency of the modulating signal is 10 kHz , so its period is 100 μs (1/10 x 10^3). The frequency of the carrier is 200 kHz, so its period is 5 μs (1/200 x 10^3). How many cycles of the carrier will fit into one cycle of the modulating signal? If your answer is 20, you're right! 100/5 =20, 10 cycles in the first 50 μs and 10 in the next 50 μs as shown below.

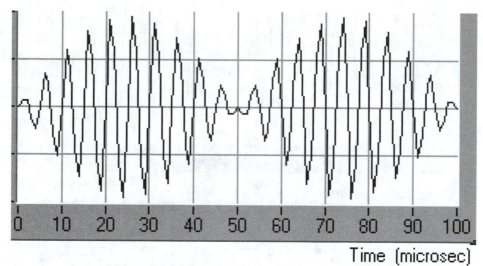

The 100 μs period of the 10 kHz modulating signal containing 20 cycles of the 200 kHz carrier

Single Sideband (SSB) AM Wave

Single sideband suppressed carrier (SSBSC or SSB for short) is widely used in modern communication technology. Citizen's Band (CB) communication, cellular, and, in general, wireless communication and other applications use SSB because it is very efficient. Less transmitted power is wasted and the required bandwidth is also less in comparison to the traditional DSBFC used by the broadcast AM radio. Let's consider the basic theory.

In Fig. 8-10, the product circuit or the Balanced Modulator produces DSBSC as described earlier. This signal is then filtered, eliminating one of the sidebands (either sideband can be removed). Removing one of the sidebands does not distort the information because the same information is contained within both sidebands.

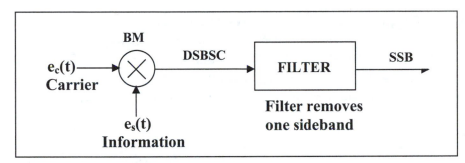

Fig. 8-10 A Balanced Modulator Followed by a Filter Produces SSB

The result of this process is shown in Fig. 8-11. The dotted lines show the elimination of the LSB and the carrier. The choice of the filter determines which sideband is to be suppressed.

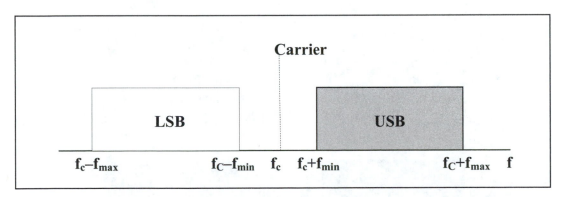

Fig. 8-11   Frequency Domain Plot of SSB Shows the Carrier and One Sideband Removed From a DSBSC signal

The numerous benefits of SSB transmission include

1. **Reduced Bandwidth**. The SSB signal occupies approximately one-half the frequency space as compared to the DSBFC.
2. **Power Saving.** The power used by the carrier and the eliminated sideband can now be channeled into one sideband.
3. **Reduced Noise.** Because SSB operates in a narrower bandwidth, its signal is less susceptible to noise than a comparable DSBFC signal.
4. **Less Fading.** Because the SSB signal operates in a narrower bandwidth than a comparable DSBFC signal, fewer frequencies are likely to travel different path resulting in spatial phase shifts, which contribute to fading at the receiver input.

Exercise 8-1: Amplitude Modulation

In this exercise you will use the provided software and experiment with AM waves including DSBFC, DSBSC, and SSB. Modulation as well as demodulation are considered. You will be able to vary parameter values and observe interactively the effect on the waveforms in the time domain (TD) and the frequency domain (FD).

Open AM.vi. (Book VIs>AM.vi)

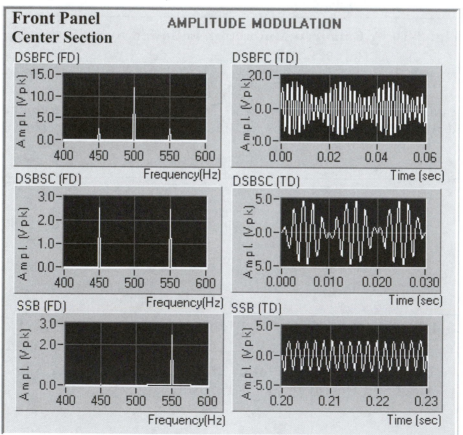

Fig. 8-12   The Front Panel and Block Diagram of Exercise 8-1

The Front Panel shown in Fig. 8-12 includes three sections: the center section, the right section, and the left section.

The **center section** displays the DSBFC, DSBSC, and SSB AM waves on Waveform Graphs in the Time Domain and in the Frequency Domain.

The **right section** is dedicated to the demodulation of the DSBSC AM wave. The sequential demodulation process is shown on three Waveform Graphs in the Frequency Domain.

The **left section** includes various controls and indicators. In the AM Parameters recessed box, digital controls set the amplitudes and frequencies of the carrier and the modulation signal. Also included here is the digital indicator for displaying the modulation index m, and the Stop switch to terminate VI execution.

The Sampling recessed box includes digital controls for setting the values of the sampling parameters. This includes the sampling frequency and the number of samples to be acquired.

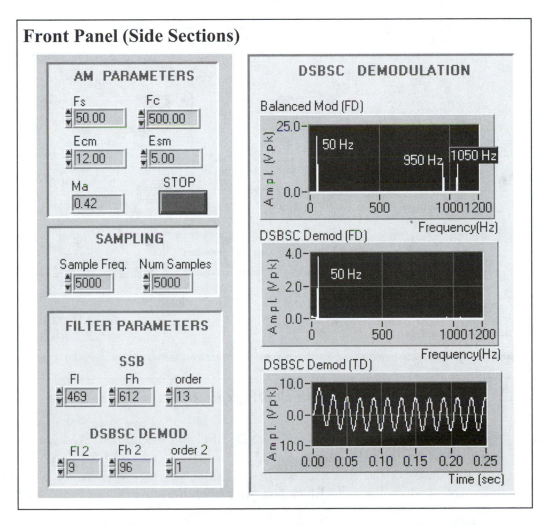

Fig. 8-12   The Front Panel and Block Diagram of Exercise 8-1 (continued)

The Filter Parameters recessed box contains controls for setting the upper cutoff frequency, lower cutoff frequency, and the order of the filter. As mentioned earlier, a filter is required in SSB generation to suppress on of the sidebands. Also, a filter is required in the demodulation process. When the DSBSC and the carrier are applied to the balanced modulator, its output includes the sum and difference frequencies. The filter rejects the sum frequencies and passes the difference frequency, which is the information or the modulating signal. In both cases a Butterworth filter is used.

The Block Diagram shown in Fig. 8-12 includes three distinct sections, each assigned a specific task. The tasks are accomplished inside subVIs. SubVIs are used to minimize diagram clutter.

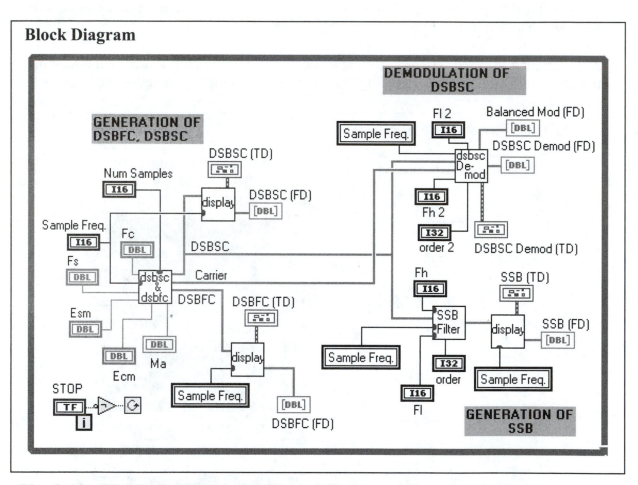

Block Diagram

Fig. 8-12   The Front Panel and Block Diagram of Exercise 8-1 (continued)

The **dsbsc & dsbfc** subVI generates the DSBFC and the DSBSC AM waves at its output together with the carrier wave. The carrier is required for demodulation. You may inspect the Block Diagram of this subVI to see software by double clicking on the icon and switching to Block Diagram when the subVI Front Panel opens.

The **SSB Filter** subVI takes the DSBSC input and, with the help of a filter, suppresses the LSB, producing USB at its output.

The **dsbsc Demod** subVI demodulates the DSBSC signal and recovers the modulating signal. To accomplish this task, the carrier is required. Inside the subVI the DSBSC signal and the carrier are multiplied (Balanced Modulator action), and the sum outputs from the product circuit are rejected by the filter that follows leaving only the modulating circuit. The Front Panel of Fig. 8-12 shows the demodulation process of a typical run in three waveform graphs. The first graph shows the output of the product circuit in the Frequency Domain, 50 Hz is the modulation, and the sums are 950 Hz and 1050 Hz because the carrier is set at 500 Hz and the DSBSC signal includes the 450 Hz and 550 Hz components. You may double click on this subVI and inspect the Block Diagram software.

Display.vi produces two outputs, one in the Frequency Domain and the other in the Time Domain. The Bundle function is used to format the X-axis scale. The "0" input to the Bundle starts the scale at 0, and the second input, the reciprocal of the sampling frequency (seconds per sample or the time between samples), provides the increment, Δt, between samples. The result is that the scale along the X-axis is in seconds. Display.vi also includes Amplitude and Phase Spectrum.vi for the Frequency Domain display.

Digital Frequency

A brief explanation is perhaps in order regarding the use of sampling frequency. It may appear odd that we are using sampling frequency in a simulation exercise where no data acquisition takes place. It must be remembered that LabVIEW is a digital environment that has nothing to do directly with the analog signals. Every signal generated or processed consists of dots or samples. For example, a sine wave shown below is made up of samples,

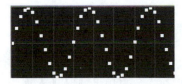

but when these samples are connected by choosing the appropriate interpolation in the Legend window, the same waveform will have the illusion of being continuous, as shown below.

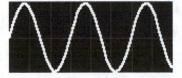

In the analog world we measure frequency in Hz or cycles per second, which has little meaning in a digital world that uses samples. Many waveform generating and filtering VIs in LabVIEW require digital frequency as input. Digital frequency is obtained by dividing the analog frequency by the sampling frequency as follows:

$$F_d = F_a/F_s \qquad \text{(8-21)}$$

where the units are: (cycles/sec)/(samples/sec) = cycles/sample. The reciprocal of F_d results in the number of samples per cycle. In the above illustration, analog frequency of 50 Hz and sampling frequency of 750 samples/sec were used. You may count the dots and find that there are 15 (750/50). If the Nyquist sampling rate rule is to be observed, which requires that the sampling rate must be at least twice the sampled frequency, then the digital frequency must be in the range of 0 to 0.5.

Frequency Modulation

When the frequency of the carrier is varied by the level of the modulating signal, the result is Frequency Modulation (FM). This is illustrated in Fig. 8-13.

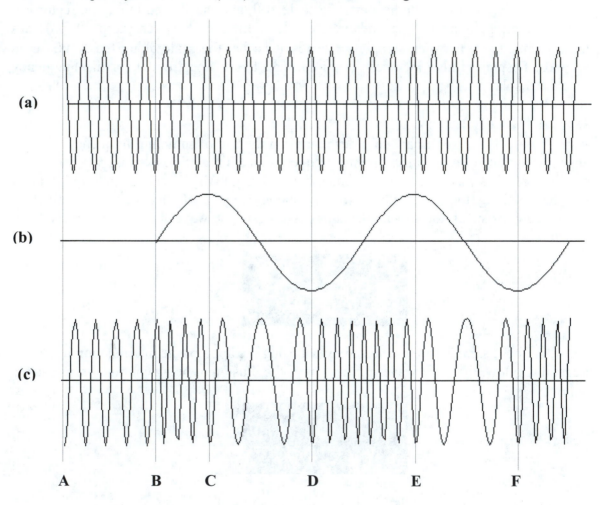

Fig. 8-13 Frequency Modulation: (a) Carrier, (b) Modulating Signal, (c) Frequency Modulated Carrier

Between points A and B in Fig. 8-13, the modulating signal is 0 V. Consequently the carrier frequency remains unchanged (period remains unchanged). Between points B and C as well as between points D and E, the voltage level of the modulating signal is increasing in the positive direction and, as can be seen in the bottom waveform, the carrier frequency is increasing (period of carrier is decreasing) as well. Between points C and D as well as between points E and F, the voltage level of the modulating signal is decreasing and so is the frequency of the carrier (period of carrier is increasing).

Throughout this process, the amplitude of the carrier remains unchanged. It should be emphasized that the change in carrier frequency, or **Carrier Frequency Deviation** as it is often called, depends upon the instantaneous voltage of the modulating signal.

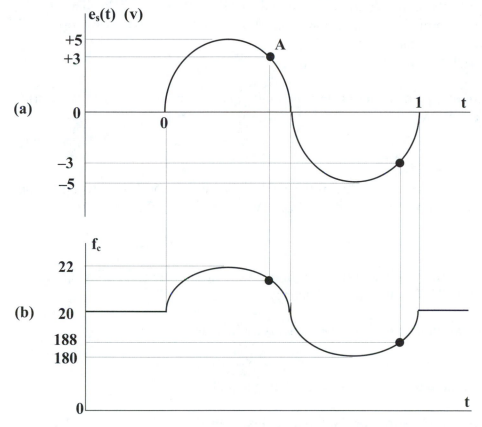

Fig. 8-14   (a) Modulating Signal, (b) Carrier Frequency Deviation

Fig. 8-14 illustrates a very important point about frequency modulation. It shows carrier frequency deviation that depends on the instantaneous voltage of the modulating signal. This is summarized in the following table:

Vs (V)	f_C (kHz)
0	200
+3	212
+5	220
−3	188
−5	180

Thus, the carrier rest frequency (no modulation) is 200 kHz. At the positive peak (+5 V) of modulation, carrier frequency swings to 220 kHz, and at the negative peak of modulation (−5 V), it swings to 180 kHz.

The frequency of the modulating signal is $(1/1 \times 10^{-3}) = 1$ kHz. This means that the frequency contractions and expansions shown in the bottom waveform of Fig. 1 are performed 1000 times per second.

The important quantity here is *maximum frequency deviation* (F_D) of the carrier, which is defined as follows:

$$F_D = kE_{sm} \qquad (8\text{-}22)$$

where E_{sm} is the peak voltage of the modulating signal and k is constant associated with the modulator. For example, if k = 4 kHz/V and E_{sm} = 5 Vpk, then according to Eq. 8-22

$$F_D = kE_{sm} = (4)(5) = 20 \text{ kHz}$$

This is the same value as illustrated in Fig. 8-13 where the carrier swing is 20 kHz above and below the 200 kHz carrier frequency.

The modulation index of the FM signal is defined as

$$m = F_D/f_s \qquad (8\text{-}23)$$

where f_s is the frequency of the modulating signal.

The percent modulation for the FM signal is defined as

$$\% \text{ Mod} = [F_D \text{ actual}]/ [F_D \text{ max}] \qquad (8\text{-}24)$$

Frequency Spectrum of FM Wave

In order to determine various spectral line voltages and powers, we must use the Bessel function table. Part of this table is illustrated below:

m	J_0	J_1	J_2	J_3	J_4	J_5	J_6	J_7	J_8 ...
0	1.00								
0.25	0.98	0.12							
0.5	0.94	0.24	0.03						
1.0	0.77	0.44	0.11	0.02					
1.5	0.51	0.56	0.23	0.06	0.01				
2.0	0.22	0.58	0.35	0.13	0.03				
2.5	−0.05	0.50	0.45	0.22	0.07	0.02			
3.0	−0.26	0.34	0.49	0.31	0.13	0.05	0.01		

Frequency spectrum determination will be illustrated by the following example

Example

Given: $E_{cm} = 1000$ V_{pk}, $f_c = 160$ kHz, $k = 2.5$ kHz/V_{pk}, $E_{sm} = 8$ V_{pk}, $f_s = 10$ kHz, RL = 50 Ω. Determine the voltage and power frequency spectra.

$F_D = kE_{sm} = (2.5)(8) = 20$ kHz
$m = F_D/f_S = 20 \times 10^3/10 \times 10^3 = 2$

The unmodulated carrier power is determined from
$P_{CU} = (E_{cm})^2/(2R_L) = (1000)^2/(100) = 10$ kW

The modulated carrier power is defined as

$$P_{CM} = (J_0{}^2)(P_{CU}) \tag{8-25}$$

Peak value of the modulated carrier voltage E_{CM} is defined in terms of the peak value of the unmodulated carrier voltage E_{cm} as

$$E_{CM} = J_0 E_{cm} \tag{8-26}$$

where J_0 is to found in the Bessel functions table. Hence the *modulatedJ carrier power and its peak voltage E_{CM}* are calculated from Eq. 8-25 and 8-26 as

$$P_{CM} = (J_0{}^2)(P_{CU}) = (.22)^2(10,000) = 484 \text{ W}$$
$$E_{CM} = J_0 E_{cm} = (0.22)(1000) = 220 \text{ Vpk}$$

The USB and the LSB that are closest to the carrier (separated from the carrier by f_s) are designated as the 1st sideband.

Its peak voltage and its power are defined as

$$E_1 = (J_1)(E_{cm}) \tag{8-27}$$
$$P_1 = J_1{}^2 P_{CU}$$

The second USB and LSB voltage and power (separated from the carrier by 2f$_s$) are defined in a similar way:

$$E_2 = (J_2)(Ecm) \text{ and} \qquad (28)$$
$$P_2 = J_2{}^2 P_{CU}$$

In a similar fashion the third USB and LSB voltage and power (separated from the carrier by 3f$_s$) are defined as

$$E_3 = (J_3)(Ecm) \qquad (29)$$
$$P_3 = J_3{}^2 P_{CU}$$

This process continues until all J values are used along the line corresponding to the modulation index m in the Bessel function table.

We next calculate the sideband voltages and powers. Since m = 2, there are only for sidebands in the Bessel function table using Eq. 8-27, 8-28, and 8-29. You should verify the results in the following table

J value	E (sideband in Vpk)	P (sideband in W)
J$_1$ = 0.58	580	3364
J$_2$ = 0.35	350	1225
J$_3$ = 0.13	130	169
J$_4$ = 0.03	30	9

This information is transferred to the frequency spectrum chart in Fig. 8-15.

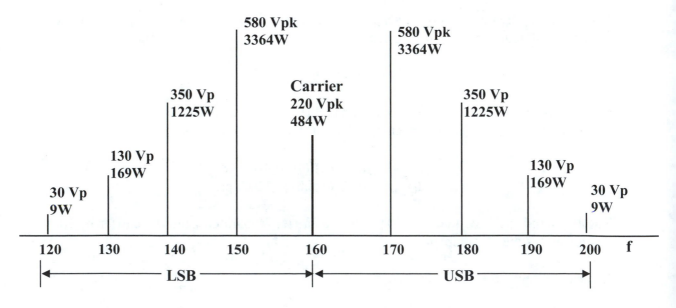

Fig. 8-15   Power/Voltage Frequency Spectrum for Above Example

The total power in an FM signal is defined as

$$P_T = P_{USB} + P_{LSB} + P_{CM} = P_{CU} \qquad (8\text{-}30)$$

The amplitude of the FM wave is constant, which is why the total carrier power is equal to the unmodulated carrier power. In frequency modulation only the frequency of the carrier varies and not its amplitude. As the modulation index changes, the power moves back and forth between the sidebands and the modulated carrier. It is possible to adjust the modulation index m for which $J_0 = 0$, thus forcing modulated carrier power to be equal to zero. This means that all of the power is in the sidebands and none in the carrier.

The bandwidth of the FM wave is defined as

$$BW = 2nf_S \qquad (8\text{-}31)$$

where f_S is the frequency of the modulating signal and n is the highest J subscript value that corresponds to a given m value. In this example m = 2, and the highest J subscript value is 4 (also the number of the upper or lower sidebands). Hence the bandwidth of the FM signal in this example is

$$BW = 2nf_S = (2)(4)(10\times10^3) = 80 \text{ kHz}$$

This value may also be calculated from the frequency spectrum in Fig. 8-15 as $200 - 120 = 80$ kHz.

Exercise 8-2: Frequency Modulation

This exercise illustrates the use of LabVIEW objects to simulate the process of frequency modulation (FM). FM.vi is included with this book. Should you decide to build it, the Front Panel and the Block Diagram are provided in Fig. 8-16.

Build FM.vi or open it (Book VIs>FM.vi).

In the Block Diagram of Fig. 8-16, the task of the For Loop is to generate the FM wave whose equation may be expressed mathematically in the time domain as follows

$$V_o(t) = E_{cm}\cos[2\pi f_c i/N + (ksE_{sm}/f_s)\sin(2\pi f_s i/N)] \qquad (8\text{-}32)$$

where

E_{sm}, E_{cm} are the peak voltages of the modulating signal and carrier, respectively
f_s, f_c are the frequencies of the modulating signal and the carrier, respectively
N is the total number of samples of $V_o(t)$ array
$F_d = ksE_{sm}$, the maximum deviation of carrier frequency
$m = ksE_{sm}/f_s$, the modulation index of the FM wave

The value of N is set to 1000 samples to be acquired. The iteration terminal **I** will, therefore, vary from 0 to 999, and on the count of 1000 the For Loop will terminate execution. At this time the 1000 elements of the one-dimensional array V_o stored in the

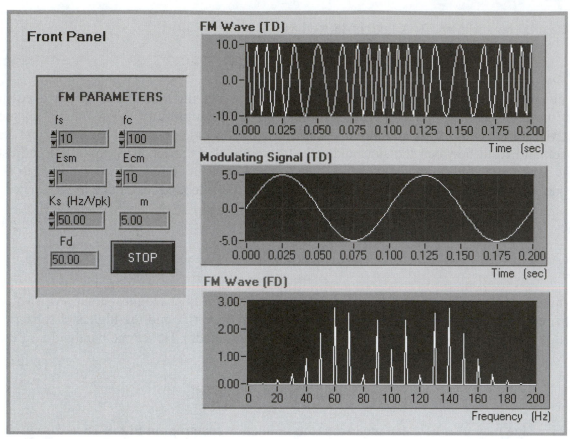

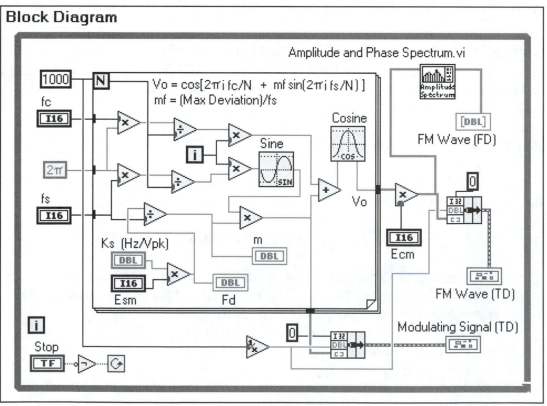

Fig. 8-16   The Front Panel and Block Diagram of Exercise 8-2

border of the For Loop will be passed outside the loop. For each iteration of the loop, the argument $2\pi f_s i/N$ of the modulating signal is calculated, the sine of this argument is calculated, and the result is multiplied by the modulation index m. The carrier's fixed argument is added to the preceding result $(k s E_{sm}/f_s)\sin(2\pi f_s i/N)$. This is now the total argument of the carrier, the first part is fixed and the sinusoidal portion provides the modulation of the carrier frequency. The cosine of this total argument is now calculated and multiplied by carrier's peak voltage E_{cm}. Despite all of this work only one point of $V_o(t)$ is calculated and placed in the array. This procedure is repeated N times (set to 1000 in Fig. 8-16).

One may wonder why the sine and cosine functions were used, requiring the loop to form the array and many other supporting objects. Why not use the sine and cosine pattern that generates the arrays and requires no loops? The answer is fairly simple. Upon inspection of the pattern VIs, one realizes that they are not set up to accept another function as an input (remember that we have to do a cosine of a sine.) Much of the FM software could have been accomplished with a Formula Node, which uses equations directly. The user may want to repeat this exercise with a Formula Node.

Outside the For Loop, we generate the frequency spectrum with Amplitude and Phase Spectrum.vi and format the scale for the X-axis so that it properly represents time in seconds. Should the Waveform Graph be used without such formatting, the resulting plot will be versus sample number.

The While Loop will repeat its iterations endlessly as long as the input to the Condition terminal is TRUE. For this to happen, the Stop switch must be OFF, or FALSE, which is its default setting. The VI execution will commence immediately after the RUN button is clicked. This provides an interactive environment where the user can instantly observe the result of a parameter change.

The VI execution will be terminated when the you click on the Stop switch. The mechanical action "Latch When Pressed" has been selected for the Stop switch. This action is shown in the adjoining illustration. When the switch (Boolean control) is depressed, the control's new value is retained until the VI reads it at least once. After the VI has read it, the switch reverts to its default value. This action, which is often used to stop the While Loops, will not be affected if you repeatedly press the button. Six types of mechanical actions are available for each switch. To see them, right-click on the switch in the Front Panel and choose the Mechanical Action option from the menu.

The recessed box labeled as FM Parameters contains digital controls for setting voltages and frequencies and two indicators for displaying maximum frequency deviation of the carrier F_d and the modulation index m.

Run this VI and experiment with different parameter values.

Transmission Lines

Any transmission line serves as a conducting medium for an electronic signal. There are many types of transmission lines. Some of the most common are shown in the illustration below.

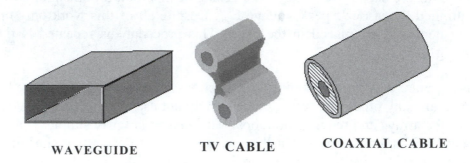

WAVEGUIDE　　　　**TV CABLE**　　　　**COAXIAL CABLE**

The waveguide is used predominantly at microwave frequencies above 3 GHz. The two wire transmission 300 Ω cable shown in the illustration above is used to connect the antenna to the television receiver. Finally, the coaxial cable has a special advantage because its outer conductor (braid) shields the inner conductor from external noise and other interference. Coaxial cables are used over a wide frequency range extending into the microwave range up to 10 GHz. They are often used to interconnect electronic equipment modules, and many cable service providers also use coaxial cables in their distribution network.

The understanding of the transmission line theory is important in high-frequency applications. The theory applies to the so-called long lines whose length is comparable to the wavelength. The wavelength is defined as the distance that a wave travels in the time duration of one cycle. Expressed mathematically the wavelength is defined as

$$\lambda = v_c/f \tag{8-33}$$

where v_c is the speed of light (3×10^{10} cm/s or 3×10^8 m/s) and f is the frequency of the signal. It is clear from the above equation that λ becomes small at very high frequencies, and the length of the line may be less than λ. When this happens, some or all of the incident signal may be reflected by the load in the manner of light striking a shiny surface. Combination of the incident and the reflected waves results in stationary or standing waves on the transmission line.

The negative effects of reflections and standing waves may be summarized as follows:
- **Loss of power at the load.** Because some of the incident power is reflected by the load, less power is absorbed by the load.
- **Cable insulation breakdown due to arcing.** In applications where voltages are large, standing wave voltage may be sufficiently large to cause arcing between the outer and inner conductors (in the case of a coaxial cable), destroying the cable.

- **Equipment damage.** In certain high-power applications, such as a transmitter delivering 50 kW of power to an antenna over a transmission line, the reflected power is travelling from load to the transmitter and when it reaches the transmitter, it may damage the final transmitter stage.

Transmission line theory applies to high-speed computers as well. Consider the PC shown in the illustration. It delivers high-speed data to a peripheral. We know from Fourier analysis that the high-frequency components that make up the pulse are contained in the leading and trailing

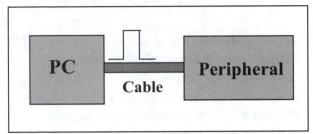

edges of the pulse. The lower the rise time, the higher the frequencies. For nanosecond pulses these frequencies extend into the microwave band, where the characteristics of the connecting cable must be carefully selected.

The transmission line theory that specifically applies to the LabVIEW software in Exercise 8-3 is presented here. This theory should provide the reader with a better understanding of the results emerging from the exercise.

Transmission Line Equations

The transmission line equations are generally derived from the T model that includes an incremental length Δx, as shown in the illustration below. The series effects of the line (R and L) are represented by Z and the shunt effects (G and C), which are due to the insulating material between the two conductors of the transmission line, are represented by Y.

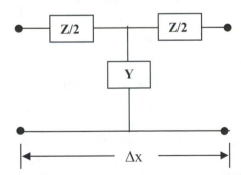

R, L, G, and C are distributed parameters given by the manufacturer on a per-unit-length basis. For example, if $C = 40$ pF/m then a 10 cm length of the line will have a capacitance of 4 pF (40×0.1).

The final form of transmission line equations for voltage, current, and impedance may be summarized as follows:

$$E(y) = (I_L/2)[(Z_L + Z_o)e^{j\beta y} + (Z_L - Z_o)e^{-j\beta y}] \tag{8-34}$$
$$I(y) = [I_L/(2Z_o)][(Z_L + Z_o)e^{j\beta y} - (Z_L - Z_o)e^{-j\beta y}] \tag{8-35}$$

$$Z(y) = E(y)/I(y)$$
$$= Z_o[(Z_L + Z_o)e^{j\beta y} + (Z_L - Z_o)e^{-j\beta y}]/[(Z_L + Z_o)e^{j\beta y} - (Z_L - Z_o)e^{-j\beta y}]$$
$$= Z_o[(1 + |\Gamma_v|e^{-j(2\beta y - \psi)}]/[1 - |\Gamma_v|e^{-j(2\beta y - \psi)}] \tag{8-36}$$

An explanation regarding the properties and characteristics of these equations is in order.

Transmission Line Equations Depend on Distance

Time is one of the implicit independent variables because these waves are sinusoidal in nature, and distance y along the transmission line is the other independent variable. This is unusual if one considers low frequency applications (short line relative to the wavelength) where measured voltage is the same at all points along a conductor, but not at microwave frequencies where standing waves are involved.

Incident and Reflected Waves

In the illustration below, a load on the right is connected to the generator on the left by a transmission line. Because it is generally easier to measure currents on the load side, the transmission line equations are expressed in terms of load current and load impedance. There are instances when this is not done. For the zero distance, reference is taken at the load. Hence, y = 0 is at the load and y increases in the positive direction from right to left.

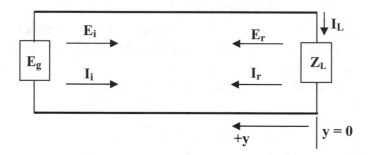

Eq. 8-37 expresses total voltage on the line as a function of distance y. This equation is in two parts.

$$E(y) = E_i + E_r = (I_L/2)(Z_L + Z_o)e^{j\beta y} + (I_L/2)(Z_L - Z_o)e^{-j\beta y} \tag{8-37}$$

Notice that the first part of Eq. 8-37 is the incident voltage that is launched by the generator and is travelling at slightly less than the speed of light toward the load. The second part of Eq. 8-37 represents the voltage reflected by the load (in the manner of light that is reflected by a surface) that is traveling toward the generator.

Eq. 8-38 expresses total current on the line as a function of distance y. This equation is also in two parts, the sum of the incident and reflected waves.

$$I(y) = I_i + I_r = [I_L/(2Z_o)](Z_L + Z_o)e^{j\beta y} - [I_L/(2Z_o)](Z_L - Z_o)e^{-j\beta y} \tag{8-38}$$

The first part represents the incident current, and the reflected current is expressed by the second part.

The incident and the reflected waves combine to form a stationary wave referred to as a standing wave. Standing waves include current and voltage when their respective incident and reflected components combine. The standing wave amplitude can be measured at any point on the line. Slotted lines are designed for exactly that purpose. Because the incident and reflected components have a magnitude and phase, they must be added as vectors in determining the resultant.

Voltage Reflection Coefficient and VSWR

The voltage reflection coefficient Γ_v is defined as

$$\Gamma_v = E_r/E_i = |\Gamma_v|e^{j\psi} = [Z_L - Z_o]/[Z_L + Z_o] \qquad (8\text{-}39)$$

It is also a vector whose magnitude is $|\Gamma_v|$ and whose phase is ψ. For a given transmission line Γ_v depends only on the value of the load impedance Z_L and perhaps for that reason it is measured at the load ($y = 0$). As shown in the illustration, ψ is the angle between the incident and reflected voltages at the load. The magnitude $|\Gamma_v|$ represents the percentage of incident voltage reflected by the load. Therefore its value is in the range of 0 to 1. A value $|\Gamma_v| = 0$ means no reflection or all of the incident voltage is absorbed by the load, and $|\Gamma_v| = 1$ signifies that all of the incident voltage is reflected by the load.

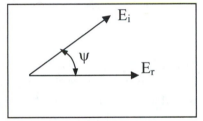

The Voltage Standing Wave Ratio (VSWR) is expressed as follows:

$$\rho = E_{max}/E_{min} = [1 + |\Gamma_v|]/[1 - |\Gamma_v|] \qquad (8\text{-}40)$$

This is a real quantity representing largeness of the standing waves as a ratio of maximum voltage of the standing wave to its minimum. Because VSWR is a function of Γ_v, it is indirectly a function of the load impedance. The value of ρ ranges from 1, corresponding to $|\Gamma_v| = 0$, to infinity when $|\Gamma_v| = 1$. $\rho = 1$ is an ideal, never attained in practice.

Characteristic Impedance

Characteristic impedance Z_o of the transmission line is defined in terms of Z and Y of the transmission, the series and shunt effects discussed earlier, as

$$Z_o = [Z/Y]^{1/2}$$

After substituting for Z and Y we get

$$Z_o = [(R + j\omega L)/(G + j\omega C)]^{1/2}$$

If the real terms in the above equation are smaller than the corresponding imaginary terms, which is often the case at microwave frequencies, then we get an approximate form:

$$Z_o = [L/C]^{1/2} \qquad\qquad (8\text{-}41)$$

Maximum and Minimum Values

From a purely intuitive standpoint, when the numerator in $Z = E/I$ is as large as possible and the denominator is as small as possible, the value of Z must be a maximum. Conversely, when the numerator is as small as possible and the denominator is as large as possible, we must be referring to the minimum value of Z. As a result of this reasoning we conclude that

$$Z_{max} = E_{max}/I_{min} \qquad\qquad (8\text{-}42)$$
$$Z_{min} = E_{min}/I_{max} \qquad\qquad (8\text{-}43)$$

According to Euler's identity, $e^{\pm j\theta} = \cos(\theta) \pm j\sin(\theta)$, then
$$e^{-jk\pi} = -1, \text{ k odd}$$
$$e^{-jk\pi} = +1, \text{ k even}$$

Substituting $2\beta y - \psi = k\pi$ in Eq. 8-36, we see that when k is even, the numerator of $Z(y)$ is maximum and its denominator is minimum, resulting in Z_{max}. At this point on the line we also have E_{max} and I_{min}. And when the value of k is odd, the numerator of $Z(y)$ is minimum and its denominator, maximum, resulting in Z_{min}. At this point on the line we also have E_{min} and I_{max}. Solving for y we get a set of equations that are used to determine the distance to voltage maxima and minima:

$$y_{max} = (k\pi + \psi)/(2\beta) \quad k = 2, 4, 6, \ldots \text{ (if } E_{min} \text{ occurs first)} \qquad (8\text{-}44)$$
$$y_{max} = (k\pi + \psi)/(2\beta) \quad k = 0,2, 4, 6,\ldots \text{(if } E_{max} \text{ occurs first)} \qquad (8\text{-}45)$$

$$y_{min} = (k\pi + \psi)/(2\beta) \quad k = 1, 3, 5, 7, \ldots \qquad\qquad (8\text{-}46)$$

Eq. 8-44 and 8-45 are used to locate voltage maxima and minima. Hence the first voltage minimum occurs at $k = 1$, the second minimum at $k = 3$, the third minimum at $k = 5$ and so on. Regarding the distance from the load to the various maxima, it depends on whether a voltage maximum or a voltage minimum is closest to the load. If a voltage minimum occurs first, then $k = 2$ locates the first voltage maximum in Eq.8-44. But if voltage maximum is closest to the load, then $k = 0$ locates its position in Eq. 8-45.

The values of Z_{max} and Z_{min} are deduced from the preceding discussion as

$$Z_{max} = \rho Z_o \qquad\qquad (8\text{-}47)$$
$$Z_{min} = Z_o/\rho \qquad\qquad (8\text{-}48)$$

As stated earlier, the total voltage on the line is a vector sum of the incident and reflected components, and the maximum value of standing wave voltage on the transmission line must occur when the two components are aligned so that they aid each other ($\psi = 0°$).

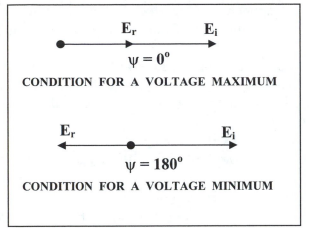

Using similar reasoning, the voltage minimum must occur when the incident and the reflected voltage components oppose each other ($\psi = 180°$). This is illustrated in the adjoining diagram.

In conclusion, at the point on the transmission line where voltage is a maximum, the impedance is also maximum and the current is minimum; and the point on the transmission line corresponding to a voltage minimum also corresponds to the impedance minimum and current maximum. The table below provides a summary.

All standing waves repeat their waveform pattern every $\lambda/2$.

PARAMET	$y_{max} = (k\pi + \psi)/2\beta$ k even	$y_{min} = (k\pi + \psi)/2\beta$ k odd
Z	$Z_{max} = \rho Z_o$	$Z_{min} =$
E	$|E_{max}| = |E_i| +$	$|E_{min}| = |E_i| -$
I	$|I_{min}| = |I_i| - |I_r|$	$|I_{max}| = |I_i| + |I_r|$

Propagation Factor

The exponential terms, $e^{-j\beta y}$, that appear in Eqs. 8-34, 35 and 36 are somewhat different in their original form. The $-j\beta$ term is derived from the propagation factor γ. It is defined as

$$\gamma = \alpha + j\beta \qquad (8\text{-}49)$$

where α is the attenuation constant, and β is the phase factor. The attenuation constant specifies the amount of attenuation of the incident or the reflected waves per unit distance. In this section we are considering only the lossless lines. Short lines, for example, are likely to behave like lossless lines. The phase factor defined by

$$\beta = 2\pi/\lambda \qquad (8\text{-}50)$$

The phase factor specifies the amount of phase change of the incident and the reflected components as they move along the line. Finally, by equating $\alpha = 0$ for a lossless line, $\gamma = j\beta$, hence, the form of Eqs. 8-34, 8-35, and 8-36.

Four Cases of Infinite VSWR

The four types of load impedance that result in an infinite VSWR are shown in Fig. 8-17: short circuit, open circuit, pure L, and pure C.

(a) Short Circuit ($Z_L = 0$)

$|\Gamma_v| = 1$, $\psi = 180°$, $E_{min} = 0$, y_{min} at load, $\rho = \infty$, y_{max} is $\lambda/4$ from load

(b) Open Circuit ($Z_L = \infty$)

$|\Gamma_v| = 1$, $\psi = 0°$, $E_{min} = 0$, y_{max} at load, $\rho = \infty$, y_{min} is $\lambda/4$ from load

(c) Pure Inductance ($Z_L = j\omega L$)

$|\Gamma_v| = 1$, $E_{min} = 0$, $\rho \infty$, y_{max} is $< \lambda/4$ from load

(d) Pure Capacitance ($Z_L = 1/[j\omega C]$)

$\Gamma_v| = 1$, $E_{min} = 0$, $\rho \infty$, y_{max} is $< \lambda/4$ from load

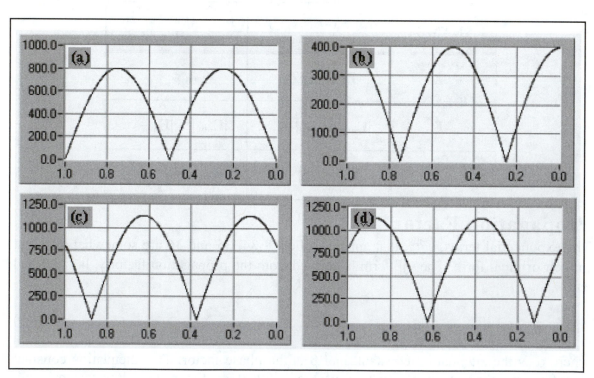

Fig. 8-17   Four Cases of Infinite VSWR: (a) $Z_L = 0$, (b) $Z_L = \infty$, (c) $Z_L = $ Pure L, (d) $Z_L = $ Pure C

Five Cases of Finite VSWR

The five types of load impedance that result in finite VSWR are shown in Fig. 8-18:
Z_L pure R > Z_o, Z_L pure R < Z_o, Z_L = R-L combination, Z_L = R-C combination, Z_L = Z_o.

(a) Z_L pure R > Z_o

$\psi = 0^\circ$, y_{max} at load, y_{min} is $\lambda/4$ from load

(b) Z_L pure R < Z_o

$\psi = 180^\circ$, y_{min} at load, y_{max} is $\lambda/4$ from load

(c) Z_L = R + jωL

y_{max} is < $\lambda/4$ from load

(d) Z_L = R − j(1/ωC)

y_{min} is < $\lambda/4$ from load

(e) Z_L = Z_o

$\Gamma_v = 0$, $\rho = 1$, $E_r = 0$

Case (e) represents a **matched condition** between the load and the transmission line, resulting in no reflections, $E_r = 0$. The incident wave is completely absorbed by the load. Note that in the illustration in Fig. 8-18, the RMS value of the incident voltage is the same everywhere on the line.

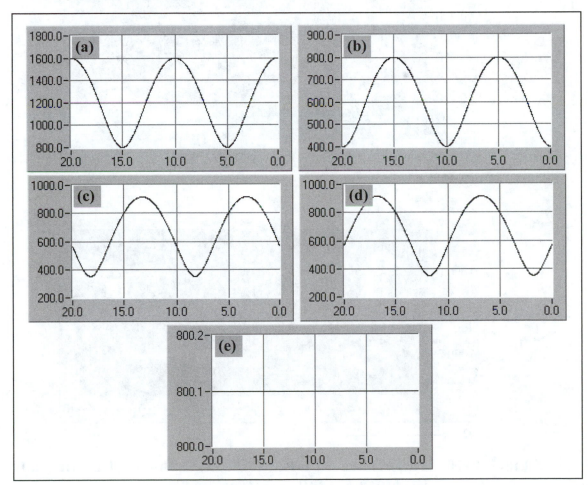

Fig. 8-18   Cases of Finite VSWR: (a) Z_L pure R > Z_o, (b) Z_L pure R<Z_o, (c) Z_L = R + jωL, (d) Z_L = R − j/ωC, (e) Z_L = Z_o

Open Circuit and Short Circuit Loads

Although open circuit (infinite impedance) and short circuit are not the type of load impedances that one is likely to use, their properties should be briefly considered.

When infinity is substituted for Z_L in Eq. 8-36 an indeterminate situation results because infinity is divided by infinity. The indeterminacy may be easily resolved, and the final form of impedance is

$$Z(y) = -jZ_o\cot(\beta y) \qquad (8-51)$$

Similarly, the final expression for short circuit load is

$$Z(y) = jZ_o\tan(\beta y) \qquad (8-52)$$

Simulation runs done on LabVIEW produce the impedance plots shown in Fig. 8-19.

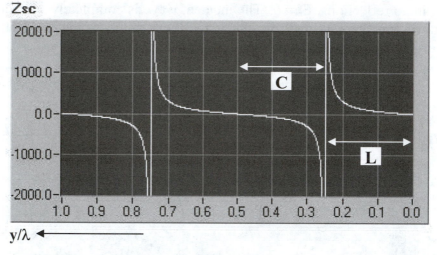

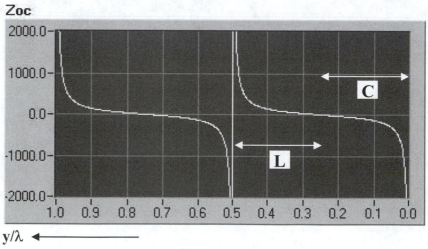

Fig. 8-19 Impedance Plots for (a) Short Circuit and (b) Open Circuit

The following properties of the open and short circuited lines may be deduced from these plots. As shown in the illustration here, the line may behave capacitive or inductive at its input, depending on its length.

A λ/4 length of a line whose load is a short circuit and a λ/2 length of a line whose load is an open circuit both have infinite impedance (open circuit) at their input. Recall that the impedance of an ideal (no resistance) parallel resonant circuit at resonant frequency is infinity.

On the other hand, a λ/4 length of a line whose load is an open circuit or a λ/2 length of a line whose load is a short circuit both have zero impedance (short circuit) at their input. Recall that the impedance of an ideal (no resistance) series resonant circuit at resonant frequency is zero.

Discrete passive components such as an inductor or a capacitor have parasitic effects, and they don't behave the way they should at microwave frequencies. For this reason, at microwave frequencies the inductive and the capacitive effects are synthesized with sections of transmission lines.

As an example, consider the following requirement. Determine the length of short circuited 100 Ω transmission line at 750 MHz, so that the input impedance to the line behaves like a 10 pF capacitor.

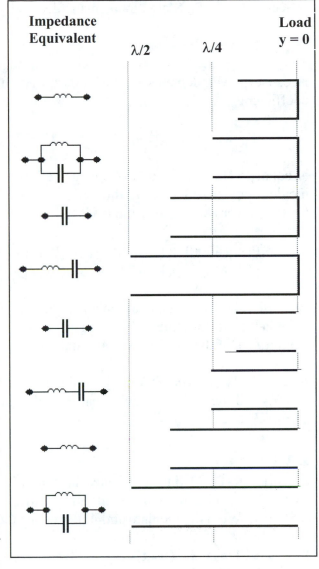

Substituting $Z(y) = -j10.6$ Ω, the capacitive reactance of the 10 pF capacitor at 750 MHz, in Eq. 8-52

$$-j10.6 = jZ_o \tan(\beta y)$$
$$\beta y = 2\pi y/\lambda = 3.04$$
$$y = 19.3 \text{ cm} = 7.7 \text{ in.}$$

$$Z_0 = 100 \text{ Ω}$$
$$10 \text{ pF}$$
$$Z_L = 0 \text{ Ω}$$
$$\longleftarrow 19.3 \text{ cm} \longrightarrow$$

Power Distribution on the Transmission Line

The power absorbed by the load is the difference between the incident power P_i and the reflected power P_r. Stated mathematically

$$P_L = P_i - P_r \tag{8-53}$$

Because power depends on the square of voltage, it follows that the power reflection coefficient

$$|\Gamma_p| = P_r/P_i = |\Gamma_v|^2 \tag{8-54}$$

is equal to the square of the voltage reflection coefficient. It represents the percent of incident power reflected by the load. Using Eq. 8-54 in Eq. 8-53 we get the expression for the power absorbed by the load.

$$\begin{aligned} P_L &= (|E_i|)(|I_i|)(1 - |\Gamma_v|^2) = (I_L^2)(Re(Z_L)) \\ &= (|E_i|^2/Z_o)(1 - |\Gamma_v|^2) = (I_L^2)(Re(Z_L)) \end{aligned} \tag{8-55}$$

Because incident voltage and incident current are in phase, their ratio is the characteristic impedance of the line, a real quantity, and the incident power is a product of the magnitudes of the incident voltage and the incident current as expressed by Eq. 8-55.

Alternatively, power absorbed by the load may be calculated as the product of load current and the resistive or the real part of load impedance, $Re(Z_L)$, because the reactive part of Z_L does not dissipate power.

Example 8-1

Consider a 100 Ω line terminated in $Z_L = 60 - j40$ Ω. The frequency of the generator signal is 1.5 GHz and the load current $I_L = 8$ A. Load current is taken as the phase reference. We will calculate and illustrate the theory previously discussed.

1. **Voltage Reflection Coefficient and VSWR**
   From Eq. 8-39 and 8-40
   $$\Gamma_v = [Z_L - Z_o]/[Z_L + Z_o] = [60 - j40 - 100]/[60 - j40 + 100] = (0.34)(\angle{-121°})$$
   Hence $|\Gamma_v| = \mathbf{0.34}$
   $$\psi = \mathbf{-121°}$$
   $$\rho = [1 + |\Gamma_v|]/[1 - |\Gamma_v|] = (1 + 0.34)/(1 - 0.34) = \mathbf{2.03}$$
   ....

2. **Maximum and Minimum Impedance on the Line**
   From Eq. 8-42 and 8-43

   $$Z_{max} = \rho Z_o = (100)(2.03) = \mathbf{203\ \Omega}$$
   $$Z_{min} = Z_o/\rho = (100)/(2.03) = \mathbf{49.3\ \Omega}$$

3. Incident and Reflected I and E

From Eq. 8-37

$E_i = (I_L/2)(Z_L + Z_o)e^{j\beta y} = (8/2)(60-j40 + 100)(1 \angle \beta y) = 659.7 \angle \beta y - 14°$

Hence

$$|Ei| = \textbf{659.7 V}$$
$$\angle Ei = \boldsymbol{\beta y - 14°}$$

where

$$\beta = 2\pi/\lambda = 2\pi/20 = 0.314 \text{ rad/cm}$$
$$\lambda = v_c/f = 3\times10^{10}/1.5\times10^9 = 20 \text{ cm}$$

$E_r = (I_L/2)(Z_L - Z_o)e^{-j\beta y} = (8/2)(60-j40 - 100)(1 \angle \beta y) = 226.3 \angle 225° - \beta y$

Hence

$$|Er| = \textbf{226.3 V}$$
$$\angle Er = \boldsymbol{225° - \beta y}$$

The phase change due to the fact that waves travel is expressed by βy. Taking the value of $\beta = 0.314$ rad/cm, we can see that the incident wave undergoes a phase change of $180°$ $[(0.314)(10)]$ as it travels toward the load. The reflected wave also undergoes the same change in phase except that it travels toward the load.

The incident and reflected currents are taken from Eq. 8-38:

$I_i = [I_L/(2Z_o)](Z_L + Z_o)e^{j\beta y} = (8/200)(60-j40+100)(1 \angle \beta y) = 6.6 \angle \beta y - 14°$

where

$$|Ii| = \textbf{6.6 A}$$
$$\angle Ii = \boldsymbol{\beta y - 14°}$$

$I_r = -[I_L/(2Z_o)](Z_L - Z_o)e^{-j\beta y} = (1 \angle 180°)(8/200)(60-j40-100)(1 \angle -\beta y) = 2.3 \angle 45 - \beta y$

where

$$|Ir| = \textbf{2.3 A}$$
$$\angle Ir = \boldsymbol{45° - \beta y}$$

....

4. Maxima and Minima of I and E

As discussed previously, the condition for voltage maxima and minima is shown in this illustration. Using the values from part 3 of this example

$|E_{max}| = |Ei| + |Er| = 659.7+226.3$
$\qquad\qquad = \textbf{886 V}$

$|E_{min}| = |Ei| - |Er| = 659.7-226.3$
$\qquad\qquad = \textbf{433.4 V}$

CONDITION FOR A VOLTAGE MAXIMUM

$\psi = 0°$

$E_r \qquad E_i$

CONDITION FOR A VOLTAGE MINIMUM

$\psi = 180°$

$E_r \qquad\qquad E_i$

Applying the same procedure for the currents

$|I_{max}| = |Ii| + |Ir| = 6.6+2.3 = \textbf{8.9 A}$
$|I_{max}| = |Ii| - |Ir| = 6.6-2.3 = \textbf{4.3 A}$

5. **Distances to the Minima and Maxima**

Because the given load impedance is capacitive, the voltage minimum of the standing waves is closest to the load; therefore, Eq. 8-46 is used in calculating the distance to the first voltage minimum (k = 1) from the load as

$$y_{min1} = (k\pi + \psi)/(2\beta) = [(180 - 121)/(4 \times 180)] = .08\lambda = (.08)(20) = \textbf{1.6 cm}$$
or $\quad (y/\lambda) = \textbf{0.08}$

$$y_{min2} = (k\pi + \psi)/(2\beta) = [(3 \times 180 - 121)/(4 \times 180)] = 0.58\lambda = (0.58)(20) = \textbf{11.6 cm}$$

Since standing waves repeat their pattern every half wavelength, we can predict that the third minimum will be at $11.6 + 10 = 21.6$ cm.

Because the voltage minimum is closest to the load, we may use Eq. 8-44 in calculating the distance from the load to the first voltage max

$$y_{max1} = (k\pi + \psi)/(2\beta) = (2 \times 180 - 121)/(4 \times 180)\lambda = 0.33\lambda = 0.33 \times 20 = \textbf{6.6 cm}$$
or $\quad (y/\lambda) = \textbf{0.33}$

It can be verified the second voltage maximum occurs for k = 4 at $\textbf{0.83}\lambda = \textbf{16.6 cm}$. Another way to calculate the second voltage maximum is to realize that the successive maxima and minima are always separated by $\lambda/4$. This is true for voltage, current and impedance.

The minima and maxima for the current are deduced from the fact that current is maximum at the point on the line where voltage is minimum and vice versa. Thus

$$y_{max1} = 0.33\lambda \qquad y_{min1} = 0.08\lambda$$
$$y_{max2} = 0.83\lambda \qquad y_{min2} = 0.58\lambda$$
$$y_{max3} = 1.33\lambda \qquad y_{min3} = 1.08\lambda$$

The impedance maxima and minima are the same as those for voltage.
The first three minima and maxima for E, I, and Z are summarized in the table below.

	y_{min1}	y_{min2}	y_{min3}	y_{max1}	y_{max2}	y_{max}
E	0.08λ	0.58λ	1.08λ	0.33λ	0.83λ	1.33λ
I	0.33λ	0.83λ	1.33λ	0.08λ	0.58λ	1.08λ
Z	0.08λ	0.58λ	1.08λ	0.33λ	0.83λ	1.33λ

Fig. 8-20 shows a LabVIEW simulation run using the given values in Example 8-1. The six cursors in each of the standing waves provide accurate readings. The simulation values agree with the calculations of the example.

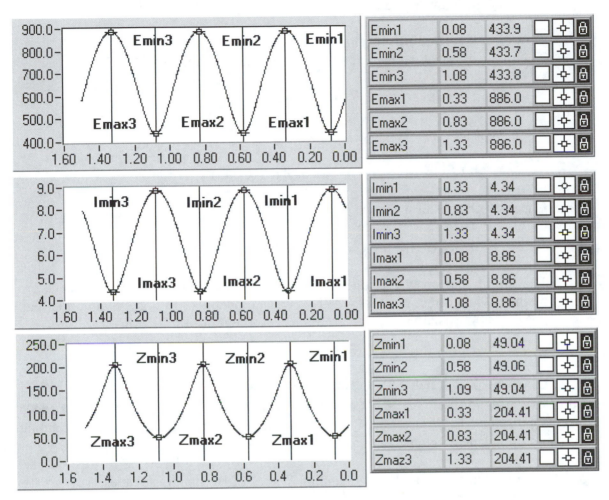

Fig. 8-20   E(y), I(y), Z(y) Standing Waves for Example 8-1

6.   E, I, and Z at Arbitrary Point on the Line
Let's calculate E, I and Z 4 cm from the load.
From Eq. 8-34
$$E(y) = (I_L/2)[(Z_L + Z_o)e^{j\beta y} + (Z_L - Z_o)e^{-j\beta y}] = E_i + E_r$$
Using the values for E_i and E_r from step 3 of this exercise and substituting for
$\beta y = (2\pi/\lambda)y = (360/20)(4) = 72°$

$$E(4\ cm) = 659.7 \underline{/58°} + 226.3 \underline{/153°} = 349.6 + j559.5 - 201.6 + j102.7$$
$$= 148.0 + j662.2 = 678 \underline{/77.4°}\ V$$

$$I(4cm) = I_i(4) + I_r(4) = 6.6 \underline{/58°} + 2.3 \underline{/-27°} = 3.5 + j5.6 + 2.0 - j = 5.5 + j4.6$$
$$= 7.2 \underline{/39.9°}\ A$$

$$Z(4) = E(4)/I(4) = [= 678\underline{/77.4°}]/[7.2\underline{/39.9°}] = 94.2\underline{/34.5°}\ \Omega$$

The reader should use the following form of impedance from Eq. 8-36
$$Z(y) = Z_o\,[(1 + |\Gamma_v|e^{-j(2\beta y - \psi)}]/[1 - |\Gamma_v|e^{-j(2\beta y - \psi)}]$$
to verify the above result.

7. **Power Distribution on the Line**
    The incident power on the line is calculated using values from step 3 as

$$P_i = |E_I|^2/Z_o = (659.7)^2/100 = 4352\ W$$

From Eq. 8-55

$$P_L = (|Ei|^2/Z_o)(1 - |\Gamma_v|^2) = (4352)[1 - (0.343)^2] = 3834\ W$$

Checking this result with

$$P_L = (I_L{}^2)[\mathrm{Re}(Z_L)] = (8)^2(60) = 3840\ W$$

The slight discrepancy is due to rounding of Γ_v. If more decimal places are retained, the correct result of 3840 W will be obtained. In this example the load reflects 4352 − 3840 or 512 W, which is travelling toward the generator.

Exercise 8-3   The Standing Waves

In this exercise you will use the LabVIEW simulation VI to test and verify various concepts presented in this section. The VI provided with this book will plot standing waves and impedances, and perform a variety of calculations.

Open Standing Waves.vi (Book VIs>Standing Waves.LLB>Standing Waves.vi).

We will examine the Front Panel and the Block Diagram of Standing Waves.vi and explain its features.

As shown in Fig. 8-21, the Front Panel includes four recessed boxes.

The **Standing Waves** recessed box includes four X-Y graphs. They are as follows:
   The **E(y) and I(y)** waveform graphs display voltage and current standing waves respectively.
   The **Regular Z** waveform graph displays impedance plots for all load conditions except two: the open circuit and the short circuit.
   The **Special Z** waveform graph displays the impedance variation along the line for two load conditions: the open circuit or the short circuit. The Special Z slider control in the recessed box labeled *Special Settings* can be set to display the open circuit case or the short circuit case. The string indicator above the Special Z graph will decide which type of Z_L is being processed, the open circuit or the short circuit.

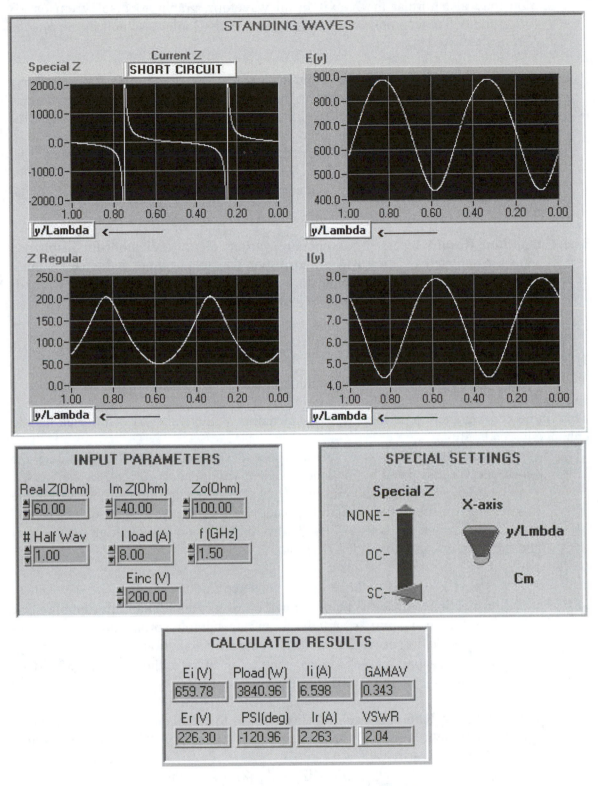

Fig 8-21 The Front Panel of Standing Waves.vi

You may also format the X-axis in all waveform graphs to display cm or y/λ (y/Lambda). To accomplish this switch the X-axis Boolean switch in the *Special Settings* recessed box.

The **Input Parameters** recessed box contains various digital controls for specifying various parameter values, such as real and imaginary components of load impedance (load current, frequency, etc.). There is a digital control for E_{inc}, the incident voltage. This value is used only for the case of the open circuit load, where the incident voltage is undefined; it will not be used by the software for all other load conditions. To specify the open circuit load condition, enter "0" in the Im Z digital control, and any value greater than 50,000 in the Real Z digital control.

The **Calculated Results** recessed box contains various digital indicators for displaying values calculated by the software corresponding to the input parameter values. The names of most of these indicators are intuitive; for example, GamaV stands for Γ_v, and Psi represents ψ, the angle of the voltage reflection coefficient.

Block Diagram

The Block Diagram shown in Fig. 8-22 includes a Sequence structure with five frames.
Frame 0
This frame is designated to perform various calculations. Several subVIs are used here to enhance its readability.

Pow_Inc_Ref.vi is a subVI whose designated task is to calculate the power dissipated by the load, incident voltage, and current, as well as the reflected voltage and current. The calculated values may be viewed in the Front Panel on the digital indicators inside the *Calculated Results* recessed box. You may examine the software inside this subVI by double-clicking on the icon and then switching to the Block Diagram.

Psi.vi is a subVI whose task is to determine the angle ψ of the voltage reflection coefficient Γ_v. Examination of its Block Diagram (double-click on the icon) reveals a tedious and meticulous procedure that checks all four quadrants to ensure that the trigonometric calculations are done properly. If the imaginary part of Z_L is zero, then ψ must be either 0° or 180°. This assignment is done by the nested Boolean Case structure.

Gamav_Rho.vi is a subVI whose task is to calculate the voltage reflection coefficient and VSWR by implementing into software the equations presented earlier:

$$|\Gamma_v| = |Z_L - Z_o|/|Z_L + Z_o| \quad \text{and}$$
$$\rho = [1 + |\Gamma_v|]/[1 - |\Gamma_v|]$$

Block Diagram

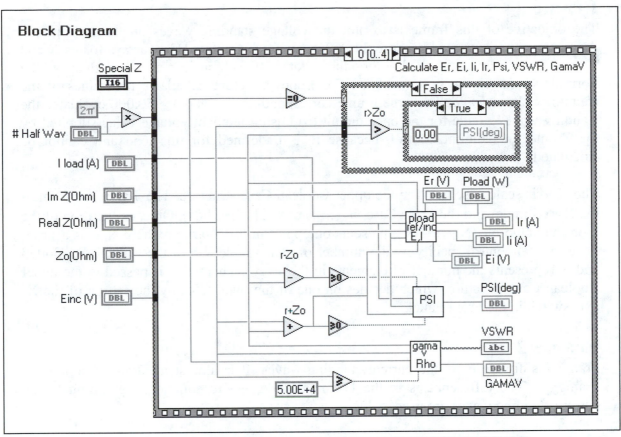

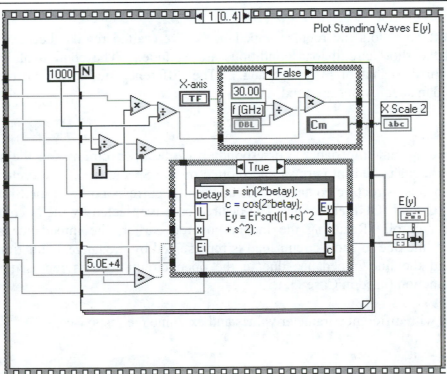

Fig 8-22 The Block Diagram Frames 0, 1, 2, 3, 4 of Standing Waves.vi

Frame 1

The objective of this frame is to plot the voltage standing waves on the X-Y graph. Arrays must be generated because the X-Y waveform graph requires arrays for its X and Y axes. This is done with the For loop that creates 1000 element arrays for both axes. The Formula Node (at the bottom of the For Loop) structure calculates the values of the standing voltage wave. The True Frame corresponding to $Re(Z_L) > 50,000$ evaluates the standing wave voltage for the open circuit load using incident voltage value supplied by the Front Panel digital control, because E_i is undefined for this load and cannot be calculated.

The X-axis scale is selected in the upper Boolean Case structure. The True Case formats the X-axis in cm and the False Case, in terms of y/λ (y/Lambda). On each iteration of the loop the value of $(N_{HW})(i)/N$ represents one point in the X-array, where N_{HW} is a Front Panel digital control specifying the number of half wavelength of the wave to be plotted and N represents the number of elements in the arrays. This value is passed to the upper Boolean Case structure, which decides on one of the two scales by the setting of the X-axis switch in the Front Panel.

Frame 2

Frame 2 software plots the current standing waves. It is almost identical in content to Frame 1. The difference is in the Formula Node, where equations pertaining to the current standing waves are used.

Frame 3

Frame 3 software plots the regular impedance curve (called regular because the open circuit and the short circuit load conditions are excluded). The content of Frame 3 is almost identical to that of Frames 1 and 2. The difference lies in the Formula Node, where impedance equations are used.

Frame 4

In this frame special impedances corresponding to the open circuit and the short load conditions are calculated into arrays and plotted on the Special Z X-Y waveform graph. The Boolean Case structure contains cases 0, 1, and 2 corresponding to the Short Circuit (SC), Open Circuit (OC), and None. The desired case is selected by the *Special Z* Front Panel slide control. Recalling the theory covered earlier, the impedance along the transmission line due to short circuit load is represented by the tangent function (used in Case 0), and the line impedance due to the open circuit load is represented by the cotangent function (used in Case 1).

Experiment with different parameter values and examine the responses.

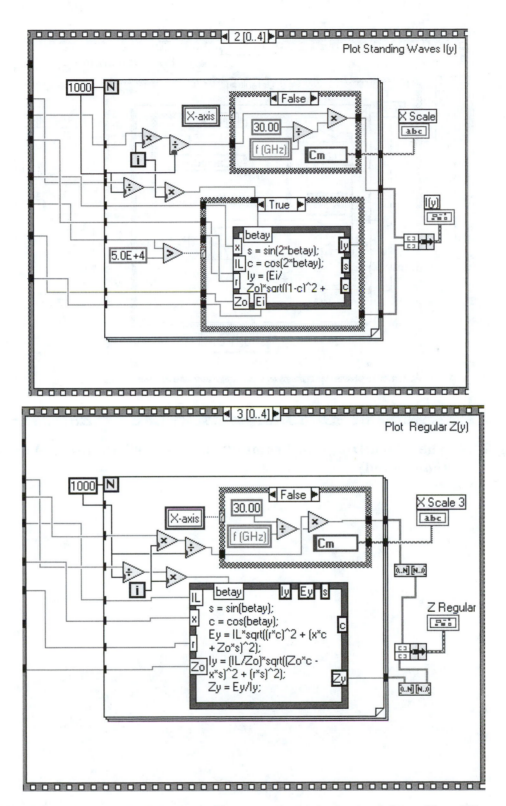

Fig 8-22 The Block Diagram Frames 0, 1, 2, 3, 4 of Standing Waves.vi

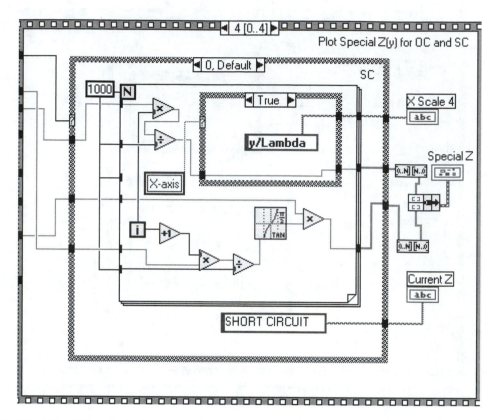

Fig 8-22 The Block Diagram Frames 0, 1, 2, 3, 4 of Standing Waves.vi (continued)

Chapter 9
About Data Acquisition

The Analog Signal and the Digitizing Process

In the world of modern electronics we are often confronted with the task of controlling a physical variable, processing a physical variable, verifying the behavior of a physical variable, or perhaps doing something else with a physical variable.

All physical variables, including motion, pressure, and temperature are rooted in the physical manifestation of nature and have nothing to do with electronics, as their behavior is characterized by molecular behavior in the case of temperature, or variations in the density of air in the case of air pressure.

All electronic equipment is made up of boards, chips, and other devises. The operation of chips and electronic devices is based on voltage and current, and so cannot act directly on physical variables such as temperature or pressure. In order to process a physical variable we must first convert its behavior parameters from physical form to electronic form in terms of voltage and current. If, in addition, we wish to use a computer as part of the process, we then have additional obstacles to overcome. The operation of a computer is based on voltages and currents as is the case with all electronic devices, but in addition the computer operation uses a code made up of 0's and 1's.

It can be seen from the above discussion that special steps must be taken before the parameters of a physical variable are in the proper form suitable for computer processing. The first step involves parameter conversion from the physical form to the electronic form. As shown in the illustration below, a transducer (also called a sensor) converts the physical variable, temperature, to a voltage signal. We call this type of signal an analog signal because its waveshape is continuous, and not broken or discrete as is the case with digital signals.

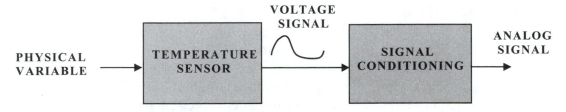

The output from the temperature sensor is generally small and not suitable for processing. Consequently, additional signal conditioning is in order. A typical signal conditioning circuit is an amplifier, which provides voltage gain to the weak signal, thus making it larger in amplitude and suitable for processing. Signal conditioning may be this simple, or more involved including substantial signal shaping and filtering. In any case, with the help of the signal conditioning circuit, the physical variable, temperature in this case, is now ready for processing.

This signal is an analog signal that can be processed by an analog circuit. It should be stressed that this signal is an electronic representation of a physical variable, made possible through the use of a temperature transducer. There are many other transducers that convert acceleration, position, velocity, and a variety of other physical variables to electronic form.

In today's world, however, computers are often used to process analog signals. This involves an additional procedure to convert the analog signal to digital form.

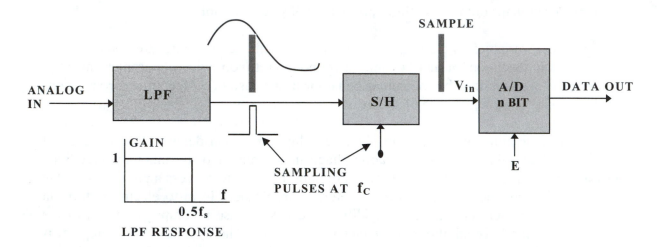

Fig. 9-1 A Typical Digitizing Circuit

A typical digitizing circuit is shown in Fig. 9-1. The heart of the digitizing circuit is the analog-to-digital converter (A/D, sometimes called ADC). It accepts a constant input voltage and produces a unique n-bit code at its output. Some of the important ADC parameters include the following:

Resolution

Resolution is defined as the smallest analog input that the converter can detect. It is defined mathematically as

$$R = [\text{Voltage Range}]/[G \cdot 2^N] \qquad (9\text{-}1)$$

where: Voltage Range is the full scale (FS) voltage at output of ADC
G is the gain of the instrumentation amplifier preceding the ADC
N is the number of bits of the ADC

For example, a 12-bit converter with a voltage range of 0–10 V (dynamic range) has a resolution of 2. 44 mV ($10/2^{12}$). Hence, if the analog input to the ADC changes by less than 2.44 mV, the output code of the converter will remain unchanged. In this example a

12-bit ADC divides the 10 V dynamic range of the converter into 4096 (2^{12}) discrete levels and represents each of the levels by a unique 12 bit code.

Fig. 9-2 illustrates a simple 3-bit ADC with an 8 V dynamic range that digitizes a triangular waveform. The resolution in this case is 1 V ($8/2^3$). Notice that each whole

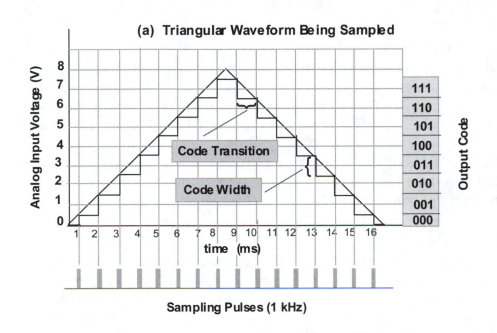

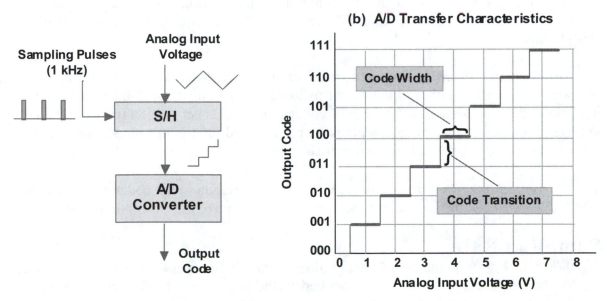

Fig. 9-2 ADC Converter Characteristics: (a) Sampling Triangular Waveform ADC Transfer Characteristics

number representing the input analog voltage occurs in the middle of a code. For example, the code 011 represents all input voltages between 2.5 V and 3.5 V. This is, of course, within the resolution of 1 V but at the same time one can't help but realize that an error is associated with the digitizing process, specifically in this case; namely, that all voltage variations over a range of one volt are represented by the same code. The price that must be paid for digitizing a signal is the loss of some of the input information. This error is called the *Quantization Error* and its value is $\pm \frac{1}{2}$ LSB. Since the output changes by 1 LSB when the input changes by 1 V, the quantization error in this case is 1/2 V.

Differential Linearity

In practice, code width (see Fig. 9-2) varies from one interval to the next. The differential linearity specification places limits on the width of the code. If the manufacturer specifies a differential linearity of $\pm \frac{1}{2}$ LSB, the width of the code can be anywhere from 0.5 LSB to 1.5 LSB ($1 \pm \frac{1}{2}$ LSB). Excessive differential non-linearity can lead to missed codes.

The errors that can distort the ideal picture of A/D conversion shown in Fig. 9-2 include *offset error, gain error, linearity error,* and *differential non-linearity.*

Accuracy

The accuracy of an ADC is often specified in terms of linearity errors: differential and integral. The ideal converter has the conversion response shown in Fig. 9-2, where points at the top of each transition, if connected, fall on the straight line. Any deviation from this straight line represents an ADC error. In an ideal converter the difference between the midpoint of any code width and the midpoint of an adjacent code width is exactly 1 LSB. In a practical converter this difference can vary, resulting in deviation from a straight line.

The manufacturer specifies this type of an error as the difference between the ideal midpoint of a code width and the actual measured midpoint of the code width. These deviations are expressed as a % of the FS voltage, or in parts per million (ppm) of the FS voltage. (Remember that % represents parts per hundred, and ppm represents parts per million, so that 0.01 percent is the same as 100 ppm.)

The *differential linearity error* is concerned with the variation in the code width, and the *integral linearity error* is concerned with the overall shape of the conversion curve.

Sampling Rate

As shown in Fig. 9-1, the signal in a digitizing circuit first passes the low-pass filter (LPF), the anti-aliasing filter, then is applied to the sample and hold circuit (S/H) and finally to ADC. For each sampling pulse applied to the S/H circuit a sample of analog input waveform is created and, as shown, applied to the ADC.

The sampling rate specifies the frequency at which the analog waveform is to be sampled. According to Nyquist, your sampling frequency must be at least twice the highest frequency component in the waveform being sampled. When sampling a sine wave whose frequency is 25 kHz, for example, the minimum sampling frequency should be 50 kHz (see Fig. 9-1).

Nyquist's sampling theorem sets the theoretical minimum sampling rate. However, in practical applications the actual sampling rate is higher.

Although the actual sampling rate can be set by the user in a particular application, the maximum sampling rate is set by the design of the board. In fact, the maximum sampling rate depends on the A/D converter settling time specification. The A/D converter experiences a transient response after each conversion, and the next conversion cannot be done until the transients subside to an acceptable level, as determined by the settling time.

For example, the MIO-16E-10 is a multipurpose I/O data acquisition board with 16 analog input channels and a settling time of 10 μs. The maximum sampling rate for this board is 100 kbps (1/10 μs) if data is acquired on one channel. If data is acquired on two analog channels, the maximum sampling rate is 50 kbps; if on four channels, 25 kbps on four channels, and so on.

In conclusion, when a continuous analog signal is digitized, the digitizing circuit represents the signal by a sequence of dots or samples, as shown below, and each sample is represented by an n-bit code. Should the original signal exhibit some unusual behavior between samples, the digitized signal will never see it. The sampling signal rate can easily be adjusted if that is a concern.

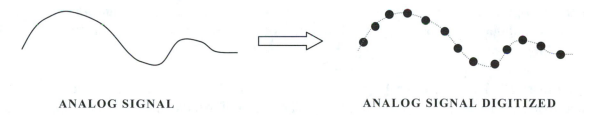

ANALOG SIGNAL   **ANALOG SIGNAL DIGITIZED**

The digitized signal can now be processed by software or transmitted to another location. The LabVIEW software, for example, will process the digitized signal.

If there is a need to return to the analog world, the digital-to-analog converter (D/A or DAC) can perform this task as shown below.

Analog signal multiplexing allows many analog signals to be processed almost simultaneously. As shown in Fig. 9-3, the multiplexer (MUX) is shown symbolically as a rotating switch that makes brief contact with each of the analog input channels. The

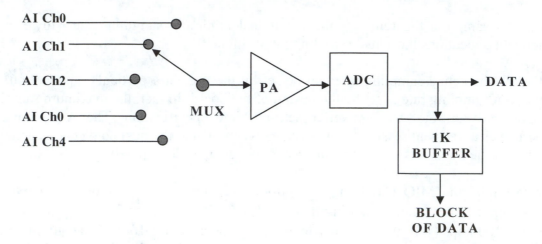

Fig. 9-3 A Multiplexed Data Acquisition Scheme

Programmable Amplifier (PA), whose gain is software controlled, provides signal conditioning, and the resulting signal is applied to the A/D converter. The LabVIEW data acquisition environment supports this kind of arrangement, although many of their data acquisition boards, such as LabPC$^+$ and MIO–16E–10, have eight analog input channels and the ADC circuit has 12-bit resolution. The dynamic range is typically 10 V, which can be configured as ± 5 V or 0 to 10 V, and the maximum sampling rate for the above boards is about 100 kbps. In general, the cost of the board increases as the speed of the board increases.

At times the data rate is so high that software can't keep up. LabVIEW offers data acquisition VIs for just such a situation. The data is buffered in FIFO, and the VI, as shown in Fig. 9-3, retrieves blocks of data at regular intervals. Buffered data acquisition that uses timers on DAQ board offers a realistic alternative to synchronized high speed data acquisition. In this book we will be using this method.

Data Acquisition Hardware and Software

The **DAQ Board** plugs into your PC's expansion slot. It serves as the interface between the real world and the computer, allowing data to enter or leave the computer (see Fig. 9-4).

The DAQ board contains many components that are necessary for data acquisition. A typical board has 8 analog input data channels that are multiplexed and applied to the instrumentation amplifier. The A/D converter digitizes the analog data. The on-board FIFO (**F**irst **I**n **F**irst **O**ut memory) provides a temporary storage of data in buffered data acquisition applications.

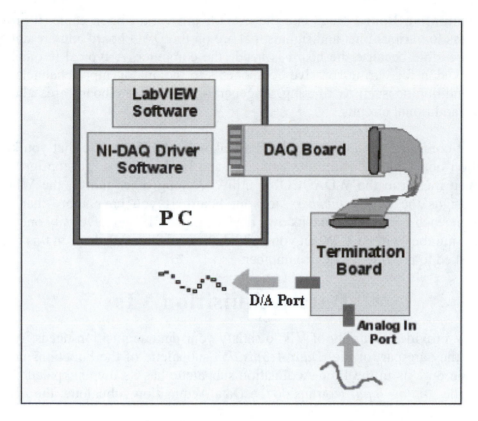

Fig. 9-4 Data Acquisition Hardware/Software

There are also two D/A converters that convert digital data to analog form and pass it to the analog output ports for use by external devices.

Digital I/O ports allow exchange of digital data between the computer and external devices.

The on-board counter/timer chips provide interaction with the hardware. In many applications, such as motion control, software overhead is excessive, resulting in software being unable to keep up with a control application. In that case the use of timed waveforms can be quite effective in speeding up the application.

National Instruments (NI) offers several DAQ boards that vary in speed and resolution.

The **NI-DAQ Driver Software** is a high-level software for the plug-in DAQ boards. NI-DAQ takes care of all low-level hardware and system programming, providing transparent DMA (direct memory access) and interrupt services. It has routines for analog and digital I/O, counter/timer, calibration, and configuration. It also provides support for RTSI (Real-Time System Interface bus, developed by National Instruments for DAQ board products), where multiple DAQ boards are synchronized in a data acquisition process.

The **Termination Board** plugs into the DAQ board via a ribbon cable. It provides you with access to various pins and signals that are on the DAQ board and are not otherwise easily accessible because the board is inside the computer. A typical termination board has a provision for wiring passive components to the analog input channels to provide signal conditioning such as filtering. A general-purpose breadboard area allows you to construct additional circuits.

You must configure the DAQ board. This makes LabVIEW aware that you have a data acquisition board plugged into the computer and ready for use. To configure the DAQ board, you must run the WDAQCONF utility. Among other things, the WDAQCONF utility assigns the *device number* to your board. If you have more than one board plugged in, each board will have its own device number. If you have only one board, then its device number will be 1. When you build data acquisition VIs later in this chapter, you will be asked to provide the device number.

Data Acquisition VIs

LabVIEW provides a variety of VIs to satisfy your data acquisition needs. As shown in Fig. 9-5, they are part of the *Data Acquisition* subpalette of the Functions palette. Each of the icon options in the Data Acquisition subpalette has its own subpalette. When you click on the *Analog Input* icon inside the Data Acquisition subpalette, the Analog Input subpalette opens with various analog input VIs.

The Data Acquisition subpalette also includes *Analog Output*, *Digital I/O*, *Counter*, and other subpalettes.

Easy VIs   perform most common I/O operations. They are easy and simple to use because the complexity of configuring and setting up the data acquisition VI is designed into the Easy VI. They usually include the Intermediate VIs, which in turn are made up of advanced VIs. The Easy VIs, however, lack flexibility and power. *AI Sample Channel.vi* and *AI Acquire Waveform.vi* are examples of Easy VIs.

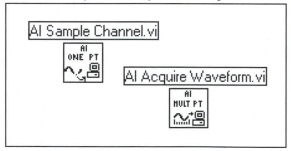

Intermediate VIs are made up of Advanced VIs. They have more power and flexibility. Examples of Intermediate VIs include AI Read.vi and AI Single Scan.vi, shown in this illustration, and others. The three VIs in the Input Utility Analog Input subpalette, shown in Fig. 9-2, are also Intermediate VIs and are examples of special-purpose applications that offer solutions to common analog input problems.

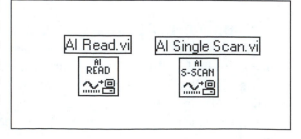

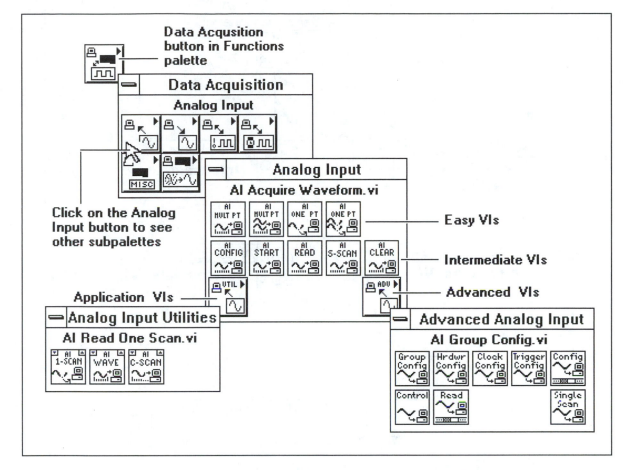

Fig. 9-5 Analog Input Data Acquisition VIs

Advanced VIs are the fundamental building blocks for all data acquisition VIs. They have maximum power and flexibility and are not more difficult to learn than the Intermediate VIs. Although you get more power and flexibility using Advanced VIs, you may also find that your Block Diagram has a lot more blocks and various other objects.

Consider *AI Sample Channel.vi,* which is an Easy VI. This VI will occupy only one block in your Block Diagram if you use it. But if you decided to accomplish the same thing with Intermediate and Advanced VIs, then instead of one block, you will have eight blocks: five Advanced VIs, two intermediate VIs, and General Error Handler.vi from the Time & Dialog palette. See the hierarchy diagram of Fig. 9-6.

Of course, you will have the opportunity to provide your own parameter values rather than accepting the default values programmed into the Easy VI. This is where the power and the flexibility come in when you use the Advanced and the Intermediate VIs. In any case, there is more work in using the Advanced VIs, and your Block Diagram may be a lot more cluttered. But you will have the opportunity to fine-tune various parameters and thus optimize the performance of your VI.

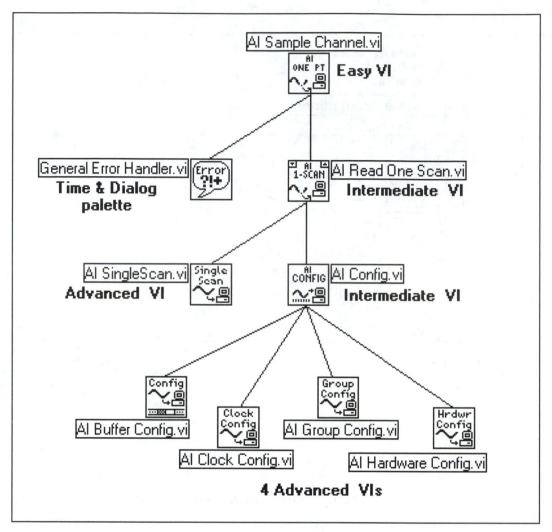

Fig. 9-6 Hierarchy Diagram for AI Sample Channel.vi

Data Acquisition Programming Options

As mentioned earlier, the Intermediate and the Advanced data acquisition VIs give you greater flexibility in programming. If you decide to use the Intermediate or the Advanced VIs, you should be aware of several programming options. You should also be familiar with some terminology that will be used in this section.

A ***Scan*** is one acquisition or one reading from each analog or digital input channel used in a data acquisition process.

An ***Update*** is one write to each of the analog output channels. Typically, a DAQ board has two analog output channels or ports. Each of these ports is driven by a D/A converter. Remember that data coming from the computer is in digital form and must be converted to analog form before being applied to the analog output port. The digital output data is applied to the digital ports.

Scan Rate and Sampling Rate One scan is one sample per channel. The scan rate is number of scans per second. Consequently, the sampling reate is the product of the scan rate and the number of channels (N) as expressed below:

Sampling Rate = (Scan Rate)(N) samples/s

The sampling rate limit for a general purpose DAQ board is 100 ksamples/s. This limits the maximum scan rate and the number of channels in a data acquisition process. Using the above sampling rate limit, four channels may be scanned at 25,000 scans/s or two channels may be scanned at 50,000 scans/s.

Acquire Waveforms.vi, for example, requires the user to specify the number of channels, the scan rate, and the number of samples per channel.

Analog Input Data Acquisition Options

In this section we will consider two types of data acquisition options. As you will see, these options offer a single point immediate or hardware timed input and buffered types of data acquisition:

> **Single Point Input (Immediate or Hardware Timed)**
> **Waveform Input**

Immediate Single Point Input

In this type of application only one point is acquired at a time. You can use an Easy VI or an Intermediate VI in this type of data acquisition application. As shown in this illustration, if you run AI Read One Scan.vi, only one data point will be acquired from each analog input channel. However, if you place this VI in a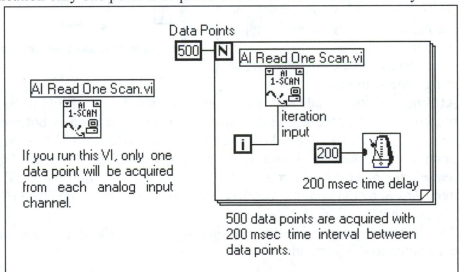

loop, many samples may be acquired and timed every 200 msec. The timing here is achieved in software. This is not hardware timing, which is faster and more accurate.

AI Sample Channel.vi, the Easy VI, could have been used in the above illustration. But as you can see from the hierarchy diagram in Fig. 9-6, *AI Read One Scan.vi,* an Intermediate VI, is included in AI Sample Channel.vi. By using AI Read One Scan.vi directly, we can save on some of the software overhead and thus speed up our data acquisition process. Notice that the iteration terminal **i** of the For Loop is wired to the iteration input of AI Read One Scan.vi.

This speeds up the operation because the VI will be initialized only on the first iteration and only data will be acquired on all subsequent iterations. If you use a While Loop instead of the For Loop, you will have a continuous data acquisition process.

Hardware Timed Single Point Input

Hardware timed single point input is faster and more precise than the Immediate Single Point data acquisition option described in the preceding section.

Hardware timing enables the *scan clock*, which times precisely the acquisition of data. The scan clock is on a DAQ board and its scan rate is selected by the user. Each scan of

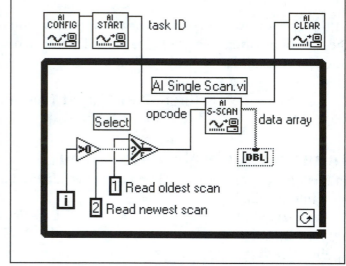

data, consisting of one data point from every channel, is placed in FIFO (first-in-first-out temporary memory storage). AI Single Scan.vi removes data from FIFO one scan at a time.

As you can see from this illustration, you need several additional VIs to configure hardware timed data acquisition.
AI Config.vi configures the hardware for an analog input operation.
AI Start.vi sets, among other things, the scan rate.

AI Single Scan.vi reads and returns one scan of data from FIFO if the acquisition is non-buffered, or from the acquisition buffer if data acquisition is buffered. Notice how the iteration terminal **i** of the While Loop is used to control S-Scan.vi; only on the first iteration (**i=0**) does the Select function apply *opcode 2* to AI Single Scan.vi, causing it to retrieve the newest scan (from FIFO if the acquisition is non-buffered). On all subsequent iterations Select.vi applies opcode 1, forcing AI Single Scan.vi to retrieve the oldest scan from the buffer.

AI Clear.vi clears the data acquisition task ID, thus releasing all resources that have been committed to this particular data acquisition.

Waveform Input

Waveform input data acquisition is a **buffered** and a **hardware timed** process. It is timed because the hardware clock is activated to guide each data acquisition point quickly and accurately. It is buffered because the data acquired during the scanning process is stored in the memory buffer and later retrieved by the VI.

AI Waveform Scan.vi, shown in this illustration, is an Intermediate Application VI that can do a waveform input type of data acquisition.

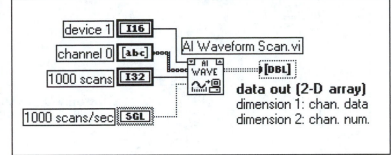

As shown in this illustration, you specify the number of scans to acquire as well as the scan rate. Recall that one scan means one data point from each channel in the list of channels. That means that we can specify other channels in addition to channel 0 if desired. For example, by entering 0:3 in the string array control in the Front Panel, you specify four channels to be scanned: channels 0 through channel 3.

As specified in the illustration, this VI will acquire 1000 data points from channel 0 in 1 second because it is scanning at 1000 scans or channels per second. The output is a two-dimensional array in which each row is one complete scan from all channels. This means that each column will include data for a particular channel. In the above illustration, column 0 will include 1000 points of data for channel 0. You can compare this one scan of data to one trace on an oscilloscope screen.

AI Acquire Waveform.vi is an Easy VI that can also do a waveform input type of data acquisition. Actually AI Waveform Scan.vi shown in the above illustration is included inside AI Acquire Waveform.vi. As mentioned earlier, Easy VIs are made up of Intermediate and Advanced VIs. They possess simplicity and lack flexibility. This VI is no exception.

Notice in the hierarchy diagram of Fig. 9-7 how many other VIs are included in an Easy VI such as AI Acquire Waveform.vi. This VI uses an application VI, which in turn uses some Intermediate VIs. The Intermediate VIs use a number of Advanced VIs. The Advanced VIs are responsible for the hardware configuration, scan clock setting, and many other chores that are transparent to the user.

To view the VI hierarchy and all the VIs that make up a particular VI, double-click on the VI icon and choose *Project>Show VI Hierarchy.*

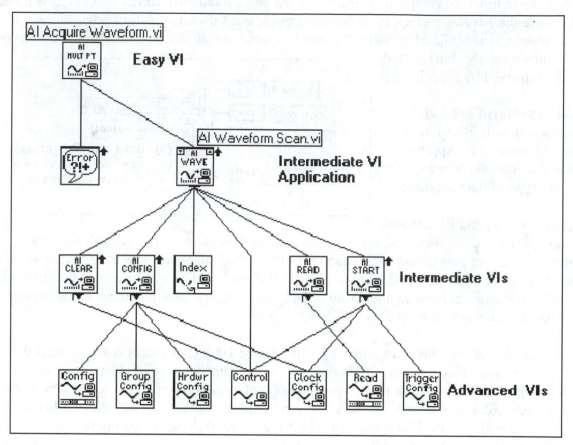

Fig. 9-7 Hierarchy Diagram for AI Acquire Waveform.vi

Digital Frequency

The frequency of analog signals is measured in Hz or cycles/sec. The digitizing process replaces the analog signal by samples. In LabVIEW, Sine Wave.vi generates an array of elements that represent a sine wave. An important point to understand is that this is not an analog signal, which is continuous, but rather a digital signal composed of samples. At its frequency input Sine Wave.vi requires a digital frequency. Digital frequency is defined as

$$f_d = f_a/f_s$$

where f_a is the analog frequency in cycles/sec and f_s is the sampling frequency in samples/sec. Consequently, the units of f_d must be (cycles/sec)/(samples/cycle) or cycles/sample. Actually the reciprocal of the digital frequency $1/f_d$, whose unit is samples/cycle, makes more sense.

The illustration shown here is the output from Sine Wave.vi where sampling frequency was set to 1500 samples/sec and the analog frequency was set to 100 Hz. Clearly the digital frequency must be

$$f_d = f_a/f_s$$
$$= 0.067 \text{ cycles/sample}$$

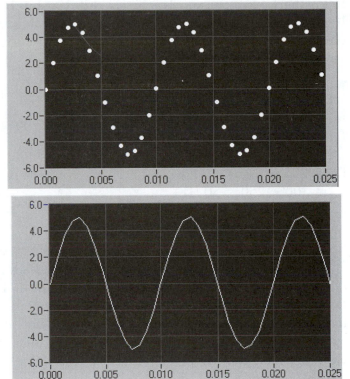

and the reciprocal $1/f_d = 15$ samples/cycle. This value can be easily confirmed by counting the number of samples contained in one cycle in the illustration.

When interpolation is restored (by right-clicking on the legend above the graph), the sine wave appears, as shown in this illustration, to be a continuous, analog-type signal.

In addition to Sine Wave.vi, other VIs such as Square Wave.vi and Saw Tooth.vi, as well as digital filters require the normalized or the digital frequency at its input. Normalized frequency ranges between 0.0 and 1.0, which corresponds to real frequency range between 0.0 and fs (sampling frequency). Suppose that signal is sampled at the Nyquist rate (fs/2); this means that it is sampled two times (two samples in one cycle). That results in the digital frequency of

$$f_d = 1 \text{ cycle/2 samples} = \frac{1}{2}$$

If we sample at twice the Nyquist rate then $f_d = \frac{1}{4}$ (4 samples per cycle). For example two cycles of a sine wave that is made up of 50 samples require an $f_d = 2/50 = 0.04$. Since the Nyquist criterion must be observed, f_d must be in the range: $0 < f_d < 0.5$.

Aliasing

Aliasing occurs when frequencies are sampled below Nyquist frequency. Aliasing frequencies can be calculated from

$$f_{al} = \left| n*f_s - f_a \right|$$

To determine the alias frequency f_{al}, choose an integer n in the above equation that will result in the smallest difference between nf_s and the analog frequency f_a of the analog signal being sampled.

Example: Given $f_s = 100$ samples/sec and the signal being sampled includes the following frequencies: 40 Hz, 80 Hz, 120 Hz, 130 Hz, 170 Hz, 210 Hz, 250 Hz, 610 Hz. Use the above equation to determine the alias frequencies. Since the analog signal must be less than $f_s/2 = 50$ Hz so that no aliasing occurs, the following alias frequencies are included in the sampled output signal:

$$f_{al1} = |100 - 80| = 20 \text{ Hz}$$
$$f_{al2} = |100 - 120| = 20 \text{ Hz}$$
$$f_{al3} = |100 - 130| = 30 \text{ Hz}$$
$$f_{al4} = |200 - 170| = 30 \text{ Hz}$$
$$f_{al5} = |200 - 210| = 10 \text{ Hz}$$
$$f_{al6} = |300 - 250| = 50 \text{ Hz}$$
$$f_{al7} = |600 - 610| = 10 \text{ Hz}$$

It can be seen that each combination of frequency components that includes (80, 120), (130, 170) and (210, 610) produces the same alias frequency. What is dramatic is shown in the illustration below. After sampling only the 40 Hz (it is $< f_s/2$) appears in the output frequency spectrum (shown by a thick spectral line), and the remainder of the input frequency components produce aliases and do not appear in the output. This represents a major distortion of the input signal caused by aliasing due to undersampling. Obviously we must raise the sampling frequency to a value above 1220 samples/sec in order to avoid aliasing.

Once the alias frequencies pass the sampling stage, it is extremely difficult to eliminate them through filtering or other processing. In practice, an analog low-pass filter (LPF) whose upper 3 db frequency is set to approximately f_s/s is placed in front of the sampling circuit as shown in Fig. 9-1. The filter ensures that all input frequency components that produce an alias because they are above the Nyquist frequency will be eliminated.

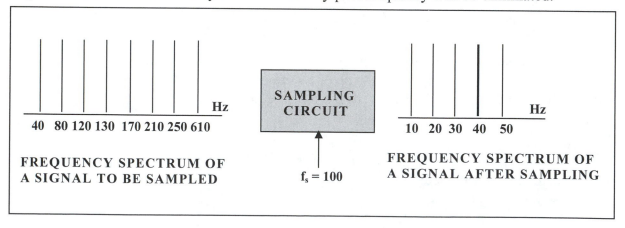

FREQUENCY SPECTRUM OF A SIGNAL TO BE SAMPLED

FREQUENCY SPECTRUM OF A SIGNAL AFTER SAMPLING

Exercise 9-1: Aliasing

The Nyquist criterion must be observed, regardless of whether an analog signal is being acquired through the data acquisition process, where it is digitized and represented by an array of numbers, or the signal is created within the LabVIEW environment through the use of a VI such as Sine Wave.vi.

As stated earlier, Nyquist criterion form sampling requires that the value of the sampling frequency must be greater than the highest input frequency component being sampled. We referred to the Nyquist frequency earlier as one-half the sampling frequency, which is the highest permissible input frequency.

Should the Nyquist criterion be violated, aliasing will occur. Aliasing is difficult to correct once it occurs, and it creates distortion of the signal being sampled because alias frequencies have been added to the output that were not a part of the original input.

It is the objective of this exercise to test and verify the aliasing phenomenon. The provided software is inside Aliasing.LLB library (Book Vis>Aliasing.LLB>Aliasing.vi).

Open Aliasing.vi. Let's examine the Front Panel and the Block Diagram of this VI.

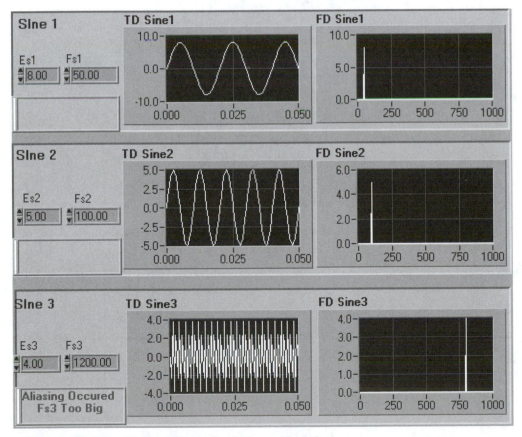

Fig. 9-8 The Front Panel of Exercise 9-1

There are three sections inside the center recessed box. Each section pertains to one sinusoid. Its frequency and amplitude are set here and its Time Domain (TD) and Frequency Domain (FD) Waveform Graph displays are included here. There is also a string indicator to alert the user that aliasing has occurred. In Fig. 9-8 the lower section shows this display.

The recessed box on the left side of the Front Panel contains Waveform Graph displays, one in the Time Domain and the other in the Frequency Domain of the composite signal (sum of three sine waves).

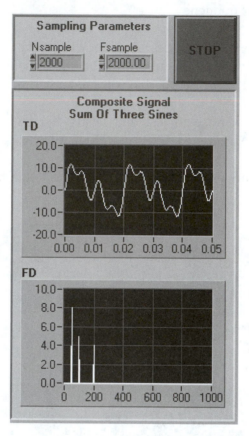

Fig. 9-8 The Front Panel of Exercise 9-1 (continued)

The Sampling Parameters section includes the digital controls for setting the sampling frequency and the number of samples. The Stop Boolean Control is used to terminate VI execution.

The **Block Diagram** shown in Fig. 9-9 includes the three sine wave function whose outputs are added in the summing block.

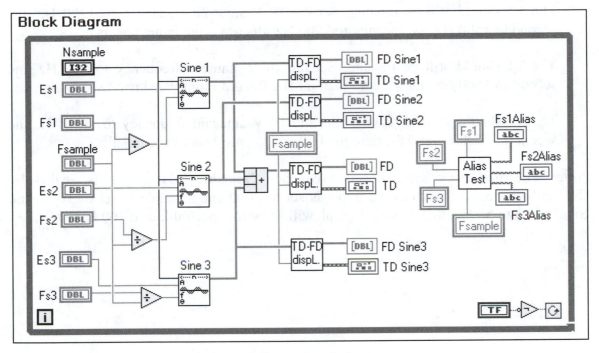

Fig. 9-9 The Block Diagram of Exercise 9-1

TD_FD Display.vi is a subVI whose task is to display the applied array in the Time Domain and the frequency spectrum in the Frequency Domain. It formats the X-axis for the Time Domain display to show true time instead of the usual sample number.

Alias Test.vi is a subVI that compares the value of the sine wave frequency to the sampling frequency. If the sampling frequency is greater than the Nyquist frequency, Alias Test.vi outputs an appropriate warning to the user that also includes the frequency that is too large.

Procedure

1. Set the sampling frequency to 2000 and the number of samples to 2000 also. Program appropriate Front Panel controls to accommodate the input signal whose frequency spectrum is shown here. Run the VI and record the Front Panel results.

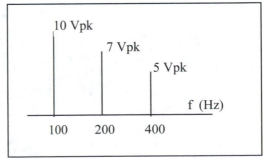

2. While the VI is running, change the frequency of the 10 V sinusoid in steps: 200 Hz, 400 Hz, 800 Hz. Record your observations. Do not stop VI execution.

3. Repeat step 2 and change frequency for the second and the third sinusoids. Record your observations.

The VI should still be running. Set the 10 V sinusoid frequency to 1200 Hz, the second to 1500 Hz and the third to 2700 Hz. Record Front Panel results.

The VI should still be running. Set the 10 V sinusoid frequency to 500 Hz, the second to 1500 Hz and the third to 2500 Hz. Record Front Panel results.

The VI should still be running. Set the 10 V sinusoid frequency to 1000 Hz, the second to 2000 Hz and the third to 3000 Hz. Record Front Panel results.

Note: The frequency scale of all Frequency Domain (FD) graphs is scaled by (Nsample/Fsample). Hence, if Nsample is set to 2000 and Fsample is set to 1000, and Sine1 is set to 50 Hz, then FDSine1 graph will show the spectral line at 100 Hz.

Analysis

1. In step 1, were the TD and FD displays supported by theory? If not, explain.

2. In step 2, were the TD and FD displays supported by theory as you changed frequency? If not, explain.

3. In step 2, were the TD and FD displays supported by theory as you changed frequency? If not, explain.

4. Explain the FD displays obtained in step 4. They are not what one would expect.

5. In step 5, the FD graphs show only one spectral line. Explain.

6. In step 6, probably the most puzzling result was obtained: all displays are gone. You can analyze any of the TD displays. Remove interpolation from the Legend window and expand the X-axis scale by setting the maximum value to 0.01. You should see a display like this:

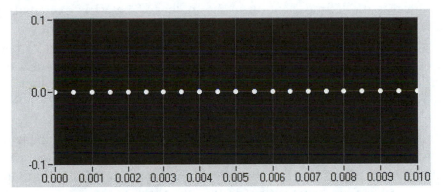

Explain why this occurred. Use the Nyquist criteria, digital frequency, its inverse, and the sketch of samples superimposed on the cycle of a sine wave as part of your explanation.

Chapter 10
Simple DAQ Experiments

In this chapter you will build and run some simple data acquisition VIs. The software for complex VIs is provided, but here you will try your own skill at finding the appropriate functions or subVIs, wiring them in the Block Diagram and finally running them.

Remember that each VI must have a Front Panel that contains various controls and indicators and a Block Diagram that includes various functions and subVIs. The Controls palette has resources necessary for the Front Panel, and the Functions palette has any object that you might require for the Block Diagram. The Controls palette can be opened by right-clicking anywhere in the Front Panel window, and in the Functions palette by clicking anywhere in the Block Diagram window.

The run button is the leftmost button in the toolbar. The VI is executed by clicking on this button. The execution of the VI proceeds in a non-traditional way. C code executes commands in the order that they were written. The execution of a VI is based on data flow. An object in the Block Diagram will execute only if each input terminal has data available. This may influence you on how you wire objects in the Block Diagram. And if priority on the execution order is important, consider using the Sequence structure.

In many data acquisition applications speed is not an issue because data becomes available at a relatively slow rate. There are situations, however, where data is fed to the data acquisition board at a rate that the software is unable to handle. In that case you may try the timed and buffered acquisition. In this arrangement the acquiring VI is configured only once instead of each time a data point is acquired. In addition, the acquiring VI retrieves a block of data instead of a single data point from the buffer.

When data is acquired from several analog input channels, it is stored in a two-dimensional array in which data from the highest channel is in the first column. Arrays in LabVIEW are zero based. Thus, if channels 0, 1, and 2 are scanned, data from channel 2 will be in column 0 of the array, data from channel 1 will be in column 1 and data from channel 0 will be placed in column 2. If this data is to be displayed on a graph, the array must be transposed. You will find the transposing VI in the Arrays subpalette of the Functions palette.

For most of your graphing needs, you will use the Waveform Chart, Waveform Graph, or X-Y Graph. The Waveform Chart accepts data on a point-by-point basis. The Waveform Graph requires the vertical (Y-axis) data to be in the array form, and the X-Y Graph requires both the vertical and the horizontal data to be in the array form.

The only way to create an array is to use a For Loop or While Loop where on each iteration of the loop a data point is stored in the border of the loop. When the loop finishes execution, the array may be retrieved by wiring an object to the border of the loop and outside the loop. The "Enable Indexing" is a default feature in the For Loop, so when you wire to the black tunnel outside the loop, the array will automatically be available to the object outside the loop. This is not the case with the While Loop; to enable indexing in the While Loop, you must right-click on the black tunnel and select "Enable Indexing."

Versions 5.1 and 6 of LabVIEW have resources for operating LabVIEW on a network or on the Internet. The Communication subpalette of the Functions palette contains TCP VIs and the Internet Toolkit subpalette contains CGI VIs, the building blocks for the CGI interface. Some experiments in this book use these VIs. Without further delay let's start with some of the simple data acquisition VI's. As mentioned earlier, you are provided with software to run the complex VIs later in this book.

Experiment 10-1: Temperature Data Acquisition (Easy VI)

In this experiment you will build a temperature data acquisition VI using AI Sample Channel.vi. This is an Easy VI. As indicated before, because Easy VIs are made up of the Intermediate and Advanced VIs, all the intricacies and complexities of the data acquisition are made transparent to the user. This Easy VI may not have the flexibility or the power required for some data acquisitions because the default configuration inside the VI may not be at its optimum setting. As far as our simple applications in this exercise are concerned, this VI is acceptable.

Hardware Requirements: DAQ board, DAQ extension board, temperature sensor (usually provided on the DAQ extension board)

Build this VI in accordance with the description of the Front Panel and the Block Diagram in Fig. 10-1 as follows:

Front Panel

1. **Waveform Chart** is in the *Graph* subpalette of the Controls palette. Open one Waveform Chart and configure it as follows (see Fig. 10-1):

    Scale:
    Vertical: **70** (min) and **90** (max)
    Horizontal: **0** (min) and **200** (max)
    Pop up on the waveform chart and choose *X Scale>Formatting...*, then choose ⌷ from the *Scale Style* palette.

    Legend:
    Type ***Temp Data*** as shown here: ⌷Temp Data ⌷

    Labels:
    Owned: ***Temperature Chart***
    Free: ***Data Points*** and ***Temperature °F***
    AutoScale X and ***AutoScale Y:*** Disable.

Thermometer is in the *Numeric* subpalette of the Controls palette. Open one thermometer and label it as shown in Fig. 10-1.

Block Diagram

2. **For Loop** is in the *Structures* subpalette of the Functions palette. Open one For Loop and resize it as necessary.
    .

You will next be opening the data acquisition VI, AI Sample Channel.vi inside the For Loop. Here is some information about its terminals and wiring requirements. This VI can be found in the Analog Input palette shown in the illustration on the following page.

The following inputs must be wired:
The *device* input refers to the DAQ board number. If you have only one board, then this value will be *1*. If you have more than one board plugged in, then each will be assigned a value during the configuration when you run the WDAQCONF utility.

The **Channel** input refers to the *Analog Input* channel that you will be using for acquiring data. This is a *string* data type input.

The high and low limit inputs need not be wired; their default values are +10 V and –10 V. These values specify the largest and the smallest input values applied to the data acquisition channel. LabVIEW uses these values to compute the gain for the programmable instrumentation amplifier whose output is used by the A/D converter.

Sample is the output terminal. It is wired to the indicator object to display the data point. The sample output, as mentioned earlier, *is only one data point*.

Open the following objects inside the For Loop:

AI Sample Channel.vi 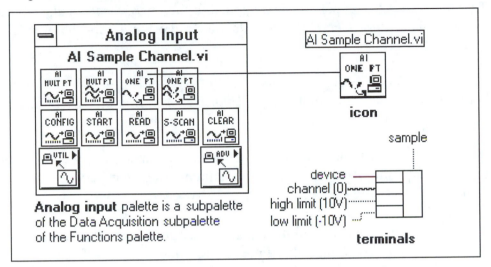 is in the *Analog Input* subpalette of the *Data Acquisition* subpalette of the Functions palette. Open one such VI.

> ***Numeric Constant*** and the ***Multiply*** functions are in the *Numeric* subpalette of the Functions palette. Open four Numeric Constants and one Multiply function. Enter the appropriate values as shown in Fig. 10-1.
>
> ***String Constant*** is in the *String* subpalette of the Functions palette. Open one string constant and type **0** into it. It is used for the channel input.
>
> ***Wait Until Next ms Multiple*** function is in the *Time & Dialog* subpalette of the Functions palette. Open one such function.
>
> ***Wire*** all objects inside the Block Diagram as shown in Fig. 10-1.

3.   The ***DAQ board*** and the appropriate ***extension board*** should have been installed at this time and configured by running the *WDAQCONF configuration utility*.

   Find out where the temperature sensor chip is located on the extension board. In some boards, including Lab PC$^+$, this chip is hard wired next to AI Channel 0 and you have to move a jumper in order to connect the chip to channel 0. If the chip on your board is connected to an analog input channel other than **0**, then *make sure that you change the 0 value stored in the channel string constant in the Block Diagram to the new value.* Also check the board number. Its value was assigned when you ran WDAQCONF. If it is *not 1*, then go back to the Block Diagram and change the value stored by the Device numeric constant.

4.   This exercise illustrates the ***Immediate Single Point Input*** type of data acquisition where AI Sample Channel.vi acquires one data point from channel 0 on device 1 (DAQ board number) and returns the voltage associated with that data point (sample output).

The temperature sensing chip outputs approximately 25 mV at room temperature. This value is scaled (multiplication by 300 as shown in the Block Diagram) to bring it approximately into the °F range. Notice that this is only an approximate calibration.

This temperature data point is next applied to the *Thermometer* indicator and also to the *Temperature Chart*. If you didn't have the For Loop, then that's all you would get, just that one point.

But the For Loop is set up to acquire 200 points with a 200 msec time delay between points provided by the *Wait Until Next ms Multiple* function. With each iteration of the loop, an additional data point is acquired.

If a *While Loop* with an ON/OFF switch were used instead of the For Loop, then the data acquisition would continue indefinitely. To stop it, you would have to click on the ON/OFF switch.

Also the time required to complete this data acquisition is 40 seconds (200 points × 0.2 seconds). To change this time, change the number of data points to be acquired or the time delay, or both.

Run this VI. As the VI is running, touch the temperature sensor chip with your fingers and note the temperature data variation on the Temperature Chart.

Save this VI as DAQ1.vi.

Front Panel

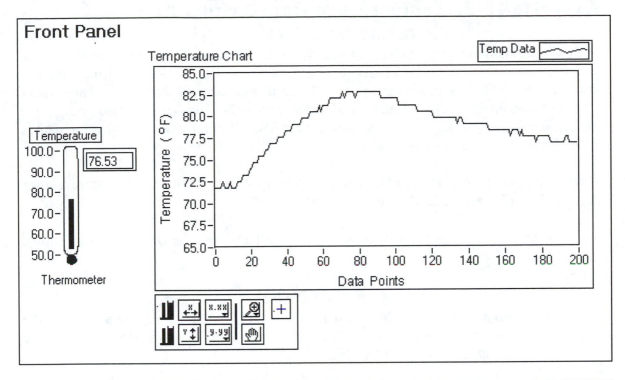

Block Diagram

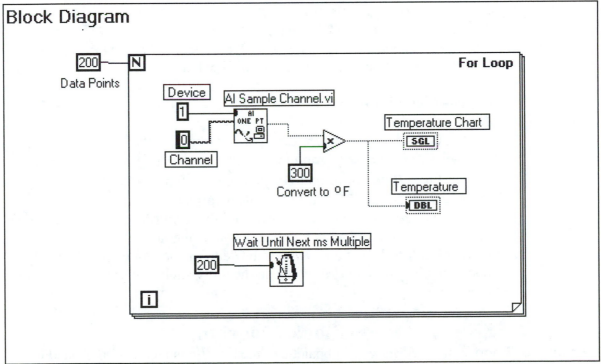

Fig. 10-1 Front Panel and Block Diagram of Experiment 10-1

Exercise 10-2: Temperature Data Acquisition (Intermediate VI)

In this experiment, you will build a VI that will acquire temperature data, except this time you will use an Intermediate VI, AI Read One Scan.vi. This VI scans all the channels wired to the DAQ board analog inputs and returns one scan of data, one measurement from each channel, as a one-dimensional array. Although in this exercise we accomplish the same type of data acquisition, we process the data differently because now the data is in the form of an array.

Hardware Requirements: DAQ board, DAQ extension board, temperature sensor (usually provided on the DAQ extension board)

You will build this VI in accordance with the Front Panel and the Block Diagram shown in Fig. 10-2 as follows.

Front Panel

1. **Waveform Chart** and **Waveform Graph** are in the *Graph* subpalette of the Controls palette. Open one Waveform Chart and one Waveform Graph and configure them as follows (see Fig. 10-2):

   Scale:
   Vertical: **70** (min) and **90** (max)
   Horizontal: **0** (min) and **200** (max)
   Popup on the waveform chart and choose *X Scale>Formatting...*,
   then choose [1.0– 0.5– 0.0–] in the *Scale Style* palette.

   Legend: Disable (optional)
   Labels:
   > Owned: ***Temperature Chart*** (Waveform Chart)
   > ***Temperature Graph*** (Waveform Graph)
   > Free: ***Data Points*** and ***Temperature °F*** (chart and graph)
   AutoScale X and ***AutoScale Y:*** Disable (chart and graph)

   Array is in the *Array & Cluster* subpalette of the Controls palette. Open one empty array shell ***and label it*** with an owned label as ***Channel.***

   String Control is in the *String & Table* subpalette of the Controls palette. Open one string control and ***drop it*** inside the array shell that you opened in the preceding step.

Block Diagram

2. **For Loop** is in the *Structures* subpalette of the Functions palette. Open one For Loop.

   Index Array is in the *Array* subpalette of the Functions palette. Open two Index Array functions. You will use one of these inside the For Loop and the other outside the For Loop. The Index Array outside the For Loop will process a two-dimensional array, so it needs an additional dimension index. To add the dimension, either drag the lower corner of the icon or pop up in the index (black rectangle) and choose *Add Dimension* from the pop-up menu. The icon should look like this: ▦□

Then disable the row index (upper black rectangle) by popping up on the row index black rectangle and choosing *Disable Index*ing. After you are done, the icon will appear as shown here: ▦▦

Numeric Constant and the **Multiply** function are in the *Numeric* subpalette of the Functions palette. Open five Numeric Constants and one Multiply function. Enter the appropriate values into the numeric constants, as shown in Fig. 10-2.

Wait Until Next ms Multiple function is in the *Time & Dialog* subpalette of the Functions palette. Open one such function.

AI Read One Scan.vi is in the *Analog Input Utilities* subpalette of the *Analog Input* subpalette of the *Data Acquisition* subpalette of the Functions palette. Open AI Read One Scan.vi.

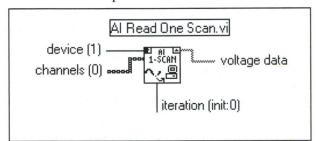

As you can see from this illustration, you need to wire only three terminals: The **Device** input is a numeric constant representing the DAQ board number. The **Channels** input is a string control array that was created in the Front Panel. The **Voltage data** output is a one-dimensional array representing one scan of data. Although in this exercise we are wiring only one analog input channel, in general many input channels are possible.

The **Iteration** input speeds up the data acquisition if you place AI Read One Scan.vi inside a loop to take more than one measurement. If you wire the iteration input of this VI to the loop iteration terminal **i**, this VI will take care of all necessary configurations such as hardware configurations on the first (**i = 0**) iteration of the loop. Then on all subsequent iterations it will only take measurements from the designated channels. This saves a considerable amount of software overhead. In this exercise we use this VI inside the For Loop, where you will wire the iteration terminal i of the For Loop to the iteration input of this VI.

Wire all objects in the Block Diagram as shown in Fig. 10-2.

3.   The DAQ board must be activated and the temperature sensing chip connected to the appropriate analog input channel. See section 3 of Exercise 10-1.

4.   This exercise also illustrates the **Immediate Single Point Input** type of data acquisition; but in contrast to Exercise 10-1, AI Read One Scan.vi returns a one-dimensional array of measurement from each wired analog input channel,

referred to as one scan. In this exercise, only channel 0 is wired, so each scan contains only one measurement. Nevertheless, it is still an array.

The For Loop acquires 200 scans or 200 temperature measurements from channel 0. Each measurement is scaled to convert it to approximate degrees Fahrenheit. The Wait Until Next ms Multiple function provides a 100 ms delay between data points for a total acquisition time of 20 seconds. If you were to use a While Loop with an ON/OFF switch instead of the For Loop, the data acquisition would continue indefinitely until you clicked on the switch to stop it.

In this exercise the Index array functions are used to extract a subset of the array. Inside the For Loop the Index Array function extracts the 0th element of each scan and displays it on the waveform chart.

Each scan (after scaling) is also applied to the border of the For Loop where a two-dimensional array containing 200 scans is created. When the For Loop completes its execution, this two-dimensional array is applied to the Index Array, which extracts the 0th column (the data for channel 0). Recall that each row of the array represents one scan and each column represents the data for one channel. The resulting one-dimensional array (0th column) is then applied for display on the Waveform Graph (Temperature Graph).

As the VI is running, you will be able to see the acquired data point by point on the Waveform Chart (Temperature Chart). Only after the VI completes execution will you see the entire temperature curve displayed on the Waveform Graph (Temperature Graph).

You may use the cursor on the Temperature Graph to read precisely the value of a point on the curve. Notice in the Front Panel of Fig. 10-2 that the cursor position as shown reads 84.96°F for data point number 50. Since the time delay inside the For Loop is 100 ms, the data point 50 will occur 5 seconds after the data acquisition begins. In this exercise the X-axis can represent time in seconds by dividing each data point by 10.

Enter 0 in the channel string array control on the Front Panel.

Run this VI. As the VI is running, touch the temperature sensor chip with your fingers and note the temperature data variation on the Temperature Chart.

Save this VI as DAQ2.vi.

Front Panel

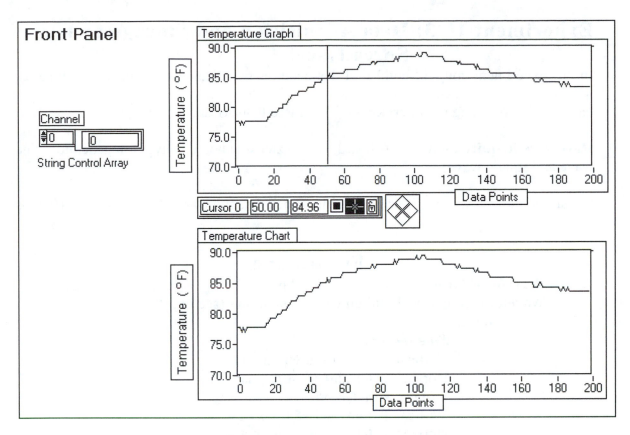

Block Diagram

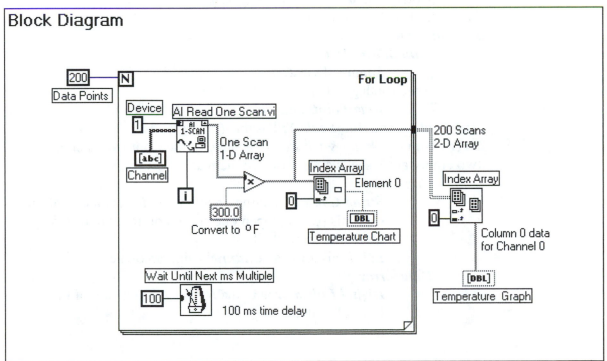

Fig. 10-2 Front Panel and Block Diagram of Experiment 10-2

Simple DAQ Experiments   259

Experiment 10-3: Data Acquisition and Storage to a Spreadsheet

In this experiment you will build a VI that acquires two channels of data and then stores the data to a spreadsheet. This is an example of an *Immediate Single Point Input* type of data acquisition using the Intermediate VI, AI Read One Scan.

Hardware Requirements: DAQ board, DAQ extension board, waveform generator (sine wave and square wave)

Construct the VI whose Front Panel and Block Diagram are shown in Fig. 10-3 as follows.

Front Panel

1. **Waveform Graph** is in the *Graph* subpalette of the Controls palette. Open two Waveform Graphs and configure them as follows (see Fig. 10-3):

    Scale:

    >> **Sine Wave Graph**
    >> Vertical: **–5** (min) and **+5** (max)
    >> Horizontal: **0** (min) and **100** (max)
    >> **Square Wave Graph**
    >> Vertical: **0** (min) and **+5** (max)
    >> Horizontal: **0** (min) and **100** (max)

    Legend: Disable (optional)

    Labels: Owned:

    >> **Sine Wave Graph**
    >>> **Square Wave Graph**
    >> Free (both graphs):
    >>> **Data Points** and **Volts**

    AutoScale X and *AutoScale Y:* Disable (both graphs)

    Array is in the *Array & Cluster* subpalette of the Controls palette. Open two empty array shells and configure them as follows:

    Channel:

    >> **String Control** is in the *String & Table* subpalette of the Controls palette. Open one string control and **drop it** inside the array shell.
    >> **Label** this array as **Channel** using an owned label.

    Data Array:

    >> **Digital Indicator** is in the *Numeric* subpalette of the Controls palette.

Block Diagram

2. **For Loop** is in the *Structures* subpalette of the Functions palette. Open one For Loop and resize it as necessary.

Index Array is in the *Array* subpalette of the Functions palette. Open two Index Array functions. Add a dimension and disable the row index as you have done in Exercise 10-2. The Array Index should look like this

AI Read One Scan. vi is in the *Analog Input Utilities* subpalette of the *Analog Input* subpalette of the *Data Acquisition* subpalette of the Functions palette. Open one such VI. See Exercise 10-2 for a terminal description for this VI.

Numeric Constant is in the *Numeric* subpalette of the Functions palette. Open four numeric constants and one multiply function. Enter the appropriate values into the numeric constants, as shown in Fig. 10-3.

Write To Spreadsheet File.vi is in the *File I/O* subpalette of the Functions palette. Open one such VI. As shown below, this VI accepts at its input either a one-dimensional or a two-dimensional array and stores it to a spreadsheet file separating columns of data by tabs.

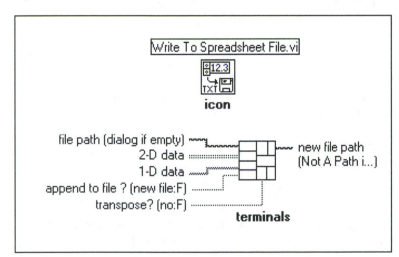

In this exercise we will apply a two-dimensional array of data as the input to this VI. This is the only input that you have to wire. The default settings for all other inputs will do just fine.

You could, however, change some of the default settings. For instance, the **append to file** input default setting is *false*. If you wire a *true* Boolean constant (from the Boolean subpalette of the Functions palette) to this input, then every time you save new data, it will be appended at the end of the old data. The *false* input (default) erases the old data. The **transpose** input switches the rows and columns of the two-dimensional input array. The default setting is *false* (no transpose), but if you wanted the input data to be transposed, then you would have to wire a *true* (Boolean constant) to this input.

Wire all objects inside the Block Diagram as shown in Fig. 10-3.

3. The DAQ board must be activated. If Ch.0 on your DAQ board can accommodate an external input or the temperature sensing chip by means of a jumper, as is the case with the Lab PC$^+$ board, make sure that Ch.0 is connected to the external source.

4. Connect a square wave generator to channel 0 and a sinusoidal source to channel 1. Adjust the amplitudes of both sources to about 4 V$_{pk}$ and set the frequency to 10 Hz.

5. In this exercise two measurements are taken from channels 0 and 1 on each scan and stored into a two-dimensional array. As the For Loop iterates from 0 to 99, a two-dimensional array of 100 rows and 2 columns is formed at the border of the For Loop. Notice that channels are scanned from highest to lowest. So if you enter 1,0 into the Channel string control array in the Front Panel, AI Read One Scan.vi will take a measurement from channel 1 first and then from channel 0. This means that the first column in the two-dimensional array (column 0) will include 100 points of data from channel 1, and the second column will contain 100 points of data from channel 0.

After the loop completes execution, the two-dimensional array is applied to two Index Array functions, Write To Spreadsheet File.vi and the Data Array.

The two Index Array functions extract the Sine Wave and the Square Wave data to be displayed on the Waveform Graphs. Notice that the numeric constant 0 extracts the 0 column (Sine Wave data) and the numeric constant 1 extracts column 1 (Square Wave data) from the two-dimensional array.

The *Array Data* displays the two-dimensional array. You can operate the two indexes with the operating tool to see the value of an element, or you can resize the array as shown in Fig. 10-3 to see several elements.

6. *Run* this VI after entering **1, 0** into the *Channel* string control array.
Note: As soon as you run the VI, Write to Spreadsheet File.vi opens the dialog, letting you specify the name of the spreadsheet file where you want to save the data.

J.U2J	4.U31
1.631	1.033
3 047	4 033
4.053	4.033
4.539	4.036
4.165	0.063
7 207	0 066
1.895	0.066
3.354	0.066
-1 204	0.066
-2 500	0.066
3 699	U.U63
-1 094	0.066

Enter *Test_Run.txt* or some other name of your choice. Notice the .txt extension, specify this file as a text file.

After running the VI, you can view your file using any text editor. In this illustration a portion of the two columns of data was viewed using the *Write* text editor, which is in the Accessories group in Windows. Any other text editor or word processor can also be used.

Open the *cursor* in both waveform graphs by popping up on the graph and choosing *Show>Cursor* from the popup menu. You may also choose the shape and color of the cursor by popping up on the cursor with the operating tool. Refer to Chapter 6 for more information on cursors.

The cursor helps you to read precisely points on the curve. Notice in Fig. 10-3 that the cursor for the sine wave is set to data point 33 and reads the value of 1.85 V and the cursor for the square wave reads the value of 4.03 V. The Data Array also shows these values.

Save this VI as DAQ3.vi and close it.

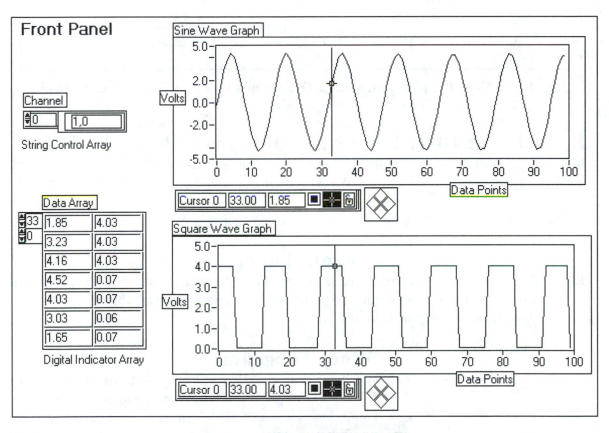

Fig. 10-3 Front Panel and Block Diagram of Experiment 10-3

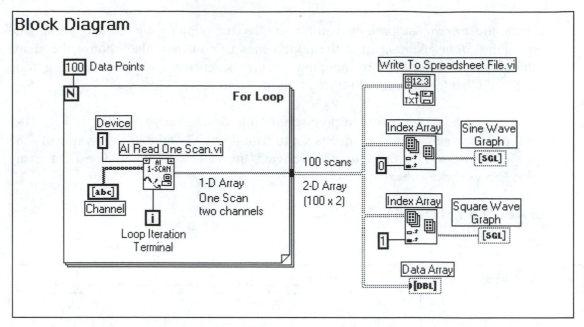

Fig. 10-3 Front Panel and Block Diagram for VI in Experiment 10-3 (continued)

Experiment 10-4: Retrieving Data from the Spreadsheet

In this experiment, you will build a VI that opens the spreadsheet file, retrieves the data, and displays the data on the graphs. We will use the data that you saved in Exercise 10-3.

Build the Front Panel and the Block Diagram shown in Fig. 10-4 as follows.

Front Panel

1. The Front Panel for this exercise is very similar to that of the preceding exercise. Copy the Front Panel from Experiment 10-3 and leave out the *String Control Array* and the *Digital Indicator Array*.

Block Diagram

2. **Read From Spreadsheet File.vi** is in the *File I/O* subpalette of the Functions palette. Open one such VI. Shown in this illustration are the icon and some of the terminals.

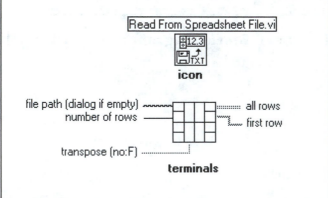

The **file path** input can be left unwired. The default is the dialog box. During execution this VI will give you access to directories and files, allowing you to open the data text file.

Number of rows is a numeric input that specifies the number of rows to be read from the file. The *all rows* output is the two-dimensional array of data that has been retrieved from the text file specified by you and is now available for processing.

Index Array is in the *Array* subpalette of the Functions palette. Open two Index Array functions. Add a dimension and disable the row index for each Index Array function, as you did in the preceding exercise.

Numeric Constant is in the *Numeric* subpalette of the Functions palette. Open three Numeric Constants and enter the appropriate values into these constants, as shown in Fig. 10-4.

3. **Wire** all objects in the Block Diagram as shown in Fig. 10-4.
4. **Run** this VI. At the beginning of execution a dialog window will open, giving you access to directories and files. Choose *Test_Run.txt* (or whatever file name you used in the preceding exercise). Observe the retrieved data plotted on the waveform graphs. These are the same two waveforms that you acquired in the preceding exercise.

Save this VI as DAQ4.vi and close it.

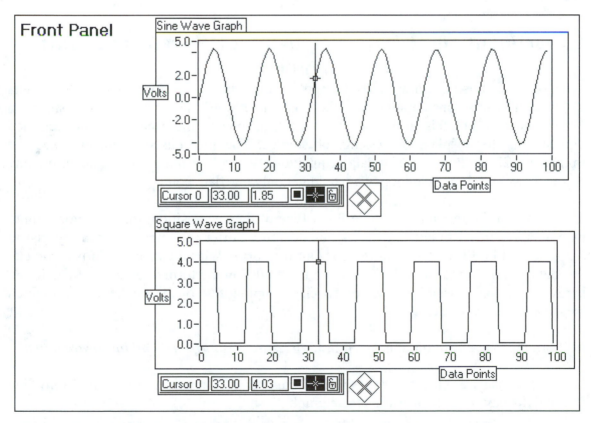

Fig. 10-4 Front Panel and Block Diagram of Experiment 10-4

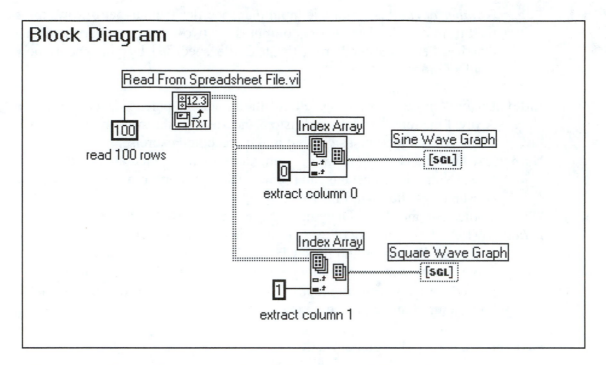

Fig. 10-4 Front Panel and Block Diagram of Experiment 10-4 (continued)

Experiment 10-5: Data Acquisition (Waveform Input: One Channel)

This exercise illustrates a Waveform Input type of data acquisition. This type of data acquisition differs considerably from the Immediate Single Point type of acquisition, in which one data point is measured at a time from one or more channels and returned immediately for display, processing, or storage. The previous experiments used the Immediate Single Point type of data acquisition. We were using a loop to acquire many points. If we didn't use the loop, we would get only one data point.

In contrast to the Immediate Single Point type of data acquisition, the Waveform Input type acquires many points and stores this data in a buffer (a temporary memory that in our case will be FIFO, the First In First Out memory). After the acquisition is complete, this data is transferred from the buffer and is returned in the form of an array. In addition, hardware timing is used because you specify sampling rate or the frequency with which measurements are made.

In short, the Waveform Input is much faster because it is ***buffered and hardware timed.***

In this exercise we will acquire a specified number of samples at a specified sampling rate and display the data on a waveform graph.

Construct the Front Panel and Block Diagram shown in Fig. 10-5 as follows.

Front Panel

1. **Waveform Graph** is in the *Graph* subpalette of the Controls palette. Open one Waveform Graph, resize it as necessary, and configure it as follows (see Fig. 10-5):

   Scale:

         Vertical: ***AutoScale Y:*** Enable

         Horizontal: Choose *Precision* as **3**

                         Enter **0.000** (min) and **0.005** (max)

                         AutoScale X : Disable

   Legend: Optional

   Labels:

         Owned: ***Sine Wave Graph***

         Free: ***Time (sec)*** and ***Volts***

   Cursor: optional

   Digital Control is in the *Numeric* subpalette of the Controls palette. Open two digital controls. Set their *Representation* to **I16** (pop up on control and choose *Representation>I16* from the popup menu).

         Label one of them as ***Num Samples*** and the other as ***Sampling Rate*** using owned labels. The samples/sec units label is a free label and is optional.

   Vertical Switch is in the *Boolean* subpalette of the Controls palette.

         Label the switch as ***STOP*** using an owned label.

         Mechanical Action: ***Latch When pressed***

         Choose *Data Operations>Make Current Value Default* from the popup menu.

Block Diagram

2. **While Loop** is in the *Structure* subpalette of the Functions palette. Open one While Loop and resize it as necessary.

   AI Acquire Waveform.vi is in the *Analog Input* subpalette of the *Data Acquisition* subpalette of the Functions palette. Open AI Acquire Waveform.vi. As shown in this illustration:

   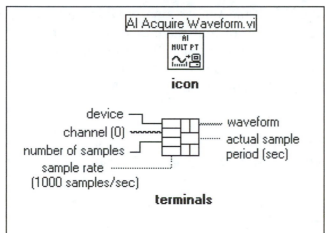

   device is a numeric constant representing the DAQ board number.

   channel is a string constant representing the channel number.

   number of samples is a numeric constant representing the number of data points to be acquired.

   sample rate is the frequency in samples/sec with which measurements or data points are taken from the specified channel.

   waveform is a one-dimensional array of data points returned from the buffer.

actual sample period is the time between samples. We will use this value to set the time interval between data points along the horizontal scale of the waveform graph.

Bundle is in the *Cluster* subpalette of the Functions palette. Open one Bundle function and resize it to three inputs.

Numeric Constant is in the *Numeric* subpalette of the Functions palette. Open three Numeric Constants, label one as **device** with an owned label and enter the value **1** (DAQ board number). Enter the appropriate values into the remaining constants, as shown in Fig. 10-5.

String Constant is in the *String* subpalette of the Functions palette. Open one String Constant, label it as **channel** with an owned label, and enter the value **0** (analog input channel number used for data acquisition).

Wait Until Next ms Multiple is in the *Time & Dialog* subpalette of the Functions palette. Open one *Wait Until Next ms Multiple* timer function.

3. *Wire* all objects in the Block Diagram as shown in Fig. 10-5.

4. At this time the DAQ board should be connected and configured, and the extension board should be connected to the DAQ board. Apply to channel 0 a sinusoidal input of 4 V_{pk} and 1 kHz frequency. Enter **500** into the *Num Samples,* and **15000** into the *Sampling Rate* digital controls on the Front Panel.

5. In this exercise 500 measurements are acquired from channel 0 into a buffer at a sampling rate of 15000 samples/sec. AI Acquire Waveform.vi returns this data as a one-dimensional array and displays it on the waveform graph, which we called Sine Wave Graph.

The fact that this is a much faster type of data acquisition is evident from the frequency of the sine wave that we are sampling here. Recall that in the previous VIs using the immediate single-point input, we used 10 Hz for the sine wave frequency.

Run this VI. Notice that the use of the While Loop provides an interactive data acquisition environment. Each block of 500 samples of data is returned by AI Acquire Waveform.vi and displayed on the Sine Wave Graph. This occurs each second with the time delay provided by the Wait Until Next ms Multiple timer function. This operation resembles the oscilloscope, where the traces occur at regular intervals.

As the VI is running you can experiment by changing the amplitude and frequency of the sine wave as well as the number of samples and the sampling rate.

To stop this VI click on the STOP switch.

Save this VI as DAQ5.vi and close it.

Front Panel

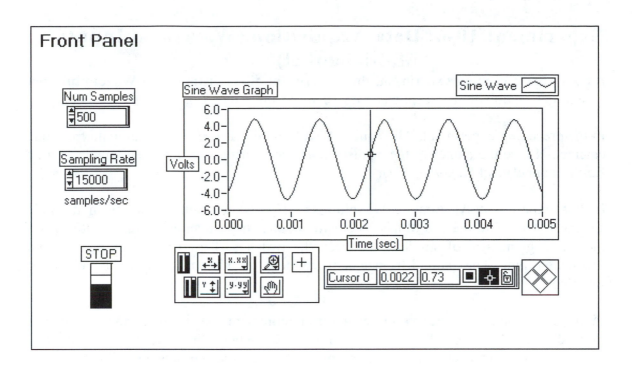

Block Diagram

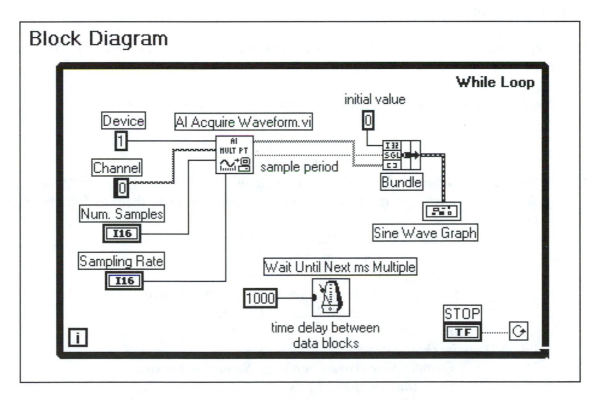

Fig. 10-5 Front Panel and Block Diagram of Experiment 10-5

Simple DAQ Experiments   269

Experiment 10-6: Data Acquisition (Waveform Input: Multichannel)

As with the preceding experiment, this experiment also illustrates the Waveform Input type of data acquisition, except this time we will acquire data from multiple channels.

In the preceding experiment, AI Acquire Waveform.vi was used to acquire data from one channel. Notice the names of the inputs: *Num Samples* (number of measurements from a single channel) and *Sampling Rate*.

In this experiment AI Acquire Waveform**s**.vi is used to acquire data from multiple channels. The 's' at the end of this VI's name implies more than one channel; otherwise, there is a great deal of similarity between the two VIs. They both use buffering and hardware timing. The main difference between them lies in the number of channels being sampled.

In this experiment we have to acquire data from more than one channel, and thus we have to deal with the idea of scanning channels. Remember that one scan means one measurement from each channel in the group. Notice the wording used in the Block Diagram in Fig. 10-6. Instead of number of samples, we use *samples/ch,* and instead of sampling rate, we use *scan rate*.

In this experiment we will acquire a specified number of samples at a specified sampling rate and display the data on a waveform graph.

Construct the Front Panel and Block Diagram shown in Fig. 10-6, as follows.

Front Panel

1. **Waveform Graph** is in the *Graph* subpalette of the Controls palette. Open two Waveform Graphs, resize them as necessary, and configure them as follows (see Fig. 10-6):

  Scale: (both graphs)
  Vertical: *AutoScale Y:* enable
  Horizontal: Choose *Precision* as **3**
  Enter **0.000** (min) and **0.005** (max)
  AutoScale X : Disable

  Legend: Optional
  Labels: (both graphs)
  Owned: *Sine Wave Graph* and *Square Wave Graph*
  Free: *time (sec)* and *Volts*
  Cursor: Optional

  Digital Control is in the *Numeric* subpalette of the Controls palette. Open two digital controls. Set their *Representation* to *I16* (pop up on control and choose *Representation>I16* from the popup menu).

Label one of them as ***Num samples/ch*** and the other as ***Scan Rate*** using owned labels. The Scans/sec units label is a free label and is optional.

Vertical Switch is in the *Boolean* subpalette of the Controls palette.

> *Label* the switch as ***STOP*** using an owned label.
> *Mechanical Action*: ***Latch When pressed***
> Choose *Data Operations>Make Current Value Default* from the popup menu.

Block Diagram

2. **Index Array** is in the *Array* subpalette of the Functions palette. Open two Index Array functions. Add dimension and disable the row index as you did in Exercise 10-2.

    While Loop is in the *Structure* subpalette of the Functions palette. Open one While Loop and resize it as necessary.

    Wait Until Next ms Multiple is in the *Time & Dialog* subpalette of the Functions palette. Open one Wait Until Next ms Multiple timer function.

    Bundle is in the *Cluster* subpalette of the Function palette. Open two Bundle functions and resize them to include three inputs.

    Numeric Constant is in the *Numeric* subpalette of the Functions palette. Open six Numeric Constants, label one as ***device*** with an owned label, and enter the value **1** (DAQ board number). Enter appropriate values into the other numeric constants, as shown in Fig. 10-6.

    String Constant is the *String* subpalette of the Functions palette. Open one string constant, label it as ***channels*** with an owned label, and enter **1, 0** (analog input channels being scanned).

    AI Acquire Waveforms.vi is in the *Analog Input* subpalette of the *Data Acquisition* subpalette of the Functions palette. Open AI Acquire Waveforms.vi. As shown in this illustration:

    device is a numeric constant representing the DAQ board number.

    channel is a string constant representing the channel number.

    number of samples/ch is a numeric constant representing the number of data points to be acquired from each channel.

    scan rate is the frequency of scanning channels in scans/sec. On each scan one measurement is taken from each channel.

waveform is a two-dimensional array of data points being returned from the buffer. Each row of the array includes data from one scan. Hence, each column row includes data for one channel.

actual scan period is the time between scans. We will use this value to set the time interval between data points along the horizontal scale of the waveform graph.

3. *Wire* all objects in the Block Diagram as shown in Fig. 10-6.

4. At this time the DAQ board should be connected and configured, and the extension board should be connected to the DAQ board. Apply to channel 1 a sinusoidal input of 4 Vpk and 1 kHz frequency and apply a 4 Vpk square wave of the same frequency to channel 0. Set the *Num samples/ch* digital control to **500** and the *Scan Rate* digital control to **15,000**.

5. In this exercise 500 measurements are acquired from each of the two channels at a scan rate of 15,000 scan/sec. The data is stored in the buffer until acquisition is complete. Then AI Acquire Waveforms.vi returns the data as a two-dimensional array.

    The two index array functions extract column 0 data and column 1 data and display this data on the two waveform graphs.

    Run this VI. As in the previous exercise, the While Loop provides an interactive data acquisition environment. 500 data points are taken from each channel and displayed as a trace on the two waveform graphs. This process is repeated every second as a result of the 1 second time delay provided by the *Wait Until Next ms Multiple* timing function. Experiment with different amplitudes, frequencies, numbers of samples/ch, and scanning rates, as well as with different waveforms.

    Save this VI as DAQ6.vi and close it.

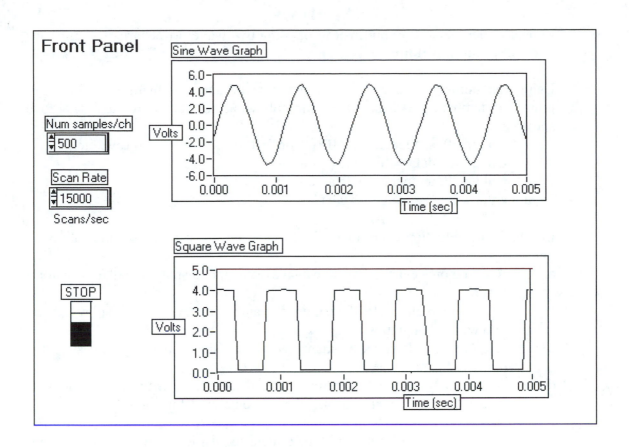

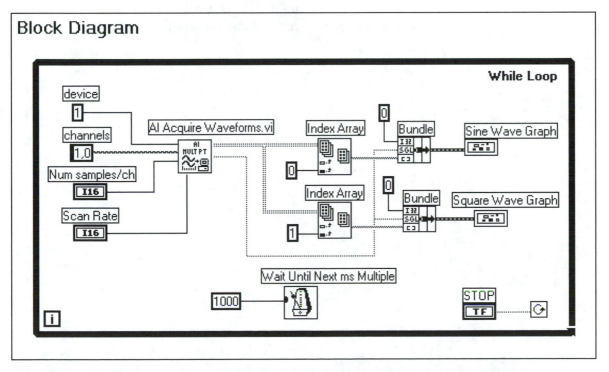

Fig. 10-6 Front Panel and Block Diagram of Experiment 10-6

Simple DAQ Experiments 273

Analysis

1. Repeat Experiment 10-6, but instead of using two waveform charts, use only one to display both waveforms.

2. Using the immediate single-point input type of data acquisition, acquire 200 points of temperature data. Using the following format, store this data to a spreadsheet:

   Column 0: sequential row ID (1 for first row, 2 for second row, etc.)
   Column 1: data point time (for a row)
   Column 2: temperature data point

   Separate the columns using two tabs.

3. In this data acquisition experiment the Front Panel must give user the following options:
   a. Acquire one channel of sine wave data and display it on a waveform graph.
   b. Acquire two channels of square wave and sine wave data. Display data on two waveform graphs.
   c. Acquire two channels of square wave and sine wave data and store the data to a spreadsheet.
   d. Acquire one channel of temperature data and determine *min* and *max* values for the data, and display the data on a waveform graph.

   Use the waveform input type of data acquisition similar to that used in Exercise 10-5 or 10-6 to obtain 500 data points per channel with a scanning rate of 15,000. Frequencies and amplitudes are to be chosen by you.

Chapter 11
Physics I Experiments

Introduction

Physics I is generally devoted to Newtonian mechanics and the study of motion. Both translational and rotational motion are considered in this chapter. The following experiments are included in this chapter:

- **Translational Motion**
- **The Pendulum**
- **Moment of Inertia**

LabVIEW software is used in all experiments to acquire and process the data. In today's world of computers and a variety of software that is readily available, it makes sense to use LabVIEW software in data acquisition and data processing. The software used in this chapter is not of commercial origin, but in a manner of speaking is home spun. Nonetheless, it is just as appropriate and perhaps even more so because it has been custom tailored to meet the needs and objectives of the experiments.

These experiments are not intended to alter the traditional approach, but rather to reinforce it in a meaningful way. In a computerized environment, the user no longer is faced with the drudgery of many hours of data collection and then with more hours of calculations and curve plotting. The computer collects the data and processes it in a manner of seconds. The implication of this is that more work can be accomplished in the allotted time, leaving the user with extra time to examine the data and draw conclusions.

So, without further delay let's begin with the first experiment.

Experiment 11-1: Linear Motion

Objective

In this experiment we will use LabVIEW to acquire acceleration analog data and save it to a spreadsheet. In the second part of this experiment, the data will be retrieved from the spreadsheet and processed. LabVIEW analysis software will use the acceleration data to determine the velocity and the distance traveled profiles. You will be expected to build LabVIEW software for data acquisition, and the analysis software will be made available to you.

Parts:

1. LabVIEW (version 5 or higher)
2. Data acquisition board (such as LabPC$^+$, or MIO-16E-10)
3. Extender board (for the DAQ board being used)
4. Acceleration sensor (PASCO Model CI-6558).
5. Wooden block, string, and weights.

Part I: Data Acquisition

The Track and the Wooden Block

One of the key components in this experiment is the track upon which the block is accelerated by the hanging weight, as shown in the illustration in Fig. 11-1 below. The track and the block are made of wood; their contacting surfaces are polished to minimize friction. As shown in Fig. 11-1, the accelerometer is attached to the wooden block. The cable connected to the accelerometer carries the +5 VDC bias and the acceleration data. For best results the weight of the wooden block should be at least 10 times that of the accelerometer.

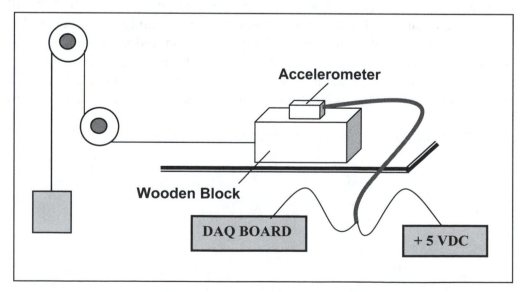

Fig. 11-1 Mechanical Setup

Procedure

1. The accelerometer uses a DIN connector whose pin-out is shown below:

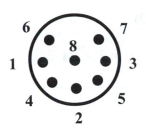

DIN Connector pins:
1. **Analog Output (+)**
2. **Analog Output (–), signal ground**
3. **NC**
4. **+5 VDC**
5. **Power ground**
6. **+12 VDC**
7. **–12 VDC**
8. **NC**

The accelerometer requires +5 VDC (pin 4), +12 VDC (pin 6), and –12 VDC (pin 7), and the power supplies ground is pin 5. The analog signal representing acceleration is connected to AI Ch0 and the signal ground (pin 2) is connected to the CH 0 ground on the extender board. The extender board is connected to the data acquisition board (DAQ board), which is inside the computer. Make these wiring connections.

2. At this time your computer should have the following operating software: LabVIEW and the DAQ board driver (NI-DAQ). The DAQ board should have been configured by the running the WDAQCONF utility, which assigns the DAQ board number (the Device in the Block Diagram of the LabVIEW program that you are about to build).

3. **Build** the VI as shown in Fig. 11-2. This is the Acceleration.vi graphical LabVIEW program that will be used in acquiring the acceleration data. It is assumed at this time that the user is familiar with the LabVIEW environment. If not, refer to Chapters 2 and 3 of *Basic Concepts of LabVIEW 4* by this author.

Front Panel

Waveform Graph (path: *Controls>Graph*):
　　Open and resize the Waveform Graph as necessary.
　　Disable *AutoScale* for both axes.
　　Set 1 digit of precision for the X-axis.
　　Set minimum and maximum values of both axes to
　　　　those shown in Fig. 11-2.
　　Labels: *Acceleration (V)* is an *owned label. Time (s)*
　　　　is a *free label*.
Digital Control (*Controls>Numeric*).
　　Open three digital controls and label them as shown in Fig. 11-2
　　using *owned labels. Device* is the DAQ board.
String Control (*Controls>String & Table)*. Open one string control.

Block Diagram

Numeric Constant (path: *Functions palette>Numeric subpalette*)
String Constant (*Function>String*)
AI Acquire Waveforms.vi (*Functions>Data Acquisition>Analog Input*)
Transpose 2D Array (*Functions>ArrayI*)
Write To Spreadsheet.vi (*Functions>File I/O*)

Wire all objects in the Block Diagram as shown in Fig. 11-2.

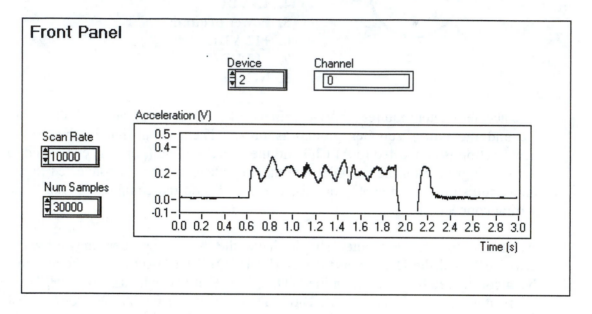

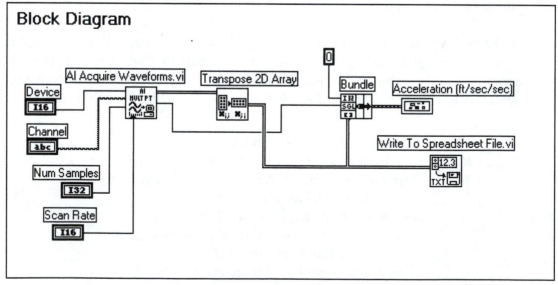

Fig. 11-2 The Front Panel and Block Diagram of Acceleration.vi.

4.   *Set* the *Scan Rate* in the Front Panel to 10000 and *Num Samples* (number of samples) to 30000.

Enter the Device number that corresponds to the DAQ board in your computer. It may not be 2 as shown in Fig. 11-2.

Enter 0 in the Channel string control. Your accelerometer should be wired to channel 0 on the extender board.

Set the SLOW/FAST switch on the accelerometer to SLOW. Just before you are ready to release the accelerometer on the block (make sure it is not moving), press the **TARE** button on the accelerometer. The TARE control zeroes out the 0 g voltage.

Run this VI and acquire several sets of data using different weights. The weights are in the neighborhood of 200 grams. When prompted, use Accel_data.txt (or a text file name of your choice) as the data file name and save it.

In the prototype setup, a wooden block shown in Fig. 11-1 traveled a distance of approximately 6.0 feet on a wooden surface. Any frictional variations between the two surfaces are reflected as oscillations in the acceleration response. A snapshot of an actual run, shown in the Front Panel of Fig. 11-2, illustrates this type of oscillation. Using an air track will provide a more stable response.

Save this VI as *Acceleration.vi* in Linear Motion.LLB inside the Book VIs folder. As you know, the .LLB extension is used to represent a library. In LabVIEW the library is not a directory, but rather a file that can contain many other VI files. There are several advantages associated with a VI Library. Any VI stored in a Library can have a name that is up to 31 characters long instead of the usual 8 characters. To save disk space, all VIs are compressed when saved and decompressed when loaded. Portability becomes simpler too because all VIs are in the same file. Refer to Exercise 9 in Chapter 2 of *Basic Concepts of LabVIEW 4* by this author for the procedure on how to create a VI Library.

Operation

Acquire Waveforms.vi, used in this experiment, provides us with *buffered* and *hardware timed* data acquisition. The clock on the data acquisition board is used to guide each acquired data point quickly and accurately.

Num Samples, the Front Panel digital control, is set for the desired number of data points, or samples. In Fig. 11-2 it is set to 30,000. The Scan Rate, also a Front Panel digital control, is used to set the scan rate, and in this experiment it is set to 10,000 scans/sec.

Acquire Waveforms.vi acquires 30,000 samples (the number of samples is set by user) at a rate of 10,000 scans/sec and stores the data in the buffer in a two-dimensional array. Each column of the array contains data from an analog input channel on the DAQ board. In this experiment, the first column contains data from Channel 0.

As shown in the Block Diagram of Fig. 11-2, the two-dimensional array is transposed so that the data can be displayed on the Waveform Graph. Notice that the *delta t* time increment (time interval between samples) used by the Waveform Graph is provided by the AI Acquire Waveforms.vi; it is equal to the reciprocal of the scan rate.

To convert the voltage samples to units of acceleration, the analog data is scaled, as shown in Fig. 11-2. The sensitivity of the accelerometer used in this experiment is 1 V/g.

The scaled data is also saved to the spreadsheet by using *Write To Spreadsheet File.vi*. When this VI executes, it will open a dialog window, giving you access to directories and files and thus allowing you to name and save this file. You can open and view this file in any text editor or word processor. You can also retrieve this data in another VI with the *Read From Spreadsheet.vi*.

Part II: Motion Analysis

In Part II of this experiment we will process the data acquired in Part I. The VI that you will use to analyze the data is called LMAnalysis.vi. It uses the acceleration data to determine the velocity and the distance profiles through repetitive integration.

The integrator that this VI uses is a home-spun algorithm that does the integration by adding the areas of many small rectangles. We know from calculus that we get velocity by integrating acceleration. Similarly, distance is obtained by integrating velocity. Hence,

$$v = \int_{t1}^{t2} adt$$

$$d = \int_{t1}^{t2} vdt$$

1.   *Open LMAnalysis.vi* (Book VIs>Physics I>Linear Motion>LMAnalysis.vi). Upon inspection of the contents of this library, you will notice that it includes other VIs. These are subVIs that LMAnalysis.vi uses during execution. The library will not operate properly without these subVIs.

The Front Panel of this VI is shown in Fig. 11-3. Upon inspection the three Waveform Graphs display the acceleration, velocity, and distance waveforms. The Menu Ring labeled *Units* allows the user to choose the metric or the English system of units. The Vertical Fill Slide labeled *Min. Value* is adjusted by the user for proper velocity and distance waveform displays. The Digital Indicators *Fin. Veloc.* and *Fin. Distance* display the terminal values of velocity and distance traveled by the block. The rectangular start/stop Button labeled *POWER* is the Boolean switch used to start or stop the execution of this VI.

2. Click on the *Units* Menu Ring with the Operating Tool and choose the units that you want to work with.

3. *Enter* the value of the Scan Rate that you used in Part 1. The analysis software will not work properly unless you enter the same value that you used for the data acquisition in Part 1.

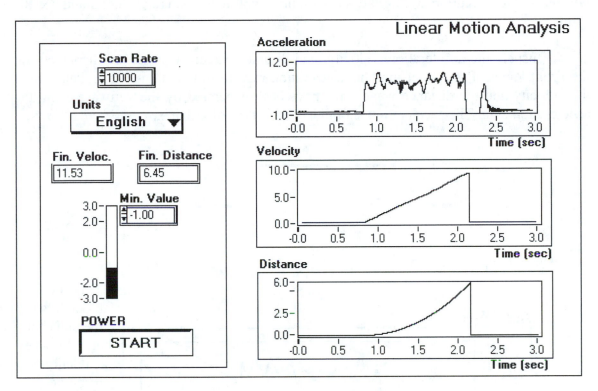

Fig. 11-3   The Front Panel of LMAnalysis.vi

4. The block with the accelerometer should now be in position and ready to slide. Push the *Tare* button on the accelerometer. This nulls out the 0 g voltage. Hold the block steady so that it doesn't move as you push the Tare button. Also, set the *Fast/Slow* switch on the accelerometer to *Slow*.

5. Click on the *Power* button with the *Operating Tool* so that the button indicates **START**.

6. *Run* the VI by clicking on the *Run* ⇨ button with the Operating Tool. A few seconds later release the block. It should slide at least 6 feet. LMAnalysis.vi is now running. You can see that it is running by observing the Run button, which looks like this: ⇨

You will be immediately prompted to open the text file where you saved the acceleration data in Part 1. Proceed to the directory where you saved the file, and open it. The acceleration waveform will appear after a few moments. If necessary, adjust the Min. Value (Vertical Slide on the Front Panel) using the Positioning Tool until the velocity and the distance waveforms appear as shown in Fig. 11-3. Fig. 11-3 shows a snapshot of a typical run.

7. The LMAnalysis VI will continue to run indefinitely. To stop the execution, click with the Positioning Tool on the Power button. The button will indicate STOP and the VI will terminate execution. While the VI is running, you may change the units and see the result.

8. Inspect the acceleration, velocity, and distance waveforms that are displayed on the Front Panel. Remember that the acceleration waveform is obtained from data, while the velocity and the distance waveforms have been generated by the software using the acceleration data. The acceleration should have this type of a shape,

but instead it appears like this:

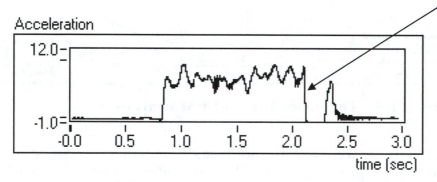

This is where the block hits the stop at the end of its motion and the acceleration suddenly becomes negative.

Notice the spikes at the end of the motion. Inspect the shapes of the acceleration, velocity, and distance profiles on the three waveform graphs in the Front Panel. Notice the rippled top on the acceleration profile; yet the velocity curve, which is derived from the acceleration curve, appears relatively clean.

Shown below are acceleration, velocity, and distance profiles from a typical run.

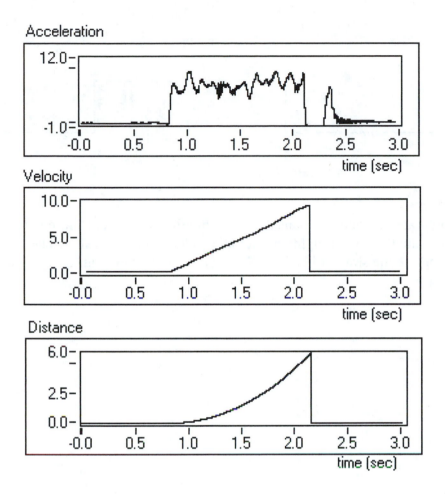

Analysis

1. What should be the shape of the acceleration motion profile? Explain.

2. The velocity profile appears to be a straight line. Should it be a straight line? Explain.

3. The distance profile curve appears to be in a shape of a parabola. Should it have a parabolic shape? Explain.

4. Integrate the acceleration manually to get the final velocity. Before you integrate, estimate the average acceleration by a solid line as shown below, and then integrate the constant between the time limits that your data shows. Compare this value with the *Fin. Veloc.* digital indicator display on the Front Panel.

5. Integrate the velocity manually to get the final distance. Before you integrate, estimate the best straight line fit to the velocity curve as shown below, and then integrate the straight line between the time limits that your data shows.

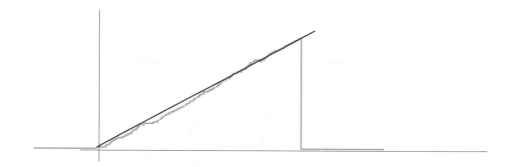

Compare the value that you calculated with the value displayed by the *Fin. Distance* digital indicator on the Front Panel. Compare these values also to the actual distance travelled by the block (measure this distance).

6. Switch to the Block Diagram of LMAnalysis.vi and open Inegadt.vi by double-clicking on the icon. When the Front Panel of Integadt.vi opens, switch to the Block Diagram. Inspect the diagram and explain how the acceleration data points are integrated. How is the resulting velocity array formed?

7. The net force that accelerates the block is the force of the hanging weight (Mg) less any frictional losses in the pulley. This force is constant. If we neglect the frictional losses in the pulley, then part of this force is used to accelerate the block and the rest of it is used to overcome the friction that exists between the surface of the block and the surface upon which it slides. Since μN is the frictional force where μ is the coefficient of dynamic friction and N is the normal force, it is possible to estimate the average value of μ by applying Newton's Second Law. Use the acquired data to determine μ.

8. We know that energy cannot be created nor destroyed. Expended energy merely changes its form. At the beginning of motion the total energy of our system is the potential energy, Mgh, where h is the distance through which the weight moves. At this time the kinetic energy is zero because the block is not in motion. And at the end of motion the energy is kinetic, $(1/2)Mv^2$. Conservation of energy requires that the initial potential energy be equal to the terminal kinetic energy plus all frictional losses. Using the experimental data estimate the total energy losses.

Experiment 11-2: The Pendulum

Objective

In the first part of this experiment we will use LabVIEW to acquire rotational motion analog data pertaining to the motion of the pendulum and save it to a spreadsheet. In the second part of this experiment the data will be retrieved from the spreadsheet and processed. LabVIEW analysis software will use the data to determine several motion profiles and calculate several parameter values. In this experiment the entire software package (data acquisition and analysis) is made available for your use.

Parts

1. LabVIEW (version 5 or higher)
2. Data acquisition board (such as LabPC$^+$, or MIO-16E-10)
3. Extender board (for the DAQ board being used)
4. Assembled vernier photogate (Vernier P/N VPG-DG).
5. Rotary motion sensor (Vernier P/N CI-6625).

Part I: Data Acquisition

The Pendulum Assembly

The pendulum assembly is shown in Fig. 11-4. The rotary motion sensor is mounted at the top of the stand. The weight is attached to the end of the metallic extension rod, which is free to swing. As it swings, it passes inside the photogate, which is clamped to the bottom of the stand.

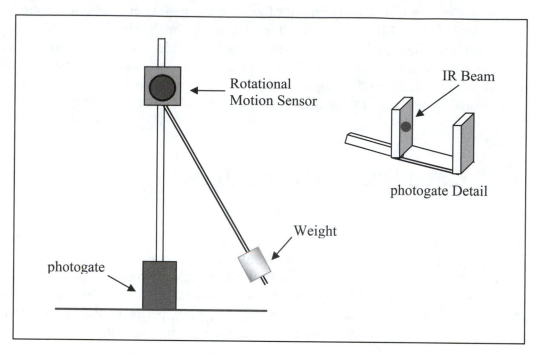

Fig. 11-4 The Pendulum Assembly

Procedure

1. The transducers that include the *photogate*, and the *rotary motion sensor* require various power and I/O connections. Because they have special plugs, you have to provide the adapters in order make various connections. The *rotary motion sensor* has a modular phone connector whose pinout is shown below.

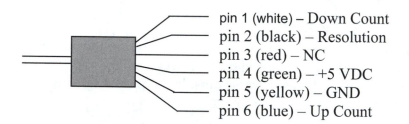

Connect +5 VDC to pin 4, GND to pin 5, and pin 1 to AI Ch. 0 (DAQ extender board).

The photogate unit uses the stereo plug shown below.

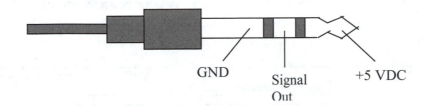

GND   Signal   +5 VDC
           Out

Connect GND, +5 VDC, and Signal Out to AI Ch. 1 (DAQ extender board).

2.   At this time LabVIEW software (version 5.1 or higher) should be in your computer and the data acquisition board (DAQ board such as Lab PC$^+$ or MIO-16E-10) should be plugged into the expansion slot inside the computer. Also, the extender board should be connected to the DAQ board, whose connector is barely visible in the back of the computer. The data acquisition driver software (NI-DAQ) should also have been loaded. Finally, the WDAQCONF utility should have been executed to configure your data acquisition board. All this is necessary before you run the data acquisition and analysis VI.

3.   Now you should become familiar with the Front Panel controls and indicators of the first rotational motion VI, **RM1.vi**. To do so, open it. The path to this VI is *Book VIs>Physics I>Pendulum.llb>RM1.vi*. The Front Panel of this VI consists of two sections that occupy an area too large for the viewing screen. Use the horizontal scroll bar to view one or the other. Fig. 11-5 shows both sections.

The left section is the ***Data Acquisition*** section. It includes the waveform graph that will display two channels of acquired data. The Menu Ring labeled ***Function*** contains two options: Data Acquisition and RM Analysis 1. Use the *Operating Tool* 🖑 to select one of the options. The section ***Scan I/O*** includes four digital controls and one digital indicator. The recommended settings are

Scan Rate = 25,000 scans/sec
Scans to read = 5,000 scans
Buffer Size = 10,000 scans
Repeat Scans to Read = 15

These are default settings, but if your panel does not show these settings, enter the values as given above. Since the *Hardware* section includes the default hardware limits, you should not change anything here. In the *DAQ Board* box, there are two controls whose values you must enter. The **Device** digital control is your board number. Enter the number assigned to your DAQ board when WDAQCONF was executed. The string control **Channels** must include the channels being scanned. Enter **1,0**.

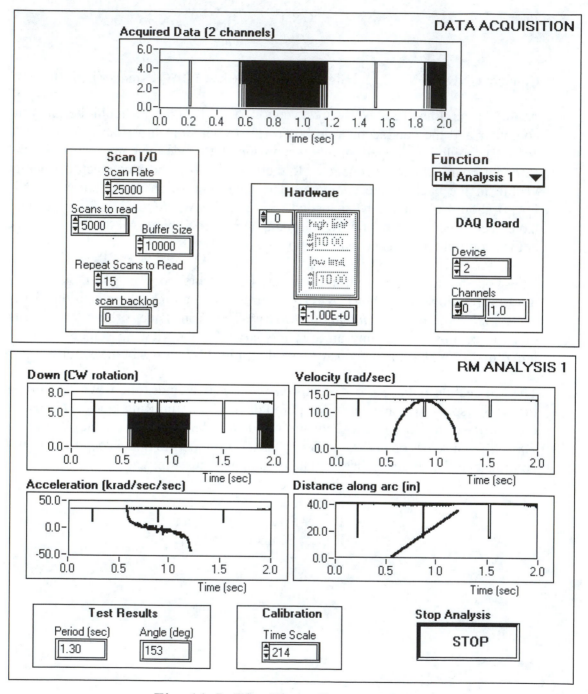

Fig. 11-5 The Front Panel of RM1.vi

The right section of the Front Panel, **RM ANALYSIS 1**, includes waveform graphs and other indicators used to display the results of data processing. The *Down (CW rotation)* waveform graph shows the two channels of data acquired during data acquisition. The *Velocity* graph is derived from the rotary motion Sensor pulses, while the *Acceleration* and the *Distance along arc* are derived from the velocity data. Because the software compresses the time scale during the formation of the velocity array, it is neccessary to restore the time scale to its original form. The **Calibration** setting (default shown in Fig. 11-5 is 214) accomplishes this task. The **Test Results** box displays the the angle travelled by the pendulum between the two extreme points, and the period.

4. Choose **Data Acquisition** from the *Function* menu in the Data Acquisition section of the Front Panel. Check the connections to the two transducers on the pendulum (the rotary motion sensor and the photogate) and turn the power ON. Displace the pendulum to the *left* by about 60 or 70° and release it, and a second or so later click on the **Run** ⬦ button. The two channels of data that you acquire will be saved to the spreadsheet. A few seconds after you click on the run button, a dialog box will appear prompting you to enter the file name for this data. Enter a name such as RMData1.txt and click on the OK button. You will see the acquired data on the waveform graph. It is important that the data that you acquire looks like this:

Pulse cluster from the rotary motion sensor. The second photogate pulse is in the middle of the cluster.

where the rotary motion sensor pulses are clustered around the the second photogate pulse that is in the middle of that cluster (the photogate pulse is not visible in this illustration). The analysis software depends on this format and will be looking for this cluster between the first and third photogate pulses. The software must also find at least three photogate pulses so that it can determine the period. This does not require any special action on your part, except to remember that the pendulum must be released from the left side so that it initially travels counterclockwise. At this time the data acquisition is complete.

Part II: Motion Analysis

Choose **RM Analysis 1** from the *Function* menu in the Data Acquisition section of the Front Panel and scroll over to the right so that you can view the analysis section. Click on the **Run** ⬦ button. You will be prompted immediately to open the text file that was saved to the spreadsheet during the data acquisition in Part I. Navigate to the directory where you saved that file and open it. It will take a few seconds for the system to load and analyze the data. When the analysis software is finished, it will present you with the results. The waveforms will appear on the waveform graphs and the period and the angle values will appear on the digital indicators. It is now time to examine and to interpret the analysis results.

Operation

This experiment demonstrates hardware timed, buffered data acquisition. Data is acquired continuously into a circular buffer (in memory) and at the same time retrieved in blocks to build a larger array of data. Buffer size is the number of scans that are held in the memory and is limited by the amount of the available RAM. By choosing a small buffer or by not withdrawing data from the buffer at a fast enough rate, an overwrite error can result where new data is written over the previous data.

In this experiment the *Scan Rate* control is set to 25,000 scans/second. With two channels being sampled, this means that each channel is sampled at 50,000 samples/sec. The *Scans to read* control is set to 5000 scans and the *Repeat Scans to Read* control is set to 15. This means that the withdrawal of 5000 scans from the circular buffer will be repeated 15 times, resulting in a two-dimensional array of 75,000 rows by 2 columns. The first column will contain the photogate data, and the second column will include the data from the rotary motion sensor.

The approach to data acquisition used in this experiment is different than that used in the Linear Motion experiment where the entire block of data was created inside the buffer and then retrieved. That approach will not work in this experiment for several reasons. First, the pulses from the rotary motion sensor are 110 μ sec wide. In order to detect these pulses, the scan rate must be sufficiently high. With 25,000 scans/sec each channel is sampled every 40 μ sec. Second, the waveform time window should be 2 to 3 seconds in order to view a sufficient amount of the pendulum motion. That means that at least 75,000 scans (150,000 samples) must be collected into an array. This would not be possible with the setup of the Linear Motion experiment because of the buffer size limitation. With the continuous data acquisition into a circular buffer and the concurrent retrieval of data blocks to form the main array outside the circular buffer, the buffer size limitation is thus removed.

RM Analysis 1.vi takes the array from the spreadsheet and separates it into two one-dimensional arrays for processing. The pulses from the photogate serve as marker pulses, indicating the midpoint of the pendulum's travel. These pulses are also used to determine the period of motion.

Fig. 11-6 illustrates, using the results from an actual run, several waveforms associated with this experiment. The top drawing shows the pendulum's position for reference points A, B, and C along the path of motion. When the pendulum is at its extreme right position, point A, and begins its motion clockwise, the rotary motion sensor outputs one 110 μs pulse for each degree of rotation. The pulses from the photogate mark the pendulum's position as it passes the midpoint, point B in the illustration. When the pendulum reaches its maximum position on the left indicated by point C, its stops and begins to move in the counter-clockwise direction. During the entire motion in the counterclockwise direction the rotary motion sensor does not output pulses. The motion repeats when the pendulum reaches point A.

The analysis software uses the pulses from the rotary motion sensor the velocity profile. As expected, the velocity is 0 at points A and C, reaching the maximum at point B, the midpoint of motion.

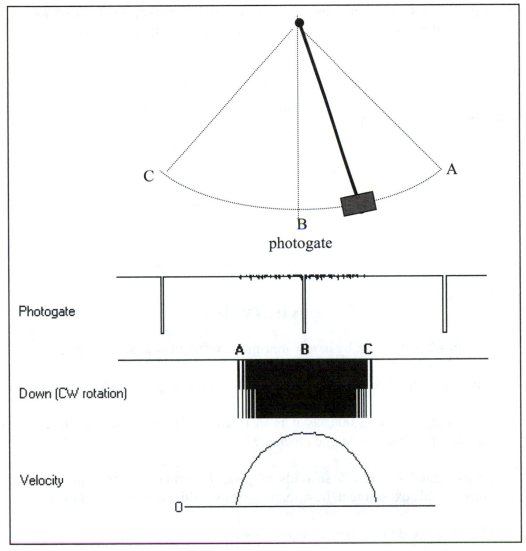

Fig. 11-6 The Analysis Waveforms from RM1.vi

The acceleration curve is derived from the velocity data through successive differentiation, using a home-spun differentiation algorithm. The acceleration is maximum and positive (positive slope) at the beginning of motion at point A, zero at point B (velocity curve has a zero slope at that point), and beyond point B the acceleration is negative (velocity curve has negative slope), reaching its maximum negative value at point C.

The distance along the arc curve is derived from the velocity curve through successive integration using a home-spun integration algorithm. The final value represents the length of arc travelled by the pendulum in one direction (Rθ).

The additional results computed by the software and displayed in the *Test Results* box, show the period of motion and the angle traversed by the pendulum between points A and C.

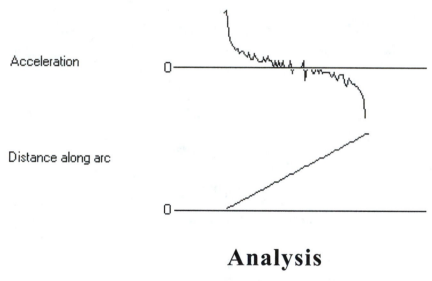

Analysis

1.   Explain what is meant by hardware timed, buffered data acquisition.

2.   Why did we have to use the circular buffer in this experiment?

3.   If the scan rate is 25,000, as it is in this experiment, how long does it take to acquire one sample from one channel?

4.   If we wanted to view 5 seconds of pendulum motion, assuming the scan rate setting in this experiment, how many scans would we have to acquire?

5.   What purpose do the photogate pulses serve?

6.   Determine the **effective** radius R of the pendulum using the classical formula for the period and the value of the period in the Test Results box on the Front Panel. Reconcile the difference between the actual R (measured) with the R predicted by data and using the classical formula.

7. Calculate the the length of the arc travelled by the pendulum between points A and C in Fig. 11-6 (in inches). Use the formula $R\theta$ for arc length and the value of the angle from the Test Results box on the Front Panel.

8. Calculate the average velocity V_{avg} of the pendulum mass along the curvature of the arc. Use the formula $V_{avg} = S/(T/2)$ where S is the arc length calculated in step 7 and T is the period in the Test Results box on the Front Panel.

9. Calculate the average angular velocity ω_{avg} in rad/sec. Since in general $V = R\omega$, use this equation and the value of V_{avg} from step 8 to determine ω_{avg}.

10. Estimate the average angular velocity, ω_{avg}, using the experimental data. Here is a hint on how you can do this. First, measure the peak value of velocity from the waveform graph on the Front Panel. Use the **Cursor Display** of the Waveform Graph to make this measurement. If you don't know how to use the cursor, see page 186 of *Basic Concepts of LabVIEW 4* by this author. The velocity that you measured is the peak angular velocity ω_{pk} in rad/sec.

Next, assume that the velocity waveform may be approximated by the half-wave rectified sinusoid shown below.

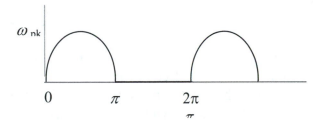

Integrate $\omega_{avg} = [\omega_{pk} \int_{0}^{\pi} \sin(\theta)d\theta]/(2\pi)$ and find the expression for ω_{avg} in terms of

ω_{pk}, then substitute the value of ω_{pk} that you measured in the beginning of this exercise. How does this value compare with the value that you calculated in step 9?

Experiment 11-3: Moment of Inertia

Objective

In the first part of this experiment we will use LabVIEW to acquire rotational motion analog data pertaining to the moment of inertia measurement and save it to the spreadsheet. In the second part of this experiment the data will be retrieved from the spreadsheet and processed. LabVIEW analysis software will use the data to determine the moment of inertia. In this experiment the entire software package (data acquisition and analysis) is made available for your use.

Parts

1.   LabVIEW (version 5 or higher)
2.   Data acquisition board (such as LabPC⁺, or MIO-16E-10)
3.   Extender board (for the DAQ board being used)
4.   Assembled vernier photogate (Vernier P/N VPG-DG).
5.   Rotary motion sensor (Vernier P/N CI-6625).
6.   Mini-rotational accessory (PASCO P/N CI-6691)

Part I: Data Acquisition

The Test Assembly

The test assembly that will be used in this experiment to acquire the data and to use this data in the moment of inertia determination is shown in Fig. 11-7. The rotary motion sensor is mounted at the top of the stand. The ring and the aluminum disk combination is attached to the top of the 3-step pulley, which is free to rotate. Several turns of string (as many turns as necessary to raise the weight slightly above photogate 1) are wound around the 3-step pulley, then the string is guided over the wheel of the super pulley and attached to the weight.

Procedure:

1.   The transducers that include the *photogate*, and the *rotary motion sensor* require various power and I/O connections. Because they have special plugs, you have to provide the adapters in order make various connections. The *Rotary Motion Sensor* has a modular phone connector whose pinout is shown below

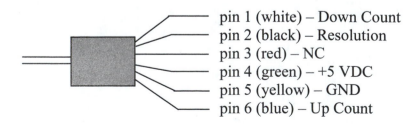

pin 1 (white) – Down Count
pin 2 (black) – Resolution
pin 3 (red) – NC
pin 4 (green) – +5 VDC
pin 5 (yellow) – GND
pin 6 (blue) – Up Count

Connect +5 VDC to pin 4, GND to pin 5, and pin 1 to AI Ch. 0 (DAQ extender board). As shown in Fig. 11-7, this experiment uses two photogate devices: photogate 1 (top) and photogate 2 (bottom). The photogate unit uses the stereo plug shown below.

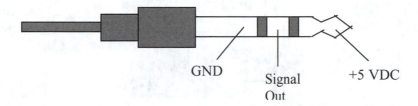

GND   Signal   +5 VDC
       Out

Connect GND, +5VDC to both photogate devices. Connect Signal Out of photogate 1to AI Ch. 1, and the Signal Out terminal of photogate 2 AI Ch. 2 of the DAQ extender board.

2. At this time LabVIEW software (version 4.0 or higher) should be installed on your computer and the data acquisition board (DAQ board such as Lab PC$^+$ or MIO-16E-10) should be plugged into the expansion slot inside the computer. Also, the extender board should be connected to the DAQ board, whose connector is visible in the back of the computer. The data acquisition driver software (NI-DAQ) should also have been loaded. Finally, the WDAQCONF.EXE utility should have been executed to configure your data acquisition board. All this is necessary before you run the data acquisition and analysis VI.

3. In this step you will be introduced to the Front Panel controls and indicators of **RM2.vi**. To do so, open it. The path to this VI is Book VIs>Physics I>Inertia>RM2.vi. The Front Panel of this VI consists of two sections that occupy an area too large for the viewing screen. Use the horizontal scroll bar to view one section or the other. Fig. 11-8 shows both sections.

    The left section is the ***Data Acquisition*** section. It includes the Waveform Graph that will display three channels of acquired data. The Menu Ring labeled ***Function*** contains two options: Acquire Data and Analysis. Use the *Operating Tool* 🖑 to select one of the options. The ***Scan Parameters*** section includes four digital controls and one digital indicator. The recommended settings are

    > Scan Rate = 25,000 scans/sec
    > Buffer Size = 15,000 scans
    > Scans to Read = 12,000 scans
    > Repeat Scans to Read = 14

    These are default settings, but if your panel does not show these values, enter them as given above. Since the *Hardware* section includes the default hardware limits, you should not change anything here. In the *DAQ Board* box, there are two controls whose values you must enter. The ***Device*** digital control includes the DAQ board number.

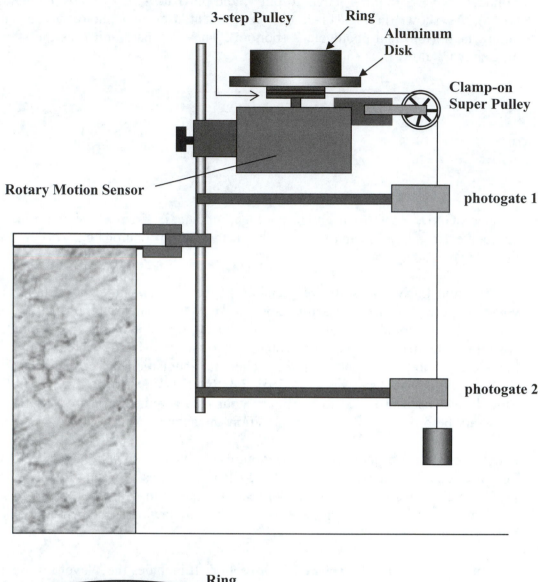

3-step Pulley Ring Aluminum Disk Clamp-on Super Pulley

Rotary Motion Sensor

photogate 1

photogate 2

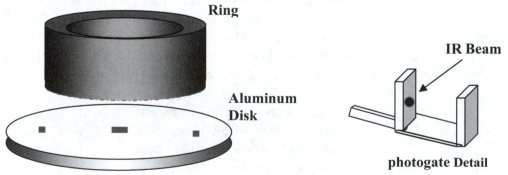

Ring

Aluminum Disk

IR Beam

photogate Detail

Fig. 11-7 The Inertia Experiment Mechanical Setup

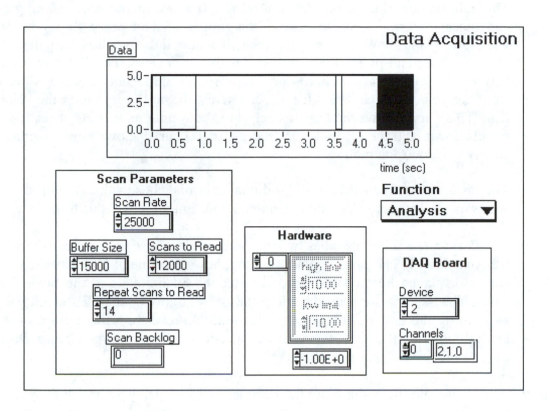

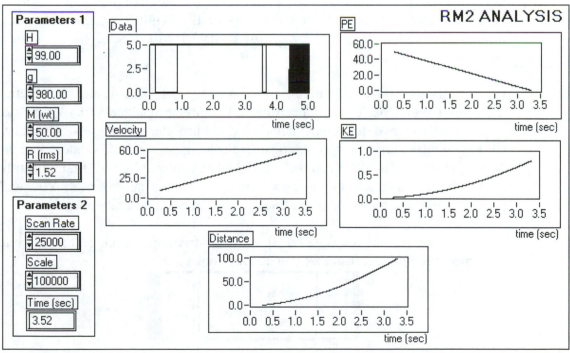

Fig. 11-8 The Front Panel's Data Acquisition and Analysis Sections of RM2.vi

The right section of the Front Panel, **RM2 ANALYSIS**, includes waveform graphs and other indicators used to display the results of data processing. The *Data* waveform graph shows the three channels of acquired data: pulses from the rotary motion sensor and pulses from the two photogate devices. The *Velocity* graph is derived from the rotary motion sensor pulses, and the *Distance* curve is derived from the velocity data. The KE (kinetic energy, $1/2mv^2$) curve uses the velocity data. This curve shows the kinetic energy of the weight as it starts at the top and travels down, and the PE (potential energy, mgH) curve shows how the potential energy varies during motion.

The **Parameters 1** box includes four digital controls whose values are used by the Analysis VI. If these values are not entered, then enter them at this time.

The **Parameters 2** box includes two digital controls whose values are also used by the Analysis VI. The Time digital indicator shows the time of travel by the weight and is displayed by the Analysis VI. Enter the Scan Rate and the Scale factor value. The Scale factor is used by the Analysis VI to divide the PE and KE values, so simplify the graph display. Any value that you read from these graphs should be multiplied by the Scale factor to get the actual value. Use the **Cursor Display** to get a precise reading from the graph.

Note: Be sure to enter the values into the Parameters 1 box using a consistent system of units. In the illustration of Fig. 11-8 the cgs system of units has been used. However, you may want to use the mks or the English system of units. The numerical results determined by the Analysis VI will depend on the system of units that you select.

4.   Choose **Acquire Data** from the *Function* menu in the Data Acquisition section of the Front Panel. Check the connections to the three transducers (the rotary motion sensor and the two photogate devices) and turn the power ON. Displace the weight slightly above the top photogate, release it, and an instant later click on the **Run** ⇨ button. The three channels of data that you acquire will first be displayed on the Data Waveform Graph and then saved to the spreadsheet. You will be prompted to enter the name of the file in a dialog box that will appear.

Enter a name such as RMData.txt, specify the drive, and click on the OK button. It is important that the data that you acquire looks like this:

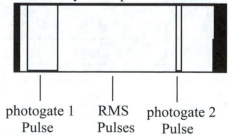

photogate 1   RMS   photogate 2
Pulse   Pulses   Pulse

The white area is made up of pulses from the rotary motion sensor (RMS). The thin white area just before the top photogate pulse represents the motion of the weight just before it passed through the top photogate. The analysis software depends on this format and will be looking for the RMS pulses between the two photogate pulses. At this time the data acquisition is complete.

5.   Repeat step 4 and acquire data, but this time *remove the ring*. Save this data to the spreadsheet.

Part II: Motion Analysis

Choose the *Analysis* option from *Function* menu in the Data Acquisition section of the Front Panel and scroll over to the right so that you can view the analysis section. Click on the *Run* ⇨ button. You will be prompted immediately to open the text file that was saved to the spreadsheet during the data acquisition in Part 1. Navigate to the directory where you saved that file and open it. Depending on the type of computer that you are using, it may take more than a minute to load the data and process it. When the analysis software is finished, it will present you with the results. The waveforms will appear on the graphs, as well as the time value on the *Time* digital indicator. We will next examine and interpret the analysis results.

Operation

This experiment demonstrates hardware timed, buffered data acquisition. Data is acquired continuously into a circular buffer (in memory) and at the same time retrieved in blocks to build a larger array of data. Buffer size is the number of scans that are held in the memory and is limited by the amount of the available RAM. By choosing a small buffer or by not withdrawing data from the buffer at a fast enough rate, an overwrite error can result where new data is written over the previous data.

In this experiment the *Scan Rate* control is set to 25,000 scans/second. With three channels being sampled, each channel is sampled at 75,000 samples/sec. The *Scans to Read* control is set to 12,000 scans and the *Repeat Scans to Read* control is set to 14. This means that the withdrawal of 12,000 scans from the circular buffer will be repeated 14 times, resulting in a two-dimensional array of 168,000 rows by 3 columns. The first column will contain the photogate 2 data, the second column photogate 1 pulses, and the last column pulses from the Rotary Motion Sensor.

The approach to data acquisition used in this experiment is different than that used in the Linear Motion experiment, where the entire block of data was created inside the buffer and then retrieved. That approach will not work in this experiment for several reasons. First, the pulses from the rotary motion sensor are 110 µsec wide. In order to detect these pulses, the scan rate must be sufficiently high. With 25,000 scans/sec each channel is sampled every 40 µsec. Second, the waveform time window should be in the neighborhood of 5 seconds in order to view a sufficient amount of the weight motion. That means that at least 125,000 scans (375,000 samples) must be collected into an array. This would not be possible with the setup of the Linear Motion experiment because of the buffer size limitation. With the continuous data acquisition into a circular buffer and the concurrent retrieval of data blocks to form the main array outside the circular buffer, the buffer size limitation is thus removed.

The RM2 Analysis portion of RM2.vi takes the array from the spreadsheet and separates it into three one-dimensional arrays for processing. The pulses from the two photogate units mark the beginning and the end of weight's motion. These pulses are also used to determine the time of travel.

This experiment uses conservation of energy to determine the inertia. In the equation

$$mgH = (1/2)mv^2 + (1/2)I\omega^2$$

mgH is the potential energy, representing the total energy at the beginning of motion. At the end of motion, this energy is converted to the kinetic energy of the weight and the rotational energy of the total inertia due to the disk, the ring, and the support. The friction in the pulley and the support as well as the weight of the string are all assumed to be negligible.

Fig. 11-8 shows the results of a typical run. The same waveform also appears in the VI's Front Panel in Fig. 11-8. The Analysis section of RM2.vi measures the time difference between successive pulses from the rotary motion sensor to determine the velocity array.

Because the mass is acted upon by the gravitational acceleration constant g, it is to be expected that its velocity should increase linearly. This is corroborated by the data in the velocity profile shown in Fig. 11-9.

The Analysis VI uses a home-spun algorithm by employing the trapezoidal rule to integrate the velocity array and produce the distance array. Its shape is a parabola, as one would expect after integrating a straight line. The final distance point, which is about 100 cm in this typical run, must match the actual distance traveled by the weight.

The Analysis VI also generates the potential energy and the kinetic energy arrays as shown in Figs. 11-8 and 11-9.

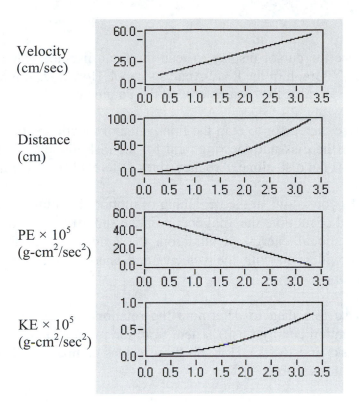

Velocity (cm/sec)

Distance (cm)

$PE \times 10^5$ $(g\text{-}cm^2/sec^2)$

$KE \times 10^5$ $(g\text{-}cm^2/sec^2)$

Fig. 11-9 Motion Data from the Analysis Section of RM2.vi.

Analysis

1. How many scans of data did you acquire in step 4 of the Procedure? Using the Scan Rate, how long should it have taken to acquire the data? How long did it actually take, judging from the Data Waveform Graph display?

2. In the Data Waveform Graph display, why is the first photogate pulse so much wider than the second?

3. Weigh the ring and measure its inner and outer diameters. Also weigh the disk and measure its diameter. Using these values, calculate the inertia, I, of the disk and the ring by using classical theory.

4. Use the data from the Front Panel to calculate the inertia based on data from:

$$mgH = (1/2)mv^2 + (1/2)I\omega^2$$

where m is the mass of the weight and H is the total distance traveled by the weight.

Calculate I for the disk and the ring. Also calculate I for the disk alone. Compare the theory-based results with those based on measurements.

5. Use the cursor display to read the final distance value from the Distance X-Y Graph in the RM2 Analysis panel. How does this value compare with the H value that you entered in the Parameters 1 box?

6. Use the cursor to read the final value of kinetic energy from the KE graph for the disk and the ring, and for the disk alone. Step 5 of the Procedure uses the disk alone. Are the values different? If so, explain why.

7. As you know, at the beginning of motion the total energy is PE and at the end of motion, the total energy is the sum of the KE of the weight and the rotational energy. In addition, the graphs in the RM2 Analysis panel provide you with the velocity and energy data over the entire span of motion.

   Use this data to determine the rotational energy (ring + disk) and the rotation energy (disk). Then, from the rotational energy calculate I(ring + disk) and I(disk). How do these values compare to those obtained earlier in this section?

Chapter 12
Physics II: Seebeck and Peltier Effects

The Seebeck Effect

The process of converting thermal energy to electrical energy demonstrates the Seebeck effect. Thomas Johann Seebeck (1770–1831) found that voltage or a difference in potential is produced when two dissimilar metals are maintained at different temperatures. The device that we call the thermocouple is a Seebeck effect device. The first experiment illustrates the Seebeck effect. But first, let's examine the theory of the Seebeck effect.

Theory

The thermocouple shown in Fig. 12-1 is an example of the Seebeck effect. Two dissimilar materials, A and B, form a test junction. The copper wires C make an electrical connection to the thermocouple wires A and B for measurement purposes. When heat is applied to the test junction, raising its temperature T above T_R, the test junction develops voltage (emf) proportional to the temperature difference $T - T_R$.

The reference junctions formed by wires A and C, and B and C are also temperature sensitive. They are usually kept at a constant reference temperature T_R. In the laboratory ice water may be used to provide the constant reference temperature. When kept at a constant temperature, the emf of the junctions does not change. In practice, the use of ice water is impractical. Instead, there are many compensating schemes to minimize the error voltage produced by the reference junctions.

The temperature range depends on type of materials used for A and B. A copper/ Constantan offers a temperature range of -200° C to $+400^\circ$ C, while platinum materials may be used up to 1500° C. Thermocouples are widely used for temperature measurement because they are fairly accurate and rugged.

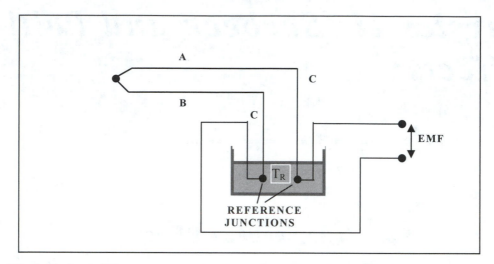

Fig. 12-1 A Thermocouple Junction Develops Voltage Proportional to the Temperature Difference

As mentioned earlier, a thermocouple can be made from two dissimilar materials. The use of semiconductors is not excluded. They too can be used to illustrate the Seebeck effect. Fig. 12-2 shows an illustration of a semiconductor converter that depends on the Seebeck effect for its operation. The illustration shows only one cell, but in order to generate enough energy to drive a motor many such cells are connected in series. A typical cell uses p-type and n-type semiconductor materials.

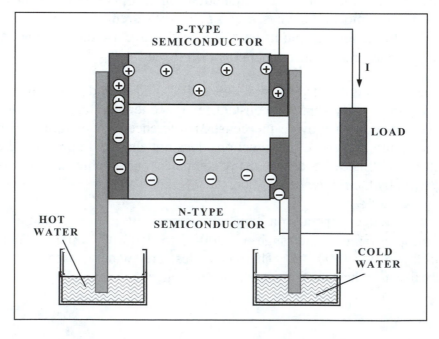

Fig. 12-2 Semiconductor Converter of Thermal Energy to Electrical Energy

As shown, a temperature differential is applied across the cell. The heat energy causes valence electrons to break away from their parent atoms and travel freely through the crystal. This occurs because the Coulomb force of attraction least binds valence electrons to the parent atom. This results in a net positive charge or a hole. The less energetic electrons do not have sufficient energy to break away from the parent atom, even with the applied heat energy; however, they can recombine with the holes. The holes thus have the appearance of moving through the crystal.

The cold end of the cell is connected to a load. This provides the path for free electrons to move through the load and re-enter the p-type semiconductor material, where they recombine with the holes. Fig. 12-2 shows the conventional current, I, but the flow of electrons is in the opposite direction.

If the load is a motor, then the current I may be sufficient to drive the motor. As current depends on the temperature difference, the hot side of the cell must reach sufficient temperature in order to develop the energy required by the motor.

Seebeck Effect and the Laws of Thermodynamics

The First Law of Thermodynamics expresses conservation of energy. According to this law, heat energy given up by the hot water is converted to mechanical work done by the motor + heat gained by the cold water + heat lost to the surroundings since the system is not thermally insulated.

The Second Law of Thermodynamics states that any irreversible process, such as the process in this experiment, progresses toward a state of greater disorder and a subsequent increase in entropy, S, which measures the disorder. A change in entropy, and thus the degree of disorder, is expressed by

$$\Delta S = Q/T$$

where Q represents the amount of transferred heat at a temperature T. In our case the hot water gives up heat energy Q_2 at temperature T_2. Because it gives up the heat, the change in entropy of the hot water is $-Q_2/T_2$. The cold water, on the other hand, gains the heat Q_1 at a temperature T_1. It therefore experiences an increase in entropy, $+Q_1/T_1$.

The Second Law of Thermodynamics requires that the total change in the entropy of the system undergoing an irreversible process increase, or in other words, become positive. Stated mathematically

$$\Delta S_{SYS} = +Q_1/T_1 - Q_2/T_2$$

According to the First Law of Thermodynamics, Q2 > Q1   because $Q_2 = Q_1$ + Work by motor + losses, and the Second Law of Thermodynamics requires that ΔS_{SYS} be positive or that

$$Q_1/T_1 > Q_2/T_2$$

which can be satisfied if $T_1 < T_2$ and it is satisfied when the fan begins to turn. As mentioned earlier, the temperature difference is very important. Imagine if the hot water and the cold water were poured into one container and the legs of the Thermoelectric Converter were immersed in that mixture. Even though the total internal energy of the water is same as before, the fan will not turn because now $T_2 = T_1$, thus violating the Second Law of Thermodynamics.

Experiment 12-1: Thermoelectric Converter

Objective

In this experiment we will observe and measure the Seebeck effect parameters. LabVIEW software is used to collect and process data. Collected data is analyzed and appropriate theory is applied.

Parts

Hardware

DAQ board (such as MIO-16E-10 or LabPC$^+$)
Extender board (associated with DAQ board)
Thermoelectric converter (PASCO TD-8550A)
Photogate (PASCO ME-9204B)
Temperature sensors (2) (PASCO CI-6505A)

Software

LabVIEW (version 5.1 or higher)
NI DAQ (DAQ board driver)
LabVIEW VIs for Experiment 12-1

Drive Sources

Fixed or variable + 12 and −12 V power supplies
Fixed DC power supply (5 V, 3 A)

Test Setup

1.   Test setup is shown in Fig. 12-3. Place the legs of the thermoelectric converter in two dishes of water. Fill one dish with cold or ice water, and place the other dish on top of a heater and fill it with room temperature water. The heater will gradually warm up the water.

2.   Connect the two terminals directly on back of the motor (attached to the thermo-electric converter) to AICh3 of the extender board. This is Seebeck voltage.

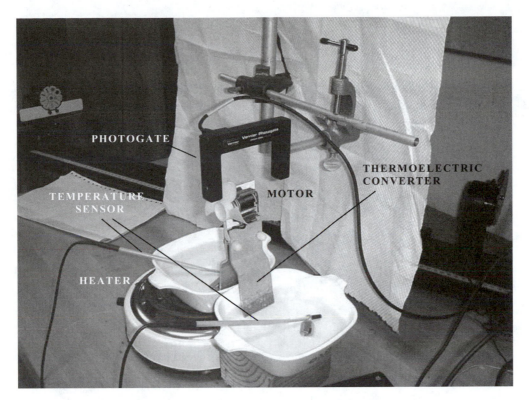

Fig. 12-3 Thermoelectric Converter Test Setup

3. Install the photogate and secure it in place by clamps. The photogate must be positioned so that the rotating fan of the motor will break its light beam. The photogate requires +5 V dc supply. Connect the photogate output to AICh0 of the extender board.

2. Immerse one temperature sensor in each dish of water. One will monitor the warm temperature and the other, the cold temperature. Connect the high temperature sensor to AICh1 of the extender board, and the low temperature sensor output to AICh2. Terminal block connections: The temperature sensors each require +12 volts to terminal 6, −12 volts to terminal 7, and terminals 4 and 5 to ground. Terminal 1 is the +signal output, and terminal 2 is the −signal output. A recommended connector interface box is shown in Fig. 12-4.

5. Flip the switch on the thermoelectric converter to the $\Delta T \Rightarrow E$ position. The experiment setup is now complete and ready to run.

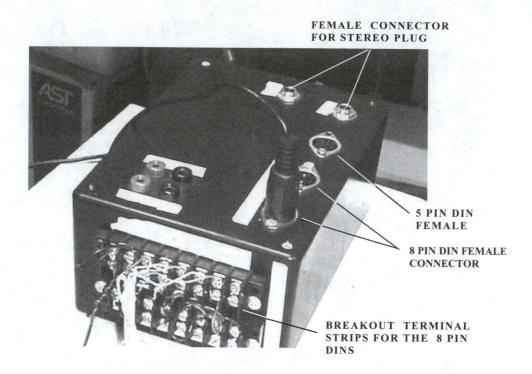

FEMALE CONNECTOR
FOR STEREO PLUG

5 PIN DIN
FEMALE

8 PIN DIN FEMALE
CONNECTOR

BREAKOUT TERMINAL
STRIPS FOR THE 8 PIN
DINS

Fig. 12-4 Connector Interface Box

Software

Fig. 12-5 shows the Front Panel of Seebeck.vi.

The Parameters recessed box includes digital and string controls that provide channels and board number information to the DAQ board. String controls provide channel information.

Delta X is a special control that sets the time interval between successive samples. In the illustration it set to 4 seconds, the default value. Because the actual test process is rather slow, 4 seconds is recommended. LabVIEW will use this value to set the X-axis for the Seebeck Effect Charts and the Temp History Overlay Charts.

The Boolean Start/Stop switch must be in the Start position before executing the VI. The Positioning tool is used to set its value. VI execution is terminated by clicking on this switch with the Positioning tool. The switch will then indicate Stop.

The Menu Ring labeled as Test Menu is used to select the test, either Seebeck or Peltier Effect.

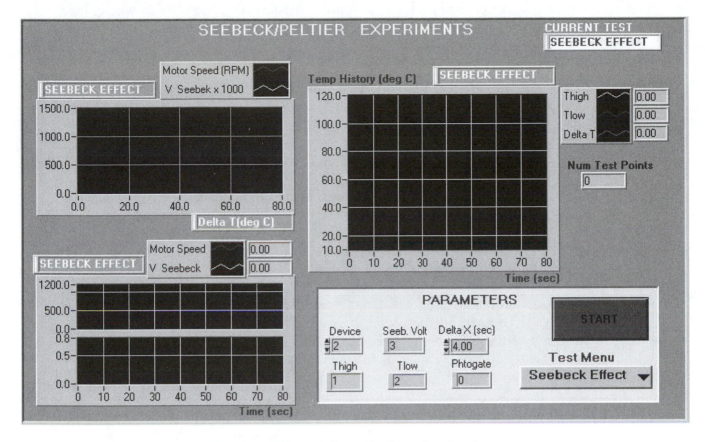

Fig. 12-5 Front Panel of Seebeck.vi

The Overlay Charts labeled as Temp History will plot three curves: THigh, TLow, and the difference, Delta T. The curves will progress point by point with 4 seconds difference (if Delta X is set to 4) between the points. Digital indicators on the right side provide values with greater precision.

The string indicator labeled as Current Test will display the name of the test currently running.

Two Stack Charts labeled as Seebeck Effect display Seebeck voltage and motor speed. These charts will vanish when the Peltier test is executed. Once again, the digital indicators to the right of the legend provide readings with better precision. LabVIEW processes the pulses from the photogate and calculates motor speed in RPM for display purposes.

The X-Y graph labeled as Seebeck Effect displays the Seebeck voltage and motor speed as a function of Delta T for the Seebeck test. It will display Delta T as a function of time in the Peltier test. LabVIEW sets the appropriate units in the string indicator along the X-axis. LabVIEW also displays the appropriate test name over each graph. Unlike the charts, which display data point by point during the test, the X-Y Graph will remain blank during the test and will display the curves when the test is terminated.

Procedure

1. Open Seebeck.vi (Book VIs>Physics II>Thermal.LLB>Thermal.vi) and switch to the Fronts Panel. The Parameters box will show the default values. You may change these values now.

2. Set the Start/Stop vertical switch to START with the Operating Tool.

3. Select Seebeck Effect from the Test Menu in the Front Panel.

4. Run the VI. Observe the data. The fan will spin faster as the temperature rises. As mentioned earlier, T high, T low, Delta T and Photogate pulse monitored on analog input channels 0 through 3 of the data acquisition board. These are displayed as a sample every four seconds (Delta X set to 4). The photogate pulses are processed during the four-second time interval and motor speed data point is displayed. To make this test work it is the difference in the temperatures of the two legs of the thermoelectric converter that is crucial. One leg is in the water being heated, while the other is in room-temperature water, or better yet, an ice bath.

5. To stop VI execution, set the Start/Stop vertical switch to STOP. When the execution stops, LabVIEW will display Motor Speed vs. Delta T and Thermal Voltage vs. Delta T curves on one X-Y Graph.

Peltier Effect

The Peltier effect, named for the French physicist Jean Charles Athanase Peltier, occurs when an external voltage is applied to two dissimilar metals. As the current flows, one of the junctions becomes hotter while the other is cooler. This effect is the reverse of the Seebeck effect.

Theory

When a battery is connected as shown in Fig. 12-6, with its negative terminal to the p-type semiconductor and its positive terminal to the n-type semiconductor, the Coulomb force of attraction forces the flow of holes in the p-type semiconductor and electrons in the n-type semiconductor to the right.

The movement of a large quantity of electrons in the n-type semiconductor is responsible for the transfer of internal energy from the left side to the right side, leaving the left side of the n-material cooler and the right side hotter. The same process occurs in the p-type semiconductor. The amount of energy transfer, and therefore, the amount of heating of the right side and cooling of the left side, is proportional to the value of current. In this experiment the current value is adjusted to 3 A.

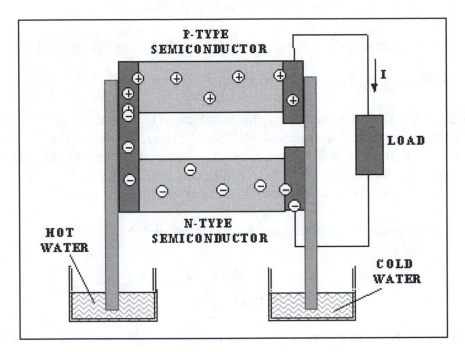

Fig. 12-6 Semiconductor Energy Converter, Electrical to Thermal

Experiment 12-2: Peltier Effect

Test Setup

1. One leg (marked "HOT") of the thermoelectric converter is immersed in a dish of water together with a temperature sensor. Connect the output of this temperature sensor to AICh1 of the extender board. The water in this dish should be room temperature.

    The other leg of the thermoelectric converter is immersed in the second dish of water with the second temperature sensor. Connect the output of this temperature sensor to AICh2 of the extender board. The water in this dish should be room temperature.

2. Connect 5 V dc, 3 A source to the banana receptacles on the thermoelectric converter. If 5 V, 3 A supply is not available, use the power amp in the external mode to get 3 A.

3. Flip the switch on the thermoelectric converter to E $\Rightarrow$ ΔT position. The experiment setup is now complete and ready to run.

Procedure

1. Open Seebeck.vi and switch to the Front Panel. Enter the appropriate values in the Parameter box. The software for this VI was described in the preceding section.

2. Set the Start/Stop vertical switch to START with the Operating Tool.

3. Select Peltier Effect from the Test Menu.

4. **Run** the VI. Observe the data. In this test Tlow, Thigh, and Delta T are monitored and displayed every 4 seconds (if Delta X is set to 4) on the Temp History Overlay Charts. The X-Y Graph, labeled as Peltier Test, will remain blank during the test and display the Delta T as a function of the time curve when the test is terminated. LabVIEW will choose the best data fit for this curve.

5. To stop VI execution, set the Start/Stop vertical switch to STOP.

Analysis

1. Explain the Seebeck effect as it applies to a semiconductor cell that is a part of the thermoelectric converter.

2. Both trays of water started with the same temperature in the Seebeck part of this experiment.
   a. What was the value of the temperature difference between the two trays after 1 minute?
   b. What was the temperature difference between the two trays at the time when the fan began to operate? What was the thermal voltage at this time?
   c. Using theory, explain why one leg of the thermoelectric converter in part b is getting warmer as time goes on.

3. Explain how the First and Second Laws of Thermodynamics apply to the first part of the experiment, which demonstrates the Seebeck effect.

4. Explain the Peltier effect in terms of the devices used in this experiment.

5. Suppose that both trays were filled in this experiment with boiling water. Will the Seebeck effect work? Explain.

6. In the Seebeck effect experiment, one graph was used to illustrate motor speed in rpm versus the temperature difference ΔT in degrees C. The approximate curve is shown in Fig. 12-7.

    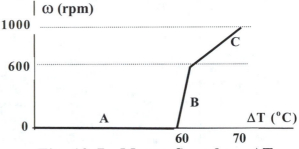

    a. Explain why segment A shows zero speed.
    b. Why is the slope of segment B greater that that of C? Explain.

    Fig. 12-7   Motor Speed vs. ΔT

7. Determine motor acceleration associated with segment B in Fig. 12-7 using this and other data collected in this experiment.

8. According to the theory for the Peltier effect in a semiconductor cell consisting of p-type and n-type material, a large quantity of electrons in the n material and holes in p material is moving from left to right as shown in Fig. 12-6. What does this motion have to do with the increase in temperature of water in the right container? Why doesn't the temperature increase in the left container?

9. In the Peltier effect experiment, how much current did you supply to the Thermoelectric Converter?

10. Regarding the Peltier effect experiment, how does the value of current applied to the thermoelectric converter affect the final temperature of the water? How much current was used and what was the final temperature?

11. Suppose that the Peltier effect experiment is done first (the switch on the thermoelectric converter is in the $E \Rightarrow \Delta T$ position). Next, the current is disconnected and the switch on the thermoelectric converter is set to the $\Delta T \Rightarrow E$ position. Will the motor rotate? Explain.

12. In the Seebeck effect experiment, water temperature was gradually increased by transferring heat from an electric heater (see Fig. 12-8 below).

    a. Calculate the heat energy in J supplied by the heater. Remember that energy = power × time.
    b. Calculate the entropy change of the hot water. Use data together with the weight of the water.
    c. Estimate the work done by the motor. This will also include various energy losses. Suggest the type of losses.
    d. Apply the First Law of Thermodynamics and estimate the overall energy loss.

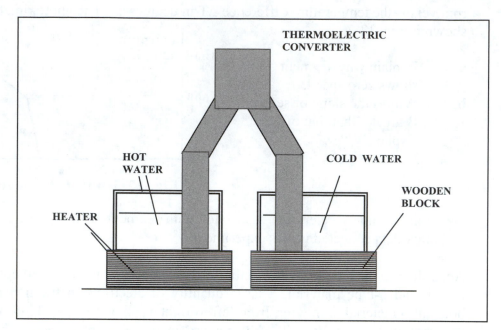

Fig. 12-8 Problem 12

13.   In the Peltier effect experiment, why is the hot water temperature increase greater than the temperature decrease of the cold water?

14.   In the Peltier effect experiment an energy of (15 W × time) Joules is supplied by the DC power sources to the thermoelectric converter. Where did this energy go? Estimate the total energy loss. Suggest the type of losses that are present in this process.

15.   Suppose that the Peltier effect experiment is conducted with both legs of the thermoelectric converter in the same tray of water. Will the temperature of the water change? Explain using theory.

Chapter 13
Physics III: Optics

Diffraction of Light

The phenomenon of diffraction of light is regarded as interference of light waves. It occurs when light passes through a slit or a small opening with sharp edges. According to Huygens' principle, every point inside the slit behaves as a source of light. All sources emit waves that travel different distances before arriving at their destination (a screen). Thus, at the screen some waves reinforce each other producing bright areas, while others are out of phase resulting in dark areas due to their cancellation. The diffraction phenomenon occurs also in radio signals that strike a sharp object in the path of their propagation.

Single Slit Diffraction Theory

Consider a monochromatic light source, such as a laser beam, incident on a single slit, as shown in Fig. 13-1. If a screen was placed at some point on the other side of the slit, our intuition suggests that a single light spot would appear on the screen. This, in fact, is not so. Instead of one spot, there are several bright spots. One in the center of the pattern is brighter and larger in diameter while the others are smaller and dimmer. The phenomenon described here is diffraction. Diffraction occurs when light strikes sharp edges as it does when it passes through a narrow slit, shown in Fig. 13-1a.

According to Huygens' principle, every point inside the slit behaves as a source of light waves. Accordingly each source emits light waves that travel to the right of the slit. Two representative point sources, identified in Fig. 13-1a as "a" and "b", are selected for the purposes of analysis. It can be seen from the diagram that the light ray emitted by source "a" travels a distance Δx further than the ray from source "b".

From the geometry in Fig. 13-1b

$$\Delta x = (d/2)\sin(\theta) \tag{13-1}$$

where d is the slit size and θ is the angle of the light rays. If the difference in path Δx traveled by rays "a" and "b" is equal to one half wavelength $\lambda/2$, then rays "a" and "b," arriving at the screen a distance D from the slit, will differ by 180° and therefore cancel. This cancellation creates an intensity minimum on the screen.

Hence

$$\Delta x = (d/2)\sin(\theta) = \lambda/2 \qquad (13\text{-}2)$$

or
$$\sin(\theta) = \lambda/d \qquad (13\text{-}3)$$

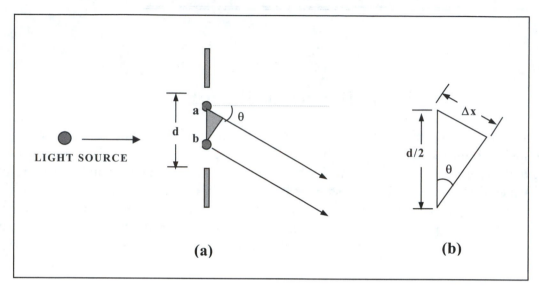

**Fig. 13-1 (a) Monochromatic Light Passes Through a Slit;
(b) Gray Triangle in (a) Enlarged For Analysis**

If four light sources inside the slit are considered, then it can be shown that

$$\sin(\theta) = 2\lambda/d \qquad (13\text{-}4)$$

with six sources
$$\sin(\theta) = 3\lambda/d \qquad (13\text{-}5)$$

With n sources inside the slit we obtain a general expression that identifies points of destructive interference where light intensity vanishes.

$$\sin(\theta) = n\lambda/d \qquad n = \pm 1, \ \pm 2, \ \pm 3, \ldots \qquad (13\text{-}6)$$

As Fig. 13-2b illustrates, each null is associated with a unique angle θ. The peaks (where light rays add) and null are a symbolic representation of the diffraction pattern. Actually, if an observer were to view the diffraction pattern on the screen, he would see a series of dots that vary in size and intensity, the center dot being largest in diameter and the brightest and the side dots decreasing in size and intensity away from the center.

If a photographic film is placed on the screen, then the diffraction pattern will appear as a series of light and dark bands as shown in Fig. 13-2c. The exposed light bands are the intensity peaks and the dark bands are the nulls.

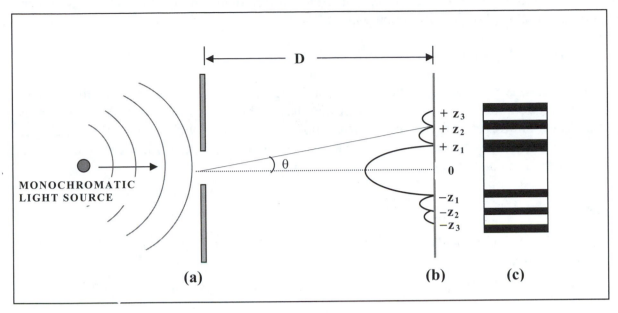

Fig. 13-2 (a) Light Source and the Single Slit; (b) Light Intensity Variation Due to Diffraction; (c) Light Intensity on Film

From the triangle shown here and from Fig. 13-2, the following equation

$$\tan(\theta) = z_n/D \approx \theta \qquad (13\text{-}7)$$

relates the distance to the nth peak z_n and the distance D between the slit and the screen. Because the ratio z_n/D is much smaller than unity the approximation $\tan(\theta) \approx \theta$ is valid. The same approximation can be made in Eq. 13-6 equating $\sin(\theta) \approx \theta$.

The two equations for θ are equated with the following result

$$z_n/D = n\lambda/d \quad \text{ or }$$
$$z_n = n\lambda D/d \qquad (13\text{-}8)$$

Eq. 13-8 relates the distance D and the size d of the slit to the nth intensity minimum z_n. In our experiment the laser diode wavelength is 660–680 nm. It is not exactly a monochromatic light source, so we will use the average wavelength of 670 nm in our calculations.

In the first part of this experiment we will produce the diffraction pattern using a laser beam as the source of monochromatic light, then manually scan the light sensor across the diffraction pattern, use the computer to collect data, and use a LabVIEW VI to process and display the diffraction pattern.

Double Slit Diffraction Theory

In Fig. 13-3 S is the source of monochromatic light, such as a laser, that ideally has one wavelength, although in practice a laser may have a line width of 20 nm. The light from the lower slit travels a distance x_2 to reach point P a distance D from the slit and the light from the upper slit travels a distance x_1. As illustrated in Fig. 13-3, the distance x_2 is longer than the distance x_1 by the amount $d\sin(\theta)$. Hence

$$\Delta x = x_2 - x_1 = d\sin(\theta) \text{ also}$$
$$\tan(\theta) = z/D$$

But $\tan(\theta) = \sin(\theta)/\cos(\theta) \approx \sin(\theta)$ because for small angles, $\cos(\theta) \approx 1$.
Consequently

$$\Delta x = dz/D$$

The difference in the paths traveled by the beams from the two slits can be converted to a phase difference by considering $\theta = \omega t$ as the basic argument of a sinusoid. Hence

$$\Delta\theta = \omega\Delta t = 2\pi f\Delta t \text{ and } \Delta t = \Delta x/v_c, \text{ where } v_c \text{ is the speed of light}$$

Therefore

$$\Delta\theta = 2\pi f\Delta x/v_c = 2\pi\Delta x/\lambda \quad \text{since } \lambda = v_c/f$$

After substituting for Δx, the final expression for the phase difference becomes

$$\Delta\theta = 2\pi dz/D$$

When the phase difference is 2π, the waves reinforce each other resulting in a peak, and when the phase difference is π, the waves cancel resulting in a null. Equating $\Delta\theta$ in the above equation to $2\pi n$, where n is an integer, and solving for z_n, the following expression for the nth intensity peak is obtained

$$z_n = n\lambda D/d \tag{13-9}$$

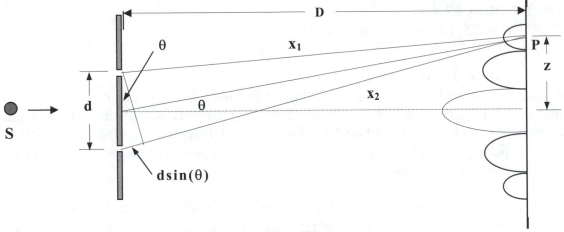

Fig. 13-3   Diffraction of Light from Two Slits

Experiment 13-1: Diffraction of Light

Objectives

- To demonstrate through measurement the properties of light diffraction.
- To expose the student to computerized data collection and data processing techniques using LabVIEW. Hardware and software are included in this process.
- To analyze the measurements and verify the diffraction theory using the measurements.

Parts

Hardware

Basic optics System (PASCO P/N OS-8515)
Rotary motion Sensor (PASCO P/N CI-6625)
Linear translator (PASCO P/N OS-8535)
Aperture bracket (PASCO P/N OS-8534)
Light sensor (PASCO P/N CI-6504A)
Connector interface box (Constructed by user)
DAQ board (such as MIO-16E-10 or Lab PC$^+$)
Extender board for the DAQ board.
RJ11 terminal

Software

LabVIEW (version 5.1 or higher)
Optics.LLB

Power Sources

+5 VDC (CI-6625, CI-6504A, and CI-6504)

Diffraction Test Setup

The transducers and sensors used in this experiment are equipped with standard connectors. Light sensor CI-6504A (used in this experiment) is provided with a 5-pin DIN connector, and light sensor CI-6504 uses an 8-pin DIN connector. The rotary motion sensor, on the other hand, is equipped with a modular phone connector. For reference, these connectors are illustrated in Fig. 13-4.

Transducers are usually equipped with a male plug. To make a connection to a specific pin, you must obtain the matching female connector. A good way to solve the connector problem is to build a connector interface box with various matching connectors wired to breakout strips that provide access to any pin on the transducer connector. An example of this type of box is shown in Fig. 13-9.

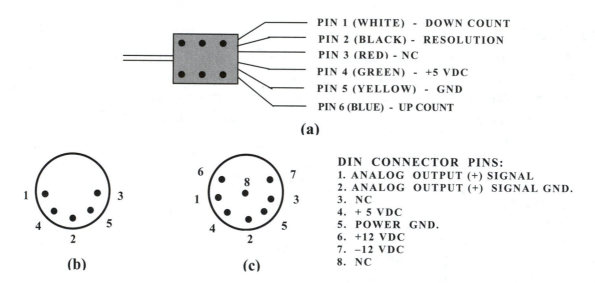

Fig. 13-4 (a) RJ11 Adapter Block; (b) 5-Pin DIN; (c) 8-Pin DIN Connectors

Fig. 13-5 shows a diagram for the diffraction experiment setup. All components are mounted on the optics bench. The laser light source is on the right side of the optics bench, and in front of it is the slit disk mounted on the slit holder. On the left side is the aperture, and behind the aperture is the light sensor mounted on the rotary motion sensor (RMS). The rotary motion sensor is mounted on the linear motion transducer, which is a toothed rod that passes through the rotary motion sensor. As the rotary motion sensor moves along the toothed rod (linear motion translator), it outputs a pulse (depending on the resolution setting) for each $1°$ or $\frac{1}{4}°$ of rotation. LabVIEW will convert these pulses to the distance scale in the diffraction graph. During the test you will manually move the rotary motion sensor across the laser beam diffraction pattern.

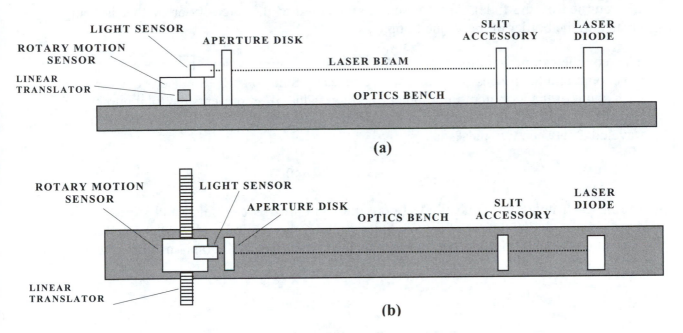

Fig. 13-5 Diffraction Test Setup: (a) Side View, (b) Top View

Software Description

In this experiment LabVIEW collects and processes data from two analog input channels. Channel 0 has the light intensity data from the light sensor and channel 1, the pulse data from the rotary motion sensor. Fig. 13-6 shows the Front Panel and the Block Diagram of the LabVIEW graphical program Diffraction.vi.

The Block Diagram consists of two frames of the sequence structure. In Frame 0, Acquire Waveforms.vi collects data from analog input channels 0 and 1 and stores it in a two-dimensional array; column 0 contains the rotary motion sensor pulse data and column 1, the light intensity data. After transposing, the Index Array function slices the pulse data and the light intensity data into two one-dimensional arrays. This data is displayed immediately on the waveform graph as unprocessed data is also passed to Frame 1 for further processing.

The processing of data is done inside the While Loop in Frame 1, which executes repeatedly until the condition terminal receives a False from the Array Size function.

H_L_I and Boolean.vi operate on pulses from the rotary motion sensor by detecting the array element's index i, which identifies a high-to-low transition of a pulse. The Array Subset 1 function uses this value of i to return a portion or a subset of the array in which the 0th element is the ith element of the original array. The subset array is recirculated by the shift register on the next iteration and tested by the Array Size function. If the array is empty, a False is passed to the condition terminal and the While Loop stops execution; otherwise the iteration process continues. The intensity value corresponding to the index i is returned by the Index Array function and stored in the intensity array that is being created in the border of the While Loop.

This completes one iteration, and the While Loop iteration counter [i] advances to the next count and the process described above repeats. Since one RMS pulse is detected for each count of the loop iteration counter, the final value of i represents the total number of detected pulses. If the length of travel of the RMS transducer is L, then the distance traveled d between two pulses is

$$d = (L)(i_{final})/1440$$

During each iteration of the While Loop, the value of d is computed and stored in the distance array in the border of the While Loop. After the While Loop stops execution, the intensity and the distance arrays are displayed on the X-Y graph as the Intensity (processed).

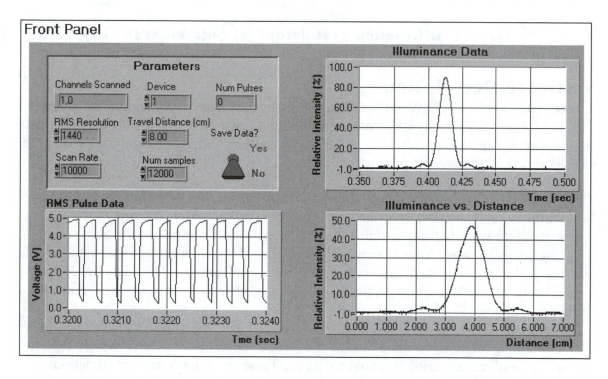

Fig. 13-6 The Front Panel and the Block Diagram of Diffraction.vi

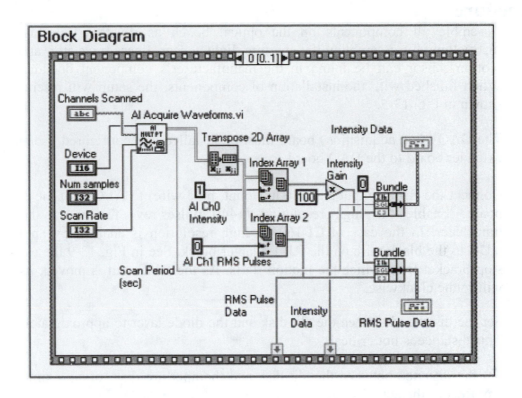

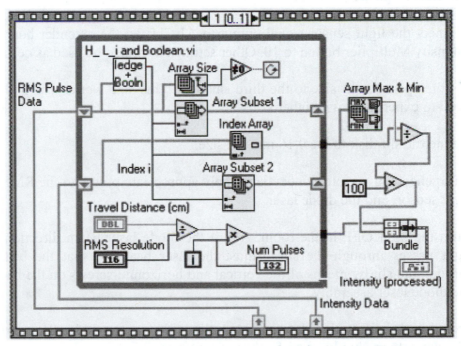

Fig. 13-6 The Front Panel and the Block Diagram of Diffraction.vi (continued)

Procedure

1. Assemble all components on the optical bench as illustrated in Fig. 13-5. Instructions for assembling the aperture, light sensor, linear translator, and rotary motion sensor can be found in the manufacturer's component documentation. When finished with the installation of components, the setup will resemble that shown in Fig. 13-5.

2. The DAQ (data acquisition) board must be installed and configured. Connect the extender board to the DAQ board.

3. Connect the RMS output signal (Up Count, blue wire) to AI Ch1 of the extender board. Enable the high resolution (1440 pulses/rev) feature of the RMS transducer. In the case of CI-1604A, high resolution is enabled by applying 0 VDC to the black wire on the RJ11 adapter block. See in Fig. 13-9 the use of the stop bracket from which the motion starts. As the RMS unit is moved, its wheel will rotate clockwise.

4. Set the distance between the slit disk and the diode laser to approximately 6 cm. This distance is not critical.

5. Set the distance between the slit disk and the aperture disk to approximately 90 cm. Record the actual value.

6. Connect the light sensor signal output to AI Ch0 on the extender board. Set the Intensity Multiplier button to 10. Other settings may also be used as desired.

7. Set the aperture slit size to the third smallest. This size works well and you may want to experiment with other sizes.

8. Rotate the slit disk to the 0.16 mm slit size.

9. After checking various connections apply appropriate power to the RMS unit, the light sensor, and the diode laser.

10. Turn the light OFF in the room. Adjust the diode laser beam direction until the light passes through the slit. Adjust the laser beam so that the beam passes through the slit by means of the vertical and horizontal screws on the back side of the diode laser assembly.

11. Examine the diffraction dot pattern on the white screen under the aperture disk. The dot pattern should be perfectly horizontal as shown here.

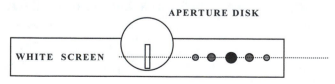

APERTURE DISK

WHITE SCREEN

It must not be skewed as shown below.

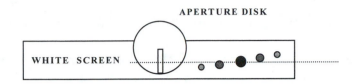

APERTURE DISK

WHITE SCREEN

If the pattern is not horizontal, adjust the outer ring on the slit disk assembly to make the dot pattern horizontal.

12. Open Diffraction.vi, (Book VIs>Physics III>Optics.LLB>Diffraction.vi) and switch to the Front Panel. Enter the required parameters in the Parameters box:

Scan Rate: 10,000 scan/sec
Num samples: 12,000 samples/chan
Channels Scanned: 1,0
Device: DAQ board number
RMS Resolution: 1440 pulses/rev
Travel Distance: Determined by user

The 10,000 scans/sec and 12,000 samples/chan result in 1.2 sec as the time in which you must slide the RMS unit over the distance of approximately 8 cm. This task can be accomplished with relative ease. In order to collect acceptable data, the motion of the RMS unit must be smooth, with no acceleration, taking approximately one second.

Two operators are needed to run this test, one to slide the RMS unit and the other to operate the VI. Set the Save Data? Boolean control in the Front Panel to Yes or No. A Yes setting will result in saving of X and Y arrays to the spreadsheet. The saved data may be recovered and plotted with Recover Spreadsheet Data.vi and used in the Analysis section. Set the switch to No in the beginning and when the experiment is running smoothly, then save the data.

Run Diffraction.vi and simultaneously slide the RMS unit so that the light sensor scans across all diffraction pattern dots on the white screen. If the Save Data? Boolean control was set to Yes, you will first be prompted to save the X-array data and then the Y-array data to a spreadsheet text file. Provide the path of your choice.

13. Repeat the procedure to obtain patterns for 0.08 mm, 0.04 mm, and 0.02 mm slits. As the slit size decreases, less light passes through the slit. To compensate for this, aperture slit size may be increased. You will be prompted to save the graph data to a spreadsheet as a text file.

14. Replace the single slit disk with the double slit disk and obtain the diffraction patterns for the following slit width/separation (in mm) combinations: 0.04/0.5, 0.08/0.25, 0.08/0.5. You will be prompted to save the graph data to a spreadsheet as a text file.

Analysis

1. If the test data was saved to a spreadsheet, it may be opened now and the graph restored. Open Recover Spreadsheet Data.vi (Book VIs>Physics III> Optics.LLB>Restore Spreadsheet Data.vi) and run it. You will be prompted twice to provide the path for the X and Y arrays.

2. Using the single slit diffraction data for 0.04 mm, 0.08 mm, and 0.16 mm slits calculate z_1 and z_2 values. Tabulate the calculated and the measured z values.

3. Compare the calculated and the measured z values. The error defined by the following equation may be used as the basis of comparison

$$E = 100(z_{nc} - z_{nm})/ z_{nc}$$

where
z_{nc} = calculated n^{th} null
z_{nm} = measured n^{th} null

Error in excess of 15% must be justified if it persists after the test is repeated.

4. Using the double slit diffraction pattern data for slit width/separation (in mm) combinations of 0.04/0.25, 0.04/0.5, 0.08/0.25, 0.08/0.5, calculate z_1 and z_2 values. Tabulate the calculated and the measured z values.

5. Compare the calculated and the measured values of z. The error defined by the following equation may be used as the basis of comparison

$$E = 100(z_{nc} - z_{nm})/ z_{nc}$$

where
z_{nc} = calculated n^{th} peak
z_{nm} = measured n^{th} peak

Error in excess of 15% must be justified if it persists after the test is repeated.

6. How does the value of z_n vary with slit size for the single slit diffraction pattern?

7. For the double slit diffraction pattern, how does z_n vary with
a. increase in slit size.
b. separation between slits.

8. Explain the effect on the diffraction pattern due to change in precision of the rotary motion sensor from 1440 pulses/rev to 360 pulses/rev.

9. What effect on the diffraction pattern can be expected when the scan rate of AI Acquire Waveforms.vi is changed from 5000 to 2000? Explain.

10. What effect on the diffraction pattern can be expected when the number of samples per channel of AI Acquire Waveforms.vi is changed from 4000 to 500? Explain.

Inverse Square Law

A source of monochromatic light, light of a single wavelength or of one frequency, generates waves that travel in nearly parallel paths and consequently do not spread out or scatter. The intensity of such light remains nearly constant as the distance from the source increases. A laser beam is an example of a monochromatic source.

On the other hand, the light emitted by the tungsten filament of an incandescent lamp includes many wavelengths that travel in all directions. The intensity of such light varies inversely as the square of the distance from the source, hence the inverse square law. An isotropic source in electronic communications behaves in a similar manner. The inverse square law applies to all situations where the energy from a radiating source scatters nearly uniformly in all directions.

Theory

We consider here the attenuation due to light beam spreading and neglect the attenuation due to absorption and other effects. Consider a light beam from a monochromatic source as shown in the illustration below.

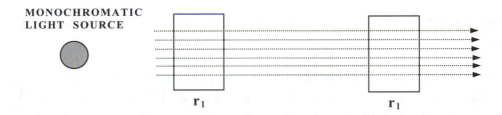

It can be seen from the illustration that the number of rays per unit area at a distance r_1 from the source is same as it is at a distance r_2. This is because the light rays travel parallel to each other, a phenomenon exhibited by a laser beam, suggesting that the intensity of light remains almost constant as it travels through space.

Let's consider now a source of light that radiates in all directions. As illustrated in the diagram below, light emitted by an ideal point source S radiates uniformly in all directions in a spherical radiation pattern. The area A on the surface of the sphere of radius R is due to a solid angle $\theta_S = A/R^2$ steradians, and the luminous flux F passes through this area. The intensity I of the point source is defined as

$$I = F/\theta_S \text{ lumens/steradian} \qquad (13\text{-}10)$$

or $F = I\theta_S = IA/R^2$, and the luminous flux intercepted by area A, the illuminance of A, is defined as

$$I_L = I/A = I/R^2 \qquad (13\text{-}11)$$

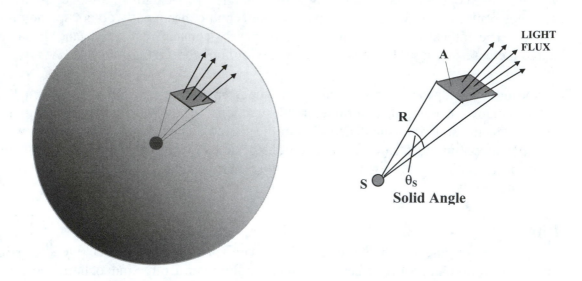

In our experiment, the area A intercepted by the luminous flux is the front area of the optical sensor.

Thus, the illuminance measures the amount of light intercepted by area A. It is evident from Eq. 13-11 that it is inversely related to the square of the distance R from the source, hence the inverse square law. This fact can be expressed in general in terms of a constant of proportionality as

$$I_L(R) = k/R^2 \qquad\qquad (13\text{-}12)$$

The ratio of I_L at r_2 to that at r_1 may be expressed as

$$I_{L2}/\,I_{L1} = (R_1/\,R_2)^2 \qquad\qquad (13\text{-}13)$$

From eq. 13-13 one can see that if the distance from the source doubles, light strength is reduced to ¼ its original value.

Experiment 13-2: Inverse Square Law

Objectives

- To demonstrate through measurement the inverse square law as it applies to light..
- To expose the student to computerized data collection and data processing techniques using LabVIEW. Hardware and software are included in this process.
- To analyze the measurements and verify the diffraction theory using the measurements.

Parts

Hardware

Basic optics system (PASCO P/N OS-8515)
Aperture bracket (PASCO P/N OS-8534)
Light sensor (PASCO P/N CI-6504A)
Connector interface box (constructed by user)
DAC board (such as MIO-16E-10 or Lab PC$^+$)
Extender board for the DAC board

Software

LabVIEW (version 5.1 or higher)
Optics.LLB

Power Sources

+5 VDC (CI-6625, CI-6504A, and CI-6504)

Software Description

In this part of the experiment only the light intensity data is acquired. When Inverse Square Law.vi completes execution, it will plot the normalized light intensity versus distance curve on the X-Y graph as shown on the Front Panel in Fig. 13-7.

The For Loop in the Block Diagram of Fig. 13-7 generates n data points determined by the Front Panel numeric controls as

$$n = Dist/Cm/Div \qquad (13\text{-}14)$$

where Cm/Div is the distance between data points. The computed value of n is applied to the loop counter N of the For Loop. The indexes of the one-dimensional array samples will range between 0 and n − 1.

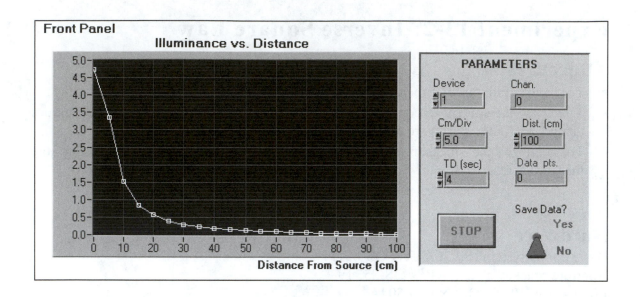

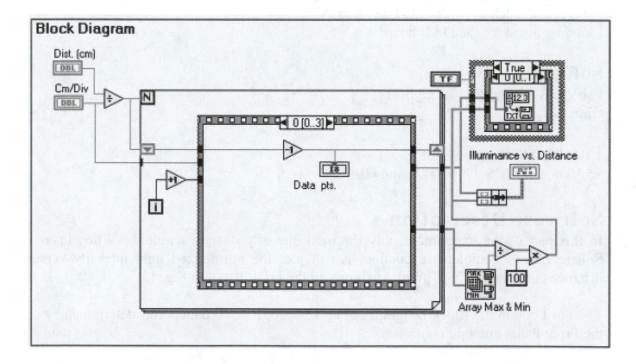

Fig. 13-7 The Front Panel and the Block Diagram of Inverse Square Law.vi

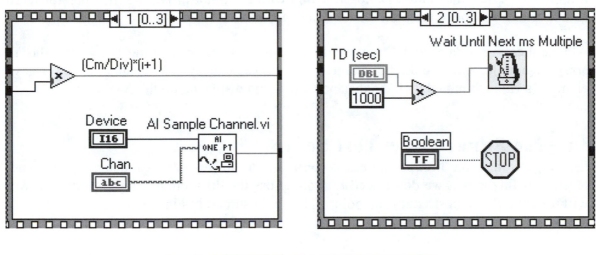

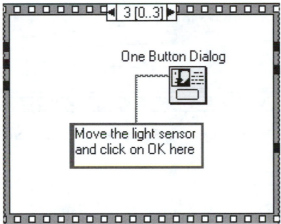

Fig. 13-7 The Front Panel and the Block Diagram of Inverse Square Law.vi (continued)

The Sequence structure inside the For Loop consists of four frames. In Frame 0 the shift register is initialized to the n value prior to the first iteration, is decremented on each subsequent iteration, and its value is displayed on the Front Panel numeric indicator Data pts., signifying to the operator the number of data points yet to be acquired.

In Frame 1 AI Sample Channel.vi acquires one illuminance data point and stores it in the one-dimensional array in the border of the For Loop to be used later for the Y-axis of the Illuminance vs. Distance X-Y graph. The corresponding X array is also formed in this frame as Cm/Div*(i+1) and stored in the one-dimensional array in the border of the For Loop.

Frame 2 provides a time delay set by the operator and an abort execution Front Panel Stop Boolean control. To set the desired time delay, the operator must enter the value in seconds into the Front Panel numeric control TD.

Frame 3 uses the One Button Dialog function with a prompt providing the operator with an opportunity to move the light sensor to a new position.

After collecting n data points, the For Loop stops execution, making the X and the Y arrays available. Outside the For Loop, the illuminance array is normalized to its largest element's value, bundled with the distance array, and displayed on the X-Y graph.

Inverse Square Law Test Setup

In comparison to the first part, the test configuration in this part is much simpler. The reason for this is that we don't need to scan across the diffraction beam, but instead we will measure the light strength at a point. This is illustrated in Fig. 13-8.

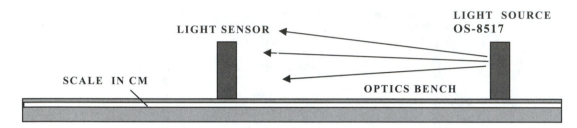

Fig. 13-8   Inverse Square Law Test Setup

Procedure

1.   Assemble the light source and light sensor on the optics bench as shown in Fig. 13-8. Cover all parts of the light source where light leakage is evident. Consult the manufacturer's manual for details.

2.   Connect the light sensor's output signal to AI Ch0 on the extender board.

3.   Set the aperture disk to full open circle.

4.   Apply appropriate power to the light sensor and to the light source.

5.   Open Inverse Square Law.vi (Book VIs>Physics III>Optics.LLB>Inverse Square Law.vi) and switch to the Front Panel. Enter the required parameter values into the Settings box.. ***You must set Cm/Div to 5 as Analysis will depend on this setting.*** Because the double-slit diffraction pattern is much finer in detail, you may want to acquire 20,000 samples/chan at a scan rate of 20,000 scans/sec. These values will allow you two seconds to move the RMS unit. You may want to change other default settings as well. Set the Save Data? Boolean control in the Front Panel to Yes or No. A Yes setting will result in saving of X and Y arrays as text files to the spreadsheet. The saved data may be recovered and plotted with Recover Spreadsheet Data.vi and used in the Analysis section.

6. Turn the room light OFF and run Inverse Square Law.vi. To preserve the integrity of the acquired data, it is very important to make the room completely dark. The VI will acquire n data points determined by Eq. 13-14. After each acquired data point you will be prompted to move the light sensor and click on the OK button in the prompt box. Move the light sensor a distance corresponding to the Cm/Div setting in the Parameters box on the Front Panel. The numeric indicator Data pts. shows the number of data points that remain to be acquired.

7. When the Data pts. indicator shows 0, the For Loop stops execution, completing the acquisition phase of the test. The VI will normalize and display the illuminance data versus distance on the X-Y graph. If the Save Data? Boolean control was set to Yes, you will first be prompted to save the X-array data and then the Y-array data to a spreadsheet text file. Provide the name and the directory for both. Record the value of Imax; you will need it in the Analysis section.

8. Replace the incandescent light source OS-8517 the diode laser. Repeat the test and save the graph data to a spreadsheet. Because the light sensor is stationary in this part of the experiment and the laser beam is very narrow, ensure that the beam is aimed directly at the light sensor's window. Make adjustments on the diode laser unit if necessary.

Analysis

1. If you wish to view the data that was saved to the spreadsheet, the files may be opened and the graph restored. Open Recover Spreadsheet Data.vi (Book VIs>Physics III>Optics LLB>Recover Spreadsheet Data.vi) and run it. You will be prompted twice to provide the path for the text files containing the X and Y arrays.

2. Open ISL Curve Fit.vi (Book VIs>Physics III>Optics.LLB>ISL Curve Fit.vi) and examine the Front Panel. This VI fits your data to the K/R^2 equation. The K Value Spread graph on the Front Panel shows the variation in the K values. The VI calculates the average K value. Enter the Cm/Div (5 was recommended in the procedure) and Dist values in the Parameters box.

Run the VI. As the VI starts execution, you will be prompted to provide the path to the saved arrays, first the X-array and then the Y-array. After the VI stops execution, the best value of K is displayed, and the ISL Data/Curve Fit X-Y graph displays the original data and the best fit to the data curve. Record all important results such as the K value, the K value spread, and the information from the ISL Data/Curve Fit graph.

3. Develop a VI that will compute and plot the error
$$E = 100|I_d - I_c|/I_d$$
where I_d = measured illuminance value, and I_c = curve fitted illuminance value computed from K/R^2 as a function of the distance. The K value was obtained in step 2. E represents the difference between the unprocessed data and the best curve fit to the data in step 2. This procedure may also be done manually. Save the results in a text file, as you may need them later.

4. The ratio of illuminance $I(r_2)$ at r_2 to $I(r_1)$ at r_1 may be expressed as
$$I(r_2)/ I(r_1) = (r_1/r_2)^2$$
and is independent of K. It is evident from the above equation that if the distance doubles; i.e, $r_2 = 2\ r_1$ the illuminance becomes ¼ its original value. In order to verify whether the data actually follows this law, create a one-dimensional array $[I(r_1)/ I(r_2)]^{1/2}$ for all data points $r_2 = 2r_1$. Clearly, $[I(r_1)/ I(r_2)]^{1/2} = 2$ for all points $r_2 = 2r_1$. If your total distance of travel in the experiment was 100 cm and cm/div was set to 5, then there are 20 data points. Of the 20 data points only about 10 points satisfy the equation $r_2 = 2r_1$. The array thus created is ideal or based on theory.

Plot the following three curves on the same X-Y graph as a function of samples that satisfy $r_2 = 2r_1$:
$$Ath = [I(r_1)/ I(r_2)]^{1/2} \text{ based on theory}$$
$$Adat = [I(r_1)/ I(r_2)]^{1/2} \text{ based on data}$$
$$E = Ath - Adat, \text{ the error}$$

5. Using the results of the previous questions,
   a. How well does the data follow the inverse square law?
   b. Comment on significant deviations (in excess of 15%) between the data and the inverse square law, taking into account the testing environment, components, and equipment accuracy.

6. Early in the analysis we were concerned with curve fitting or representing the data by an equation.
   a. Why is it necessary to represent the data by an equation that best fits the data?
   b. Suggest a practical application where a curve fitting procedure would be essential to data processing.
   c. ISL Curve Fit.vi, used in step 2 of the Analysis, is a homespun curve fitting algorithm. It works well in cases where the data is expected to fit a particular mathematical form such as K/R^2. Name other curve fitting procedures and a specific advantage that each has.

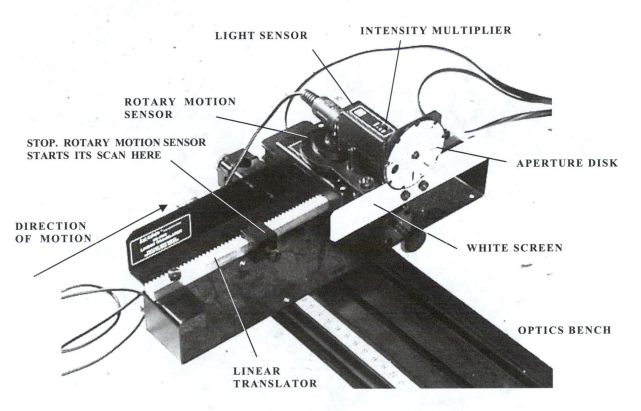

LIGHT SENSOR

INTENSITY MULTIPLIER

ROTARY MOTION
SENSOR

STOP. ROTARY MOTION SENSOR
STARTS ITS SCAN HERE

APERTURE DISK

DIRECTION
OF MOTION

WHITE SCREEN

OPTICS BENCH

LINEAR
TRANSLATOR

LIGHT DETECTION AND SCANNING ASSEMBLY

Fig. 13-9 Selected Optics Experiment Components

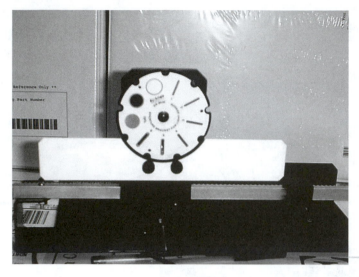

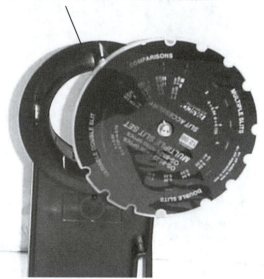

ADJUST RING FOR THE HORIZONTAL
DOT PATTERN ON WHITE SCREEN

APERTURE DISK AND THE WHITE SCREEN

THE SLIT DISK ASSEMBLY

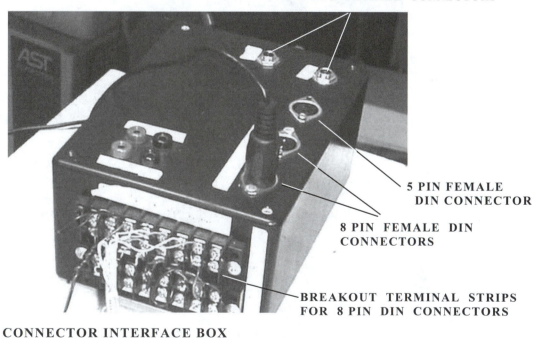

STEREO FEMALE CONNECTORS

5 PIN FEMALE
DIN CONNECTOR

8 PIN FEMALE DIN
CONNECTORS

BREAKOUT TERMINAL STRIPS
FOR 8 PIN DIN CONNECTORS

CONNECTOR INTERFACE BOX

Fig. 13-9   Selected Optics Experiment Components (continued)

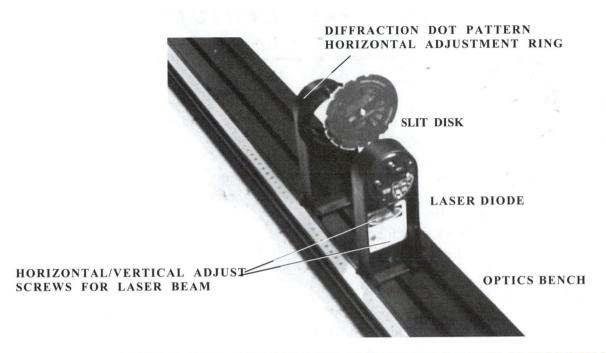

DIFFRACTION DOT PATTERN
HORIZONTAL ADJUSTMENT RING

SLIT DISK

LASER DIODE

HORIZONTAL/VERTICAL ADJUST
SCREWS FOR LASER BEAM

OPTICS BENCH

DIODE LASER AND SLIT DISK ASSEMBLIES ON THE OPTICS BENCH

LIGHT DETECTION AND SCANNING ASSEMBLY

Fig. 13-9 Selected Optics Experiment Components (continued)

Chapter 14
Motors and Generators

Introduction

In general, a second course in physics introduces students to magnetism, thermodynamics, optics, and other topics. Selected experiments included in this chapter provide the user with experimental procedures to test and verify the theory for some of the topics in Physics II. The following experiments are included:

- **DC Motor Energy Converter**
- **DC Motor Performance**
- **DC Generator**
- **AC Generator**
- **AC Motor**
- **Universal Motor with AC Drive**
- **Universal Motor with DC Drive**

LabVIEW software is used in all experiments to acquire and process the data. In today's world of computers and a variety of software that is readily available, it makes sense to use LabVIEW software in data acquisition and data processing. The software used in this chapter is not of commercial origin, but in a manner of speaking is home spun. It is nonetheless just as appropriate and perhaps even more so because it has been custom tailored to meet the needs and objectives of the experiments.

These experiments are not intended to alter the traditional approach to theory but rather to reinforce the theory in a meaningful way. In a computerized environment the user no longer is faced with the drudgery of many hours of data collection and then with more hours of calculations and curve plotting. The computer collects the data and processes it in a manner of seconds. The implication of this is that more can be accomplished in the allotted time, leaving the user with extra time to interpret the data and draw conclusions.

So, without further delay let's begin with the first experiment.

Experiment 14-1: DC Motor Energy Converter

Objectives:

- To investigate the conversion of energy from electrical to mechanical form.
- To evaluate the energy losses and motor efficiency.
- To apply LabVIEW as a virtual instrument for collecting and processing data.
- To apply LabVIEW as a controller for timing and for sequencing of events.

Parts

LabVIEW (version 5.1 or higher)
NI DAQ (DAQ board driver)
DAQ board (such as MIO-16E-10 or LabPC$^+$)
Extender board
DC Motor Converter.vi (LabVIEW software)
photogate (PASCO ME-9204B)
Variable DC power supply (to 12 VDC, 1 A)
Fixed DC power supply (5 V, 1 A)
(2) Relays (5 V, 90 mA such as SPDT Micromini by Radio Shack)
(2) Diodes (general purpose)
(2) Resistors (1 Ω, ½ W)
(2) Resistors (1 MΩ, ½ W)
(2) Darlington amplifiers (300 mA, such as NTE172A)
Current probe (PASCO CI-6556)
Motor/generator kit (PASCO CI-6513)
Weight set (300g, 400g, 500g, 600g, 700g, 800g, 900 g)

Test Setup

Fig. 14-1 shows the wiring diagram for this experiment. It includes mechanical as well as electrical requirements. The motor is placed at a height that will allow a weight on a 2.5-meter string to hang from its shaft.

The 1 Ω resistor is part of the current probe used to monitor the motor's armature current. It is wired in series with the relay's NO and NC contacts and the DC motor.

The CR1 and CR2 relays are used to start and stop the motion of the weight. CR1 provides the NO contact and CR2, the NC contact. Approximately 5 V and 80 mA are required to energize these relays. To provide the required current drive, Darlington amplifiers are used with the relay's coil wired in the collector circuit as shown in Fig. 14-1.

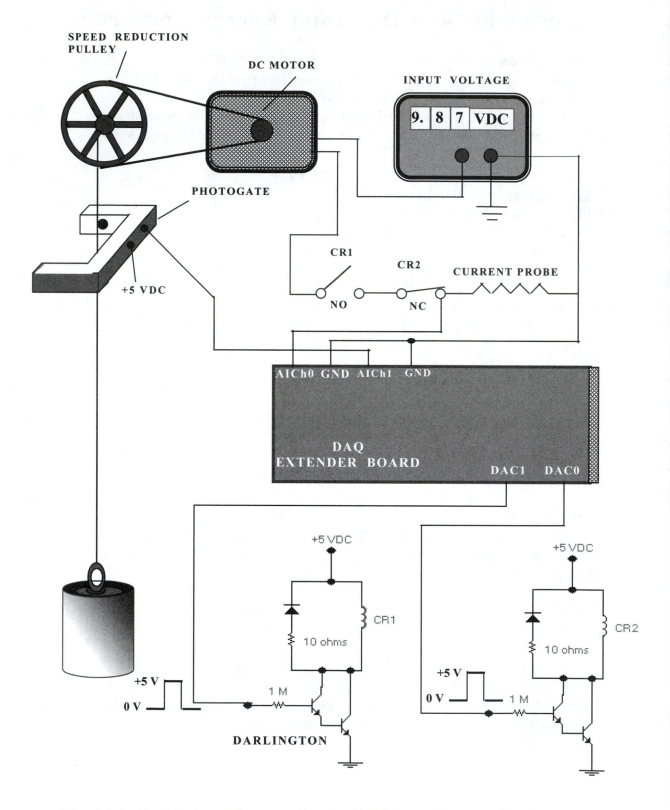

Fig. 14-1 Test Setup Diagram for the DC Motor Energy Converter

The photogate is used to sense the arrival of weight and to stop the motion. It uses a stereo plug as shown below:

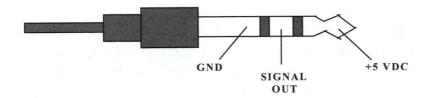

GND   SIGNAL OUT   +5 VDC

Connect GND, +5 VDC to the photogate device. Connect Signal Out of the photogate device to AI Ch. 1 of the DAQ extender board.

As shown in Fig. 14-1, the photogate is mounted close to the motor and positioned to allow the weight to move freely between its side arms.

The variable power supply may be adjusted up to 12 VDC and must provide up to 1 A of current.

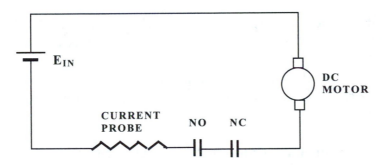

Fig. 14-2 Simplified Armature Circuit of the DC Motor

A simplified armature is shown in Fig. 14-2. The armature circuit of the DC motor does not include the photogate weight position sensor nor does it include the Darlington amplifiers that are needed to provide the proper drive for the relays.

Procedure

1. At this time LabVIEW software (version 5.1 or higher) should be on your computer and the data acquisition board (DAQ board such as Lab PC$^+$ or MIO-16E-10) should be plugged into the expansion slot inside the computer. Also, the extender board should be connected to the DAQ board, whose connector is barely visible in the back of the computer. The data acquisition driver software (NI-DAQ) should also have been loaded. Finally, the WDAQCONF utility should have been executed to configure your data acquisition board. All this is necessary before you run the data acquisition and analysis VI.

2. Build the test circuit as shown in Fig. 14-1.

3. Place the 500 g weight in the START position, approximately 2.5 meters below the motor.

4. Apply +5 VDC to the photogate and +5 VDC to the Darlington amplifier, and set Ein to 11 VDC.

5. Open DC Motor Energy Converter.vi (Book VIs>Motors and Generators>DC Motor Energy Converter.vi) and inspect the Front Panel. It has four general sections:

   The **Save Data to Spreadsheet** switch allows saving of the motor current data to a spreadsheet after the data acquisition phase.

   The **Controls** recessed box includes control buttons to process data, acquire data, or stop the motion of the weight in case of an erratic operation.

   The **Parameters** section includes digital controls for the weight, string length, and input voltage. The user will be prompted to supply these values at the beginning of the data processing phase.

   Experimental Results contains six digital indicators. They will display calculated values immediately after the data processing phase. Most of the controls and indicators will become invisible during the data acquisition phase because they are not needed at that time. They will reappear at the beginning of the data processing phase.

6. *Click* on the Run button in the Front Panel and turn ON the motor switch located on the motor housing.

7. Set the **Save Data to Spreadsheet** option as desired. If you choose YES, at the end of data acquisition phase you will be prompted to assign the file name for the data. Give the file name a .txt extension.

8. Click on the **Acquire Data** button in the Front Panel. The weight should begin to move immediately. It will travel for approximately half a minute, and as it passes through the photogate, its motion will be terminated and a message will appear on the screen indicating that the data acquisition phase is now complete. This message will give you directions to initiate the data processing phase. At this time the DC motor armature current data will be displayed (approximately 3000 samples) on the Motor Current Waveform Graph.

Note: Be sure to record the value of E_{in} as the weight is rising. You will need this value before the data processing phase.

9. Do as directed in steps 1, 2, and 3 of the message, and click on the **Process Data** button in the Controls section of the Front Panel.

10. The data processing phase takes a moment, and the numerical values are displayed in the *Experimental Results* section in the Front Panel. Record these results.

11. The final message appears asking you if you want to do another run. You must do several more runs before concluding this experiment, as follows:

Weight (g)	Input Voltage (V)
700	11
900	11
1000	11
1100	11
1200	11

Follow the directions in the final message, click on **YES,** and repeat steps 8, 9, 10, and 11, each time using a different weight. Don't forget to monitor the input voltage during the data acquisition phase. After the last run (1200 g), click on **NO** in the final message. This will terminate the data acquisition phase.

After you click on NO in the final message, the following three curves will be displayed by the X-Y graph:
Plot 0 Energy Lost.
Plot 1 Efficiency.
Plot 2 Potential Energy.

Operation

The operation of the weight lifting motor consists of a sequence of frames or events. Each frame performs a specific task.

When the operator clicks on the RUN button, the VI execution begins. The system goes into the WAIT mode until the operator clicks on the Acquire Data button in the Front Panel. When this happens, the weight begins to travel. It continues to move until it passes through the photogate and breaks the beam. The photogate outputs a signal to stop the motion of the weight. The synchronized START and STOP events generate signals, which are used to calculate the time of travel of the weight in the Process Data phase.

The START and STOP events are achieved by the NO and NC relay contacts, which are wired in series with the motor and the DC power supply as shown in Fig. 14-1 and 14-2. When one of these contacts is open, the motor's armature current is interrupted and the motor stops. The motor will operate only if both contacts are closed. This happens when the operator clicks on the Acquire Data button, energizing CR1 and causing its NO contact to close; and since CR2 is de-energized, its NC contact is closed. Since the two contacts are closed, the motor starts running and lifting the weight.

As the motor rises it will eventually pass through the photogate. The signal from the photogate energizes relay CR2, causing its NC contact to open and interrupt armature current, thus stopping the motor.

The signals that energize and de-energize the relays (5V/0V) originate at the DAC0 Out and DAC1 Out ports of the DAQ board. DAC0 Out port controls relay CR2 and its NC contact while DAC1 Out port controls relay CR1 and its NO contact.

If the weight does not stop as it passes through the photogate, the user can stop it by clicking on the *Emergency Stop Weight* in the Controls section of the Front Panel. This can happen when the photogate malfunctions, or the +5 V is not applied to the photogate or when the cable is faulty.

The execution of the DC Motor Converter.vi terminates when the operator clicks on NO in the final message at the end of the Process Data phase. The NO choice outputs a FALSE to the condition terminal of the outside While Loop, thus stopping the VI execution.

Analysis

Whenever applicable, assume W = 500 g for weight and Ein = 11 V.

1. Calculate the average velocity of the weight.

2. Verify through calculation the energy supplied to the motor by the variable power supply.

3. Verify through calculation the energy that the motor supplies to the weight.

4. Calculate the angular velocity of the motor shaft in RPM. (Hint: $V = \omega R$; use a micrometer or caliper to measure the diameter of the shaft that winds the string with weight. Then determine the speed reduction ratio provided by the pulley by counting rotations of the pulley and the motor shaft.)

5. How do power and energy differ?
   Suppose that a 100 W bulb is ON for 20 hours and another 100 W bulb is ON for 8 hours.
   How much power is supplied to each?
   How much energy is supplied to each?

6. Calculate the mechanical power in hp delivered to the load on the pulley side.

7. Identify the object(s) in the Block Diagram and explain how they make it possible to plot three curves on the same graph, with the values of weight along the abscissa? Why does this display occur just before the VI stops execution?

8. What is the main difference between the Waveform Graph and the X-Y Graph?

9. What is the main difference between the Waveform Chart and either the Waveform Graph or the X-Y Graph?

10. Could the Waveform Chart be used in this VI to plot the three curves? Explain.

11. Explain what starts the motion of the weight. Be specific. Use Block Diagram objects and explain how these objects interact with physical devices to achieve the desired task.

12. Explain what stops the motion of the weight. Be specific. Use Block Diagram objects and explain how these objects interact with physical devices to achieve the desired task.

13. Explain how the energy supplied to the motor is calculated.

14. Aside from reducing motor speed, why is the pulley used? Explain.

15. Modify the Block Diagram to allow the fourth curve (energy supplied to the motor) to be displayed on the X-Y graph. Make its color orange.

16. Modify the Front Panel and the Block Diagram to include the calculation of the mechanical power (in hp) delivered to the load, and its display on a digital indicator in the Front Panel.

17. Inspect the shape of the Efficiency curve displayed by the X-Y graph. What happens at higher loads? Explain why.

18. Why does the armature current increase at higher loads? [Hint: Consider motor speed.]

19. Consider the shape of the PE curve displayed by the X-Y graph. Prove mathematically that the displayed shape is correct.

20. **A Challenge Problem**
The data displayed by the X-Y graph is important and may be needed at a later time for inspection or analysis. Modify the Block Diagram to save this data (three curves) as a 2-dimensional array in the spreadsheet form (make sure that the data is in column form). Add to the Front Panel an object that gives the user a choice to save or not save the data.

DC Motor

Theory

We know from theory that when charge moves in a magnetic field, the magnetic field exerts force on the charge. This principle can be readily applied to the operation of a motor. First, we need a magnetic field provided by a permanent magnet or an electromagnet. In the present case we are using a permanent magnet. Then we have to produce the moving charges inside the magnetic field. This is accomplished by applying current from an external source to the armature winding. As shown in Fig. 14-3a, the current I enters the winding at pole 2 side of the armature by way of the brush terminal A and the commutator, and leaves the armature winding on pole 1 side of the armature, returning to the source E by way of the commutator brush terminal B.

A simple application of the right-hand rule on the armature winding and current I results in pole 2 of the armature being magnetized as north (N) pole and pole 1, as south (S) pole as shown in Fig. 14-3a. The opposite poles of the armature and the stator permanent magnet attract, causing the armature to rotate. Inertia carries the armature past perfect alignment of the poles. When it rotates 180°, the commutator reverses the direction of current as shown in Fig. 14-3b. The current now enters the armature winding on pole 1 side. This reverses the magnetic polarity of armature poles: pole 1 is now the north pole and pole 2, the south pole. Once again, the opposite poles of the armature and the stator attract, and the motion continues. It must be noted that if it wasn't for the commutator, which reverses the direction of armature current, and the magnetic polarity of the armature, the continuous motion beyond 180° would not be possible.

Another explanation of what causes the armature to rotate takes the physics point of view. The force on a charge q moving with velocity v in the magnetic field B is expressed by the following cross product:

$$\mathbf{F} = q\mathbf{v} \times \mathbf{B}$$

The force F is a vector that is at a right angle to the plane formed by v and B. Because the charges (electrons) that make up the armature current flow in the armature winding, the magnetic field exerts a force on the winding. The product of the force and the armature radius is the torque responsible for the rotation of the armature.

The force expression above is due to one charge. The flow of current I in a wire of length L positioned in a uniform magnetic field includes many charges (electrons). It can be shown that the force on the wire is

$$\mathbf{F} = I\,\mathbf{L} \times \mathbf{B}$$

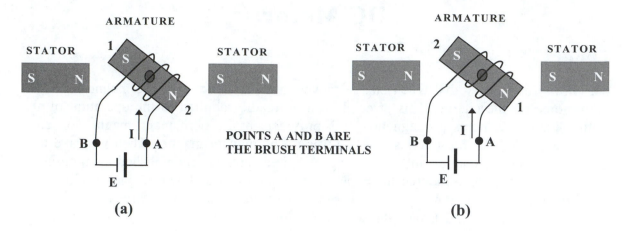

Fig. 14-3 The Commutator Reverses the Direction of Armature Current And Its Magnetic Polarity Every 180°, Making Possible the Rotation of the Armature.

Experiment 14-2: DC Motor

Objective

To examine the operation of the DC motor. Also, to demonstrate the laws of physics including electromagnetic induction, and the interaction of moving electric charge in a magnetic field. This effect is the basis of the DC motor operation. In this experiment LabVIEW software with data acquisition interface is used to collect and process the data. Acquired data is interpreted and analysis is used to verify the appropriate physics laws.

Parts

Hardware
DAQ board (such as MIO-16E-10 or LabPC+)
Extender board (associated with DAQ Board)
AC/DC motor accessory (PASCO SE-8657)
Variable gap lab magnet (PASCO EM-8641)

Photogate (PASCO ME-9204B)
Basic coil set (PASCO SF-8616)

Components/Chips
Resistors
Schmitt trigger

Software
LabVIEW (version 5.1 or higher)
NI DAQ (DAQ board driver)
Motor Generator.vi.

Drive Sources
Sine wave generator (variable frequency)

Variable DC power supply (to 12 VDC, 1 A)

Fixed DC power supply (5 V, 1 A)

Digital function generator/amplifier (PASCO PI-9587C or equivalent)

Test Setup

1. Assemble the motor accessory (armature and brushes) and the variable gap magnet. Consult the appropriate Pasco manuals for additional details. A completed test assembly diagram is shown in Fig. 14-4.

2. *Wire* the test circuit as shown in Fig. 14-5. The power amp output is connected to AICh1 of the extender board and to the two armature terminals. The photogate signal output is connected to AICh0 of the extender board. The photogate is powered with $+5 \text{ V}_{dc}$. Position the photogate so that the armature, when rotating, will break the photogate light beam. This arrangement is shown in Fig. 14-5. A clamping arrangement may be necessary to secure the photogate in the desired position.

The *Split Ring Commutator* must make contact with the brushes. The brushes must not touch the bottom of the armature coils. Apply a few drops of oil on the shaft to reduce friction.

CAUTION: The DC source that provides the armature current must be limited to 1 amp. If it is not, insert a series resistor of ½ to 1 Ω. This will protect the armature from being damaged.

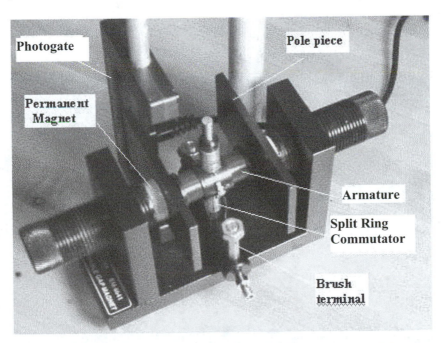

Fig. 14-4 Motor Accessory and Variable Gap Permanent Magnet Assembly. The Photogate is Positioned to Allow the Rotating Armature to Cut Its Light Beam.

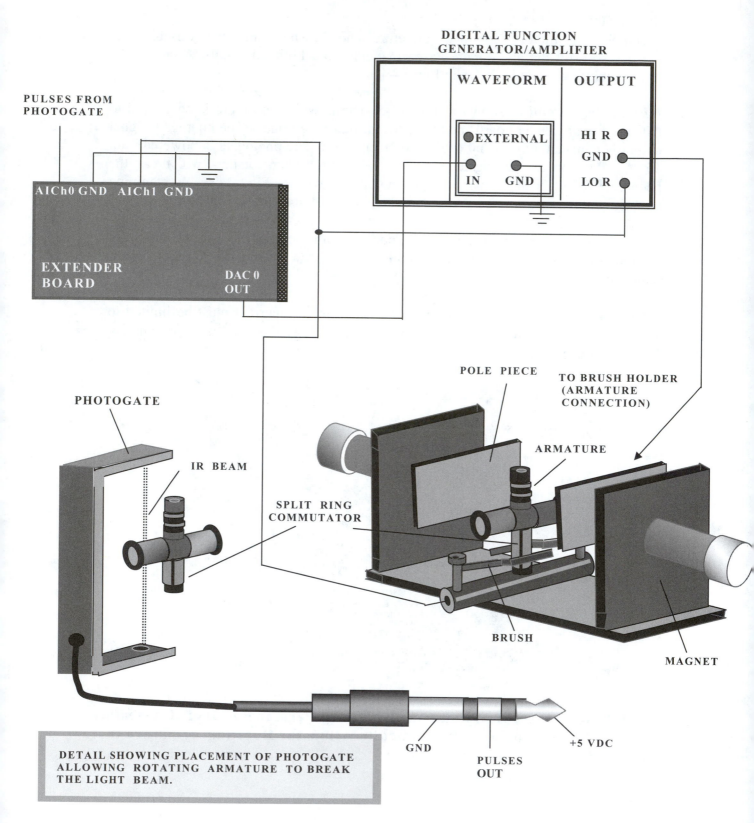

DIGITAL FUNCTION
GENERATOR/AMPLIFIER

WAVEFORM | OUTPUT

EXTERNAL

IN   GND

HI R
GND
LO R

PULSES FROM
PHOTOGATE

AICh0 GND   AICh1 GND

EXTENDER
BOARD

DAC 0
OUT

POLE PIECE

TO BRUSH HOLDER
(ARMATURE
CONNECTION)

ARMATURE

PHOTOGATE

IR BEAM

SPLIT RING
COMMUTATOR

BRUSH

MAGNET

GND

PULSES
OUT

+5 VDC

DETAIL SHOWING PLACEMENT OF PHOTOGATE
ALLOWING ROTATING ARMATURE TO BREAK
THE LIGHT BEAM.

Fig. 14-5 DC Motor Test Setup Diagram

Procedure

1. **Open** Motor Generator.vi (Book Vis>Motors and Generators>Motor Generator.LLB>Motor Generator.vi) and make the following settings in the Front Panel:

   > Choose the **DC Motor** from the Select a Test menu.
   > Set the "Save Data" Vertical Switch to **No** (you may want to save data later).
   > Set the "Calibrate" Vertical Switch to **Yes**.

   Calibration of the power amp takes place in the first part of the experiment. LabVIEW will output 2.5 Vdc to the DAC0 Out port on the extender board. This voltage, as shown in Fig. 14-5, is applied to the external input on the Digital Function Generator/Amplifier. You will see the prompt on the monitor screen directing you to adjust the power amp output to 2.5 Vdc. To do this, use the UP/DOWN buttons of the Waveform control and choose EXTERNAL. Monitor the output voltage (between the LO R and GND terminals) on a DC voltmeter and then adjust the Amplitude control button for output of 2.5 Vdc. Click on the OK button in the prompt window. The next prompt window informs you that the power amp has been calibrated. In fact, its gain has been set to 1. Click on the OK button.

2. **Click** on the Run button in the Front Panel.
   The main objective of this part of the experiment is to generate The Motor Speed vs. Motor Voltage curve from the acquired data.

   Rotate the top of the armature with your fingers to start the motor running. The motor is not self starting.

   Run the VI by clicking on the **Next** button in the Front Panel. The VI is now running. LabVIEW will increase motor voltage in small increments up to 5 V, causing the motor to accelerate. Observe on the Front Panel the Raw Data Waveform Graph. It will display something like the illustration below. This display contains data from AICh0, the photogate pulses, and the motor voltage data (the line down the center of the display) from AICh1. This display is helpful as an indication that data is being acquired.

   The data from two channels will be displayed on the waveform graph shown below.

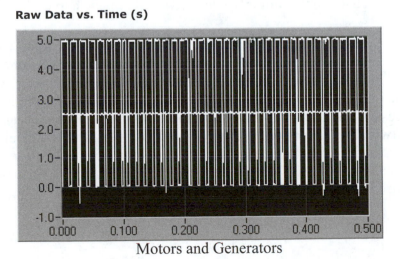

Raw Data vs. Time (s)

After all the data has been acquired and processed, the Motor Speed vs. Motor Voltage curve will be displayed on the X-Y graph shown below.

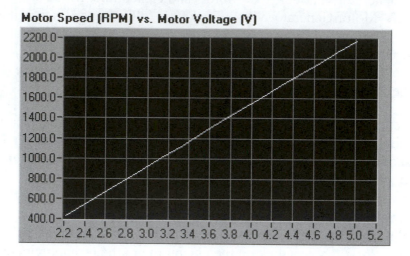

Motor Speed (RPM) vs. Motor Voltage (V)

3. This part of the experiment considers the direction of armature rotation. The direction of rotation depends on the polarity of the applied voltage.

    Click on the **Next** button in the Front Panel to start this test. LabVIEW will apply a positive voltage to the motor. Spin the armature and notice the direction of rotation.

    Click on **Next** in the Front Panel to stop this test.

    LabVIEW will now apply a negative voltage to the motor. Spin the armature to start the motor and notice the direction of armature rotation.

    Click on *Next* in the Front Panel to stop this test.

Operation

As shown in Fig. 14-5, digital function generator provides the necessary power to drive the motor. LabVIEW generates a staircase waveform of 16 steps, and each step is 0.19 V. As the ramp voltage increases, so does the speed of the motor. The main objective in this part of the test is to generate the Motor Speed vs. Motor Voltage curve.

X-Y Graph must be used in order to display the Motor Speed vs. Motor Voltage curve. Since X-Y Graph requires array inputs for both axes, we must generate a one dimensional array of motor speeds and another one dimensional array of motor voltages. To accomplish this, LabVIEW acquires a block of data at each step of the input staircase waveform, processes this data, and stores one speed and the corresponding voltage data points into the two arrays. This process is repeated 16 times and the resulting curve is displayed. Before displaying this curve, the Linear Regression VI is used to obtain a straight line. The value of the slope of this straight line (in RPM/V) may be used to predict motor speed at a desired voltage).

As the VI is running, additional Front Panel displays include the Current Motor Speed, the cumulative motor voltage displayed on the graph, and the Raw Data displayed on a Waveform Graph. The Raw Data display is particularly useful because it shows whether the data is being acquired. In case of a malfunction, the user can visually detect a problem.

In the last part of the DC motor test, the direction of rotation is tested. Since the direction of armature rotation depends on the polarity of the applied voltage, LabVIEW applies first a positive voltage to the armature, and then a negative voltage. The user should note the reversal in direction of rotation.

What have you learned from this hands-on laboratory experience?

In observing the operation of a DC motor, the user must be reminded of a powerful theory that makes possible an application such as the DC motor. This theory states that an electric charge experiences a force when moving in a magnetic field. This force results in a torque (F × armature radius) that acts on the armature that is free to rotate. The preceding statements you can read in textbooks. But here you are in a lab with a real motor at your disposal. Commercial motors have a sealed housing, but here you have access to an armature that can be easily removed.

So? What can you do manually to check some of the things mentioned?

The no-torque position of the armature occurs when it is aligned with magnetic field flux lines of the permanent magnet. If you displace the armature slightly to one side, you will feel the restoring force. Try to justify the direction of this force based on $\mathbf{F} = q\mathbf{v} \times \mathbf{B}$. Remember that this force is perpendicular to both v and B. Also note that that the two armature windings are wound in opposite directions. This means that if you pick on a turn of wire in one winding and a turn of wire in the second winding, the current will be moving clockwise in one turn and counterclockwise in the other turn of wire.

Something else you can do.

A simpler explanation of DC motor operation, offered in the Theory part of this experiment, views the armature as an electromagnet, behaving much like a solenoid. When current flows in the armature winding, it magnetizes the armature core with North and South poles at opposite ends. During the operation the armature electromagnet interacts with the North and South poles of the permanent magnet.

Remove the armature and the brushes from the variable magnet assembly. Apply a DC voltage (a few volts) across the brush terminals. Follow the path of current as it flows through the brush, the split ring of commutator, and into the armature winding. Inspect the armature winding and the direction of wires. Apply the right-hand rule and predict which pole of the armature is the North pole. Use the little magnet on a string that is included with the motor accessory unit to test your prediction.

DC Generator

Theory

Faraday's law of electromagnetic induction states

$$E_{ind} = - N(d\phi/dt)$$

The voltage E_{ind} is induced into a conductor, which moves in the magnetic field, cutting the lines of magnetic flux. This voltage is proportional to the number of turns N and to the speed at which the conductor moves. Greater voltage is induced if the conductor moves faster.

The illustration shows the edge of a loop of wire in a constant or DC magnetic field. It rotates in a counterclockwise direction with constant angular speed ω. As shown in the illustration, the angle θ is measured between the normal to the loop and the flux lines.

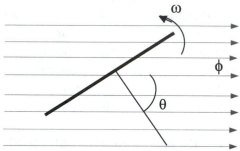

As the loop rotates with the speed ω, the angle at any time t is $\theta = \omega t$. When $\theta = 0^\circ$, the loop is perpendicular to the flux lines, resulting in a maximum number of flux lines passing through the loop. And when the plane of the loop is parallel to the field, $\theta = 90^\circ$, no flux lines pass through the loop. The cosine function best describes this behavior since $\cos(0) = 1$ and $\cos(90^\circ) = 0$. Hence

$$\phi(t) = \phi_m\cos(\omega t)$$

Since magnetic flux density $B = \phi/A$, where A is the area of the loop, the above expression for flux may be written as $\phi(t) = BA\cos(\omega t)$. Substituting this expression for flux in E_{ind} above we obtain the following

$$E_{ind} = - BAN d/dt\, (\cos(\omega t)) = BAN\omega\sin(\omega t)$$

A practical application of the induced voltage effect is the generator. As shown in the illustration below, the magnetic field is provided by a permanent magnet, and the loop of wire is attached to a cylindrical armature that is typically supported at both ends by bushings or ball bearings (not shown in the illustration). The armature is rotated at speed ω by an external device such as a motor.

The type of commutation determines the shape of voltage across the load. Shown in the illustration is a split ring commutator that produces a waveform resembling a full wave rectified type of wave, also shown below. The split ring commutator reverses the polarity of induced voltage every 180°. The resulting voltage waveform across the load resembles a full wave rectified waveform. Because it never reverses polarity, this type of a wave is referred to as the DC wave with peak value of BANω, according to the above equation for the induced voltage.

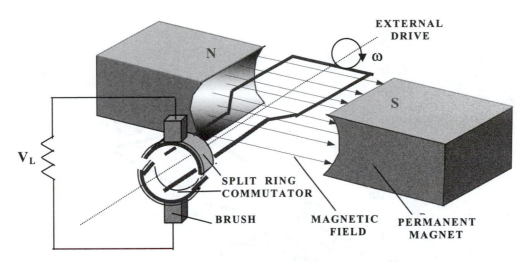

Permanent Magnet DC Generator

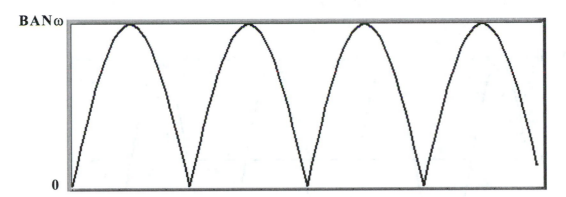

Voltage Across the Load With the Split Ring Commutator

As shown in the illustration below, the induced voltage is sinusoidal that does not reverse every 180° when a slip ring commutator is used. The illustration does not show the magnetic field.

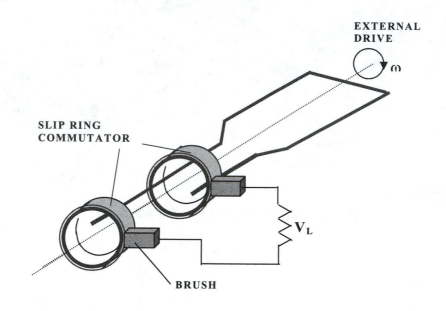

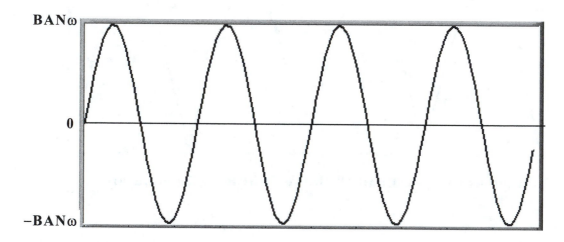

Chapter 14

Experiment 14-3: DC Generator

Objective

To examine the operation of the DC generator. Also, to demonstrate the laws of physics including electromagnetic induction, the basis of the DC generator. In this experiment LabVIEW software with data acquisition interface is used to collect and process the data. Acquired data is interpreted and analysis is used to verify the appropriate physics laws.

Parts

See Experiment 14-2.

Test Setup

See Experiment 14-2.

1. Disconnect wiring to AICh0 and AICh1 of the extender board.
2. Connect the brush terminals (armature) to AICh0 of the extender board.
3. Position the armature so that the split ring commutator makes contact with the brushes.

Procedure

1. Connect motor armature terminals (brush terminals) to AICh0 on the extender board.
2. Position the armature so that the slip ring commutator makes contact with the brushes.
3. Select **DC Generator** from the Select a Test menu in the Front Panel of Motor Generator.vi.
4. Set the "DC Gen Test" vertical switch in the Front Panel to **START**.
5. Click on the **Run** button in the Front Panel. The VI is now running.
6. *Rotate* the armature manually (one turn or more)

   a. in one direction slowly. Wait a few seconds and observe the display.
   b. in the same direction but faster.
   c. in the opposite direction slowly.
   d. in the opposite direction, but faster.

   Notice the polarity and amplitude of the induced voltage as displayed on the Waveform Graph based on direction of rotation and the speed of rotation of the armature. You may repeat the above steps as desired. Record the data.

7. *Click* on the "DC Gen Test" vertical switch with the Operating Tool and choose **STOP** to quit the test.

Operation

In this part of the experiment we demonstrate the law of electromagnetic induction. LabVIEW will acquire a block of data from AICh0 of the extender board and display it on the Waveform Graph in the Front Panel. It will continue to do so indefinitely until you set the Front Panel Vertical Switch "DC Gen Test" to STOP using the Operating Tool.

The data being acquired comes directly from the brush terminals that make electrical contact with the armature.

As you rotate the armature, inspect the display and determine whether the display is consistent with the theory in terms of the direction and the speed of the rotation. What role does the split ring commutator play in this case?

AC Generator

Theory

The theory of electromagnetic induction described above for the DC generator is exactly the same as that for the AC generator. The difference is only in the way the induced voltage is applied across the load by the commutator and the brushes. In the case of the DC generator, a split ring commutator in effect rectifies the AC armature voltage, producing DC voltage across the load.

The AC generator, on the other hand, uses a slip ring commutator. The slip ring commutator consists of two isolated rings, each connected to one side of the armature winding. As the armature rotates, the slip ring commutator does not reverse armature current every 180°. This will result in an AC waveform across the load, as shown in the illustration below:

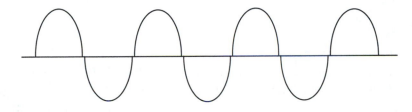

Experiment 14-4: AC Generator

Objective

To examine the operation of the AC generator. Also, to demonstrate the laws of physics including electromagnetic induction, the basis of the AC generator. In this experiment LabVIEW software with a data acquisition interface is used to collect and process the data. Acquired data is interpreted and analysis is used to verify the appropriate physics laws.

Parts

See Experiment 14-2.

Procedure

1. Connect motor armature terminals (brush terminals) to AICh0 on the extender board.
2. Position the armature so that the slip ring commutator makes contact with the brushes.
3. Select **AC Generator** from the Select a Test menu in the Front Panel of Motor Generator.vi.

4. Set the "AC Gen Test" vertical switch in the Front Panel to **START**.
5. Click on the **Run** button in the Front Panel. The VI is now running.
6. **Rotate** the armature manually (1 turn or more).

   a. in one direction slowly. Wait a few seconds and observe the display.
   b. in the same direction but faster.
   c. in the opposite direction slowly.
   d. in the opposite direction, but faster.

Notice the polarity and amplitude of the induced voltage as displayed on the waveform graph based on the direction and speed of rotation of the armature. You may repeat the above steps as desired. Record the data.

7. **Click** on the "AC Gen Test" vertical switch with the Operating Tool and choose **STOP** to quit the test.

Operation

In this part of the experiment, just as we did in the case of the DC generator, we demonstrate the law of electromagnetic induction. LabVIEW will acquire a block of data from AICh0 of the extender board and display it on the Waveform Graph in the Front Panel. It will continue to do so indefinitely until you set the Front Panel Vertical Switch "AC Gen Test" to STOP using the Operating Tool.

The data being acquired comes directly from the brush terminals that make electrical contact with the armature. In this test, however, the slip ring commutator is used.

As you rotate the armature, inspect the display and determine whether the display is consistent with the theory in terms of the direction and the speed of the rotation. What role does the slip ring commutator play in this case.

AC Motor

Theory

The theory behind the operation of the AC motor is exactly the same as that described earlier for the DC motor. The basis of this theory is that there must be a charge moving in the presence of the magnetic field. The magnetic field, as in the case of the DC motor, is provided by the variable gap magnet. The charge that moves in the magnetic field is the current that moves along the armature winding.

In order to maintain the motion of the armature in the same direction, the direction of current in the armature winding must be reversed. In the case of the DC motor, the split ring commutator accomplishes this current reversal. But the AC motor uses the slip ring commutator, which does not reverse the direction of current.

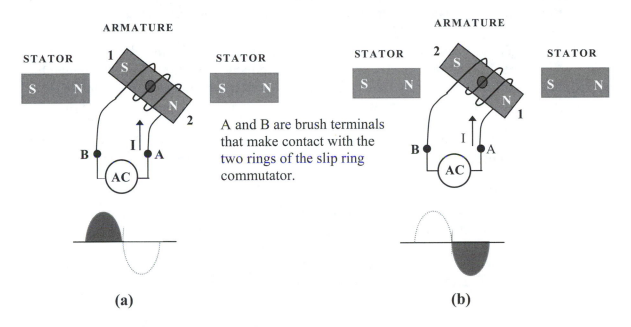

A and B are brush terminals that make contact with the two rings of the slip ring commutator.

(a)   (b)

Fig. 14-6 (a) During Positive Half Cycle, Pole 1 of Armature Is Magnetized as S and Pole 2 as N; (b) During Negative Half Cycle, Pole 1 of Armature Is Magnetized as N and Pole 2 as S.

However, the source driving the armature is a sine wave that reverses its polarity every 180°. That means that the motor makes one revolution during each cycle of the sine wave *assuming that the motor can keep up with the polarity changes of the sine wave.* If it can't, it will stop rotating. This is illustrated in Fig. 14-6. We call this type of AC motor a **synchronous** motor because it rotates in synchronism with AC, its speed being only a function of the frequency of the AC and not its amplitude.

For example, if the frequency of applied AC is 20 Hz, which means 20 cycles /s, then the armature will rotate 20 revolutions/s, or 20×60 = 1200 RPM. Only a change in drive frequency will change the speed of the motor.

Experiment 14-5: AC Motor

Objective

To examine the operation of the AC motor. Also, to demonstrate the laws of physics including electromagnetic induction, and the interaction of moving electric charge in a magnetic field. This effect is the basis of the AC motor operation. In this experiment LabVIEW software with data acquisition interface is used to collect and process the data. Acquired data is interpreted and analysis is used to verify the appropriate physics laws.

Parts

See Experiment 14-2.

Test Setup

1. *Wire* the test circuit as shown in the simplified diagram of Fig. 14-7. You may refer to the detailed diagram shown in Fig. 14-7, which shows connections that are also used here.

   V1—Variable frequency sine wave generator. Adjust the amplitude to 9 V_{rms} and the frequency to 15 Hz. Connect the output from the generator to the input of the power amplifier. The output from the power amplifier is connected to the motor armature (brush terminals).

   A1—Schmitt trigger. Converts sine wave to pulses. LabVIEW uses these pulses to determine the frequency of the AC drive. Connect the output from the Schmitt trigger to AICh1 of the extender board.

   P1—Power amplifier (PASCO PI-9587C or equivalent). Provides the necessary power to drive the motor.

   The photogate output is connected to AICh0 of the extender board. Photogate requires +5 VDC power.

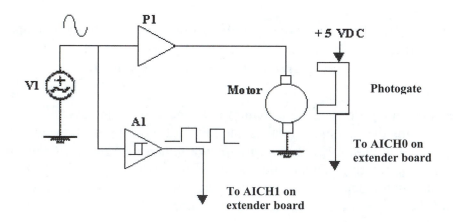

Fig. 14-7 Test Circuit For AC Motor

2. Assemble the armature and the permanent magnet as shown in Figure 14-7. Notice the position of the photogate. The rotating armature must break the light beam of the photogate.

The output of the power amplifier is connected to the **brush terminals**.

The **pole pieces** spread out and provide a uniform magnetic field in which the armature rotates.

The **slip ring commutator** must be on the bottom, making contact with the brushes.

Procedure

1. Select **AC Motor** from the Select a Test menu in the Front Panel of Motor Generator.vi.
2. *Set* the AC Motor Test Vertical Switch in the Front Panel to START using the Operating Tool.
3. *Click* on the Run button.
4. *Monitor* the amplitude of power amp output using an AC voltmeter. It should be 9 V_{rms} and the frequency should be set to 15 Hz.
   Spin the armature to run the motor. The motor should be running now. If it is not running, raise the AC drive slightly above 9 V_{rms}.
   The Motor Speed Data Waveform Graph in the Front Panel should display the acquired data.
5. Increase the frequency of AC drive slowly from 15 Hz until the motor stops. The motor will stop at some frequency less than 50 Hz. When the frequency is too high, the motor is unable to keep up with the speed at which the sine wave reverses its polarity.

   Observe the Front Panel motor speed indicators and the frequency of AC drive. The Meter indicates the ratio of revolutions/s to the frequency of AC drive.

Operation

In the AC motor test, the armature circuit is driven by a sinusoidal signal. The slip ring commutator, which does not reverse the direction of armature current, is used here. As always, the motor requires drive power, which is provided by the power amp (digital function generator/amplifier). LabVIEW acquires two channels of data: photogate pulses on AICh0 and pulses from Schmitt trigger. The photogate monitors the motor speed and the Schmitt trigger circuit converts the sinusoidal drive waveform to pulses. The pulses serve as a reference for comparing the motor speed with the frequency of the AC drive.

The operation of a DC motor, described earlier, depends on the reversal of armature current every $180°$ rotation of the armature to ensure its continuous rotation. This requirement also applies to the AC motor.

Because the split ring commutator does not reverse the armature current, the AC drive must. After one half cycle of AC drive, the armature rotates one half revolution. At this point, AC drive reverses its polarity, thus reversing the direction of armature current and allowing it to complete one rotation.

Universal Motor with AC Drive

One of the important features of the Universal Motor is that it can run on either the AC or the DC drives. In order to make this possible, the stator permanent magnet is replaced by an electromagnet using two 400 turn field windings. First, the motor operation using an AC drive is considered.

Theory

In this test the stator is not a permanent magnet as before. Now the stator uses two 400 turn windings, which are connected in series with the armature. Now the stator and the armature are both electromagnets. The same current flows through the stator and the armature windings.

As shown in Fig. 14-8, during the positive half cycle, the stator and the armature are magnetized as shown according to the right-hand rule. The attractive force of opposite poles causes the armature to rotate. After 180° of armature rotation, the magnetic polarity of stators reverses because the AC drive reverses its electrical polarity, which reverses the direction of current. However, the magnetic polarity of the armature does not reverse because we are using the split ring commutator. The reader may recall from earlier discussion that the split ring reverses the direction of armature current. The AC drive also reverses its polarity each half cycle. The two polarity reversals cancel each other with no net change in the direction of armature current during one revolution. Hence, the magnetic polarity of the armature remains constant.

This is illustrated in Fig. 14-8. During the positive half cycle of the AC drive, armature pole 1 is magnetized, according to the right-hand rule, as the North pole, while pole 2 is the South pole (Fig. 14-8a). Fig. 14-8b shows the position of the armature after 180° of rotation. As can be seen, the magnetic polarity of the stator poles reverses, but that of the armature does not.

This is a series wound motor because the stator field windings are connected in series with the armature windings. Therefore, the current is the same in both windings.

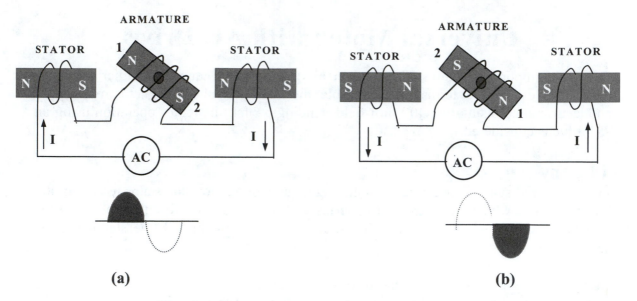

Fig. 14-8 Universal Motor Using AC Drive and Split Ring Commutator. (a) The Magnetic Polarities of the Armature and Stator are as Shown Initially. (b) After 180° of Armature Rotation, the Magnetic Polarity of the Stator Reverses, but that of the Armature Does Not.

Experiment 14-6: Universal Motor with AC Drive

Objective

To examine the operation of the universal motor driven by an AC source. Also, to demonstrate the laws of physics including electromagnetic induction, and the interaction of moving electric charge in a magnetic field. This effect is the basis of the motor operation. In this experiment LabVIEW software with data acquisition interface is used to collect and process the data. Acquired data is interpreted and analysis is used to verify the appropriate physics laws.

Parts

See Experiment 14-2.

Test Setup

1. *Assemble* the Universal Motor as shown in Fig. 14-9. You may also consult the appropriate Pasco manuals for additional details.

2. *Wire* the test circuit as shown in Fig. 14-10. The Universal Motor will be driven by an AC source in this part of the experiment.

   V1 is the variable frequency sine wave generator. Adjust the amplitude to 9 V_{rms} and the frequency to 15 Hz. Connect the output from the generator to the input of the power amplifier.

A1, the Schmitt trigger, converts sine wave to pulses. LabVIEW uses these pulses to determine the frequency of the AC drive. Connect the output from the Schmitt trigger to AICh1of the extender board.

P1, a power amplifier (such as PASCO PI-9587C), provides the necessary power to drive the motor. As shown in Fig. 14-10, the two 400 turn stator windings are connected in series. The output of the power amp is connected to this series combination. The photogate output is connected to AICh0 of the extender board. The photogate requires +5 VDC power.

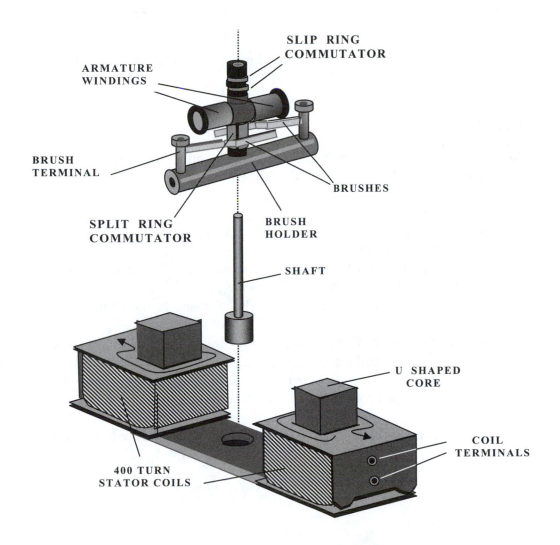

Fig. 14-9 Universal Motor Assembly

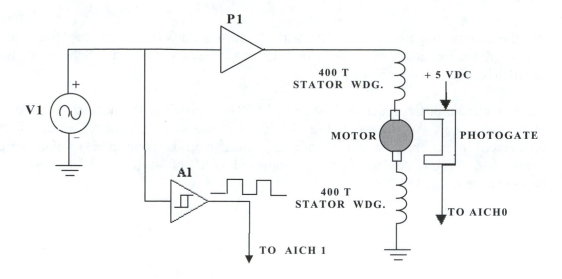

Fig. 14-10   Universal Motor Test Circuit Using AC or DC Drives

Procedure

1. *Select* Universal Motor from the Select a Test menu in the Front Panel.
2. *Set* the Universal Motor AC Test Vertical Switch in the Front Panel to START using the Operating Tool.
3. *Click* on the Run button.
4. *Monitor* the amplitude of power amp output using an AC voltmeter. It should read 9 V_{rms}. Adjust the frequency to 15 Hz.
   Spin the armature to run the motor. The motor should be running now. If it is not running, raise the AC drive slightly above 9 V_{rms}.
   The Data Waveform Graph in the Front Panel displays the acquired data from two channels.
5. Increase the frequency of AC drive slowly from 15 Hz until the motor stops. The motor will stop at some frequency less than 50 Hz. When the frequency is too high, the motor is unable to keep up with the speed at which the sine wave reverses its polarity. Observe the Front Panel motor speed indicators and the frequency of the AC drive.
6. **To end the test,** set the Universal Motor AC Test Vertical Switch in the Front Panel to STOP.

Operation

In this test, the armature circuit is driven by a sinusoidal signal. The split ring commutator, which reverses the direction of armature current, is used here. The stators consist of an electromagnet using two 400-turn coils which are connected in series with the armature windings. As the armature rotates its magnetic polarity does not reverse after one half revolution and in fact remains the same while the magnetic polarity of the stator poles is reversed each half cycle by the AC drive. The reader can probably visualize what would happen if a slip ring commutator were used instead of a split ring commutator. Both magnetic polarities now reverse each half cycle, with the result that the armature will be unable to rotate.

As always, the motor requires drive power, which is provided by the power amp (digital function generator/amplifier). LabVIEW acquires two channels of data: photogate pulses on AICh0 and pulses from the Schmitt trigger. The photogate monitors the motor speed and the Schmitt trigger circuit converts the sinusoidal drive waveform to pulses.

Universal Motor With DC Drive

Theory

The AC drive in Fig. 14-8 is now replaced by the DC drive, and the remaining setup is the same. Fig. 14-11a shows the magnetic polarity (which can be verified by applying the right hand rule) for the stator and the armature. As shown in Fig. 14-11b, after half a rotation, the armature reverses its magnetic polarity, while the magnetic polarity of the stator remains the same. Recall, the AC drive reversed the magnetic polarity of the stator, while the armature's magnetic polarity remained the same. This is the main difference in the operation. As mentioned before, in order to operate the universal motor, the drive must reverse after half a rotation the magnetic polarity of either the stator or the armature, but not both.

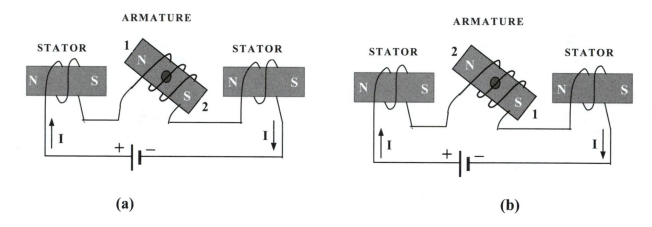

(a)   (b)

Fig. 14-11   Universal Motor Using DC Drive and a Split Ring Commutator.
(a) The Initial Magnetic Polarities of the Armature and the Stator.
(b) After 180° Armature Rotation, the Magnetic Polarity of
the Armature Reverses but that of the Stator Does Not.

Experiment 14-7: Universal Motor with DC Drive

Objective

To examine the operation of the universal motor driven by a DC source. Also, to demonstrate the laws of physics including electromagnetic induction, and the interaction of moving electric charge in a magnetic field. This effect is the basis of the motor operation. In this experiment LabVIEW software with a data acquisition interface is used to collect and process the data. Acquired data is interpreted and analysis is used to verify the appropriate physics laws.

Parts

See Experiment 14-2.

Test Setup

1. The Universal Motor assembly is same as that shown in Fig. 14-9.

2. **Wire** the test circuit as shown in Fig. 14-12. Notice the similarity between this test circuit and the AC drive test circuit shown in Fig. 14-10.

    > V1—DC source with continuously adjustable voltage. Connect the DC output to the input of the power amplifier and to AI Ch1 of the extender board.
    >
    > P1—Power amplifier (such as PASCO PI-9587C) provides the necessary power to drive the motor.
    >
    > The photogate output is connected to AICh0 of the extender board. The photogate requires +5 VDC power.

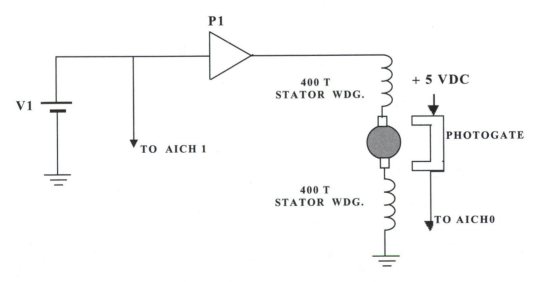

Fig. 14-12 Universal Motor Test Circuit Using DC Drive

Procedure

1. **Set** the Universal Motor DC Test Vertical Switch in the Front Panel to START.
2. **Adjust** the output of the DC source to 2.8 V.
3. **Click** on the Run button. The VI is now running.
4. At the beginning of this test you will be asked to calibrate the power amp. Adjust the power amp gain control for output of 2.8 V_{dc}.
   Spin the armature to run the motor. The motor should be running now.
   The Data waveform graph in the Front Panel displays the acquired data from two channels.
5. Vary manually the output of the DC source slowly up to 5 V_{dc} and observe the motor speed and motor voltage indicators on the Front Panel.
6. **To end the test**, set the Universal Motor DC Test Vertical Switch in the Front Panel to STOP.

Operation

The motor in this test is driven by a DC source. As stated in the Theory part, the magnetic polarity of the armature reverses each half rotation of the armature, but the magnetic polarity of the stator does not.

LabVIEW acquires data from two channels of the DAQ board. Photogate pulses are from AICh0 and the motor voltage is from AICh1. LabVIEW displays the raw data on a waveform graph, giving you an indication that data is being acquired. Then it processes the data and displays motor speed in RPM and the motor voltage on gauges in the Front Panel.

Analysis

1. State the law that makes a DC motor work.

2. Consider a simplified diagram of a DC motor. As shown in Fig. 14-13, the armature winding consists of one turn of wire made up of conductors 1 and 2. The conductors make contact with two segments of a split ring commutator. The two segments are part of the armature, and consequently they rotate with the armature. The two brushes, however, are stationary as they make a pressure contact with the commutator. The external DC power source provides current, by way of the brush and the commutator segment, to the armature winding. As shown in this illustration, the armature current I_a enters conductor 1.

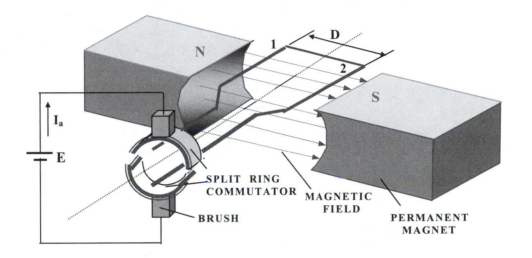

Fig. 14-13   Permanent Magnet DC

   a. Show on a Cartesian coordinate system the magnetic field B along the x-axis, the armature current I along the y-axis, and the force F_B on conductor 1 along the z-axis. Repeat the same for conductor 2.

   b. In the DC motor experiment, a variable gap neodymium magnet was used as a source of DC magnetic field. But, returning to the configuration of Fig. 14-13, the magnetic field is 15 kG, the diameter D of the loop is 4 inches, and the length of each conductor is 10 inches. The current I_a = 1.5 A. Calculate the torque, both magnitude and direction, exerted on conductors 1 and 2.

   c. Use the torque calculated in part b above to determine the power generated by the motor (in hp), assuming that the motor speed corresponding to this torque is 2000 RPM. Refer to Chapter 15 for the power equation.

   d. If the external source E = 12 Vdc, calculate the efficiency and power lost.

e. As mentioned earlier, the split ring commutator reverses current in the armature winding every 180° of rotation. Use the results of part a above to show what would happen if the current was not reversed after 180° of rotation.

3. State the law that makes the generator (AC or DC) work.

4. As shown in the illustration below, the armature of the DC generator rotates in the CW direction, conductor 1 moving up and conductor 2 moving down at the instant the plane of the loop is parallel to that of the magnetic field.

a. Calculate the induced voltage between terminals a and b. Determine and label the polarity of terminals a and b.

b. Repeat part a above when the loop advances 60° clockwise.

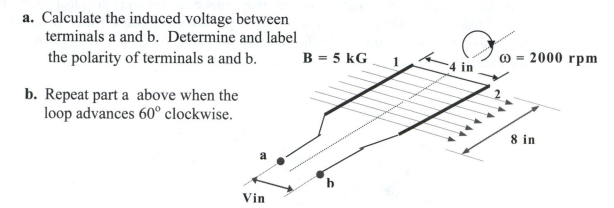

5. Consider the permanent magnet DC generator shown in Fig. 14-14.
a. Sketch the voltage V_L across the load, assuming that the motor speed is 1200 rpm. Show the time scale.

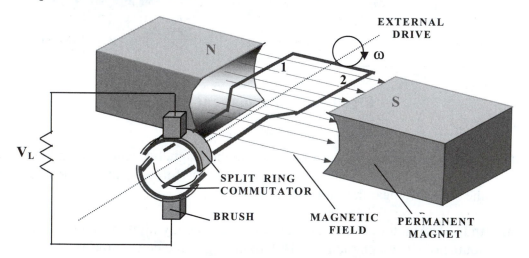

Fig. 14-14   Permanent Magnet DC Generator

b. Why is the voltage across the load DC? Explain. Determine the DC value of this wave.

c. Suppose the speed of the motor driving the generator shaft doubles. How much does the voltage V_L across the load change? Explain.

d. State all of the factors that determine the peak value of the generator voltage V_L. Which of these factors are totally under the control of the generator's manufacturer? Can the user control any of these factors? Explain.

e. Not all practical generators use a split ring or two-segment commutator; instead they use many segments and many coils wound on the armature, with each coil connected to two segments on the commutator. Imagine a commutator having four segments and two turns of wire, with each turn connected to two segments and the coils positioned at $90°$ relative to each other. Sketch the generator voltage V_L assuming that the motor driving the shaft of the generator operates at 1200 rpm. Show the time scale, the DC value, and the peak-peak ripple voltage.

6. Suppose the split ring commutator shown in Fig. 14-14 is replaced by the slip ring commutator shown below. Conductor 1 is connected to slip ring 1 and conductor 2 is connected to slip ring 2.

a. Sketch the voltage V_L across the load.

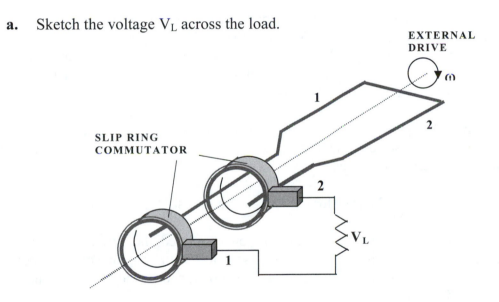

b. Explain why in this case the voltage across the load is AC.
c. Suppose that $\omega = 3600$ rpm. What is the frequency of the load voltage V_L?

7. In principle, the operation of the AC motor is the same as that of the DC motor. However, there are differences in commutation. Instead of the split ring commutator, we use the slip ring commutator. The input now is not DC, but rather AC sinusoidal waveform, which reverses its polarity every 180 electrical degrees. The magnetic polarity of the permanent magnets remains the same, but the armature magnetic polarity changes every 180° of rotation. Explain why the split ring commutator will not work here and the slip ring commutator would.

8.   The Universal Motor does not use a permanent magnet for the production of the magnetic field; instead it uses an electromagnet to produce magnetic field by the field windings, as they are called, and the current flowing in the field windings (same as the armature current) is called the field current. The electromagnet windings and the armature winding are connected in series. The split ring commutator is used in this case. The Universal Motor is used for both the DC or AC input source. Explain how the AC and the DC inputs cause the armature of the Universal Motor to rotate, paying special attention to what makes it possible to rotate past 180°.

9.   In the DC motor part of this experiment, determine the slope of the Motor Speed vs. Motor Voltage computer generated curve and predict the speed of the motor at 3.8 V.

10.   In the DC motor part of this experiment, what happened to the direction of the rotation when the polarity of the input voltage was reversed? Explain.

11.   In the DC motor part of this experiment, notice that the Motor Speed was displayed on a graph versus Motor Voltage. This is somewhat unusual because the x-axis is normally either a time or a sample number. You may examine the Block Diagram and determine how it is possible to plot such a graph. (Hint: consider the requirements of an X-Y graph.)

12.   In the DC generator test you had to move the armature clockwise and counterclockwise fast and slow. Show here the data that you obtained. Use the theory to verify your data.

13.   Repeat problem 12 above for the AC generator test.

14.   In the operation of the AC motor, you examined the Front Panel display of motor speed and the square wave associated with the sinusoidal motor drive. What was the relationship between the motor speed and the frequency of the sinusoidal drive? Explain why that is so.

15.   The AC Motor uses the sinusoidal drive and the slip ring commutator. Explain why the split ring commutator would not work.

16.   The Universal Motor in this experiment was configured with stator windings connected in series with the armature winding. In industry this is called a series wound DC motor. There is a particular problem with this type of configuration as opposed to the shunt wound or compound wound type of DC motor. Research the literature and explain what this problem is.

Chapter 15
Motor Control Experiments

Introduction

The DC motor theory was presented in detail in Chapter 14 with illustrations in Fig. 14-3 and Fig. 14-13. Some of the important points are repeated here. The motor is a transducer that converts the applied electrical energy to mechanical energy in the form of work that the motor can do by driving or rotating a mechanical load.

The fundamental principle upon which the operation of the motor is based is the interaction between a moving charge and a magnetic field. This interaction produces a force that moves the armature. The magnetic field is produced in one of two ways: by a permanent magnet, or an electromagnet, both being a part of the stationary part of the motor called the stator. If an electromagnet is used, then the stator must have windings called field windings, and a current called the field current must be applied to these windings in order to magnetize the stator poles.

The second ingredient in the operation of a motor is a moving charge that makes its way through the magnetic field. This is accomplished by applying electric current (current is defined as the movement of electrons) to the turns of the armature. Because the armature rotates and the external source supplying the current does not, we use a commutator to accomplish this task. In Fig. 14-13 a commutator consists of split rings that rotate with the armature and the stationary brushes that make pressure contact with the rings.

The commutator serves another very important function. It reverses the direction of current each half revolution of the armature. This reversal of current also reverses the direction of force (due to reaction between the magnetic field and the charge) on the armature conductors. This makes possible the continuous rotation of the armature. If this reversal of armature current did not occur, the armature would not be able to go beyond 180° and would simply rock back and forth. Reversing the polarity of the external source will reverse the direction of armature rotation.

Fig. 15-1 shows the armature circuit of a DC motor. The lumped resistor R_a represents the distributed resistance of the armature wire, typically a few ohms. E is the DC source supplying the armature current I_a. The figure also shows the effects of a load, represented by the load torque T_L rotating with the speed ω.

When the motor is operating and the armature is rotating with the speed ω, a voltage V_b called *back emf* (sometimes called counter emf) appears across the input terminals and opposes the external source that supplies the armature current as shown in Fig. 15-1.

This is a generator effect that occurs when voltage is induced into a conductor moving in a magnetic field. In accordance with Lenz's law, V_b will always oppose the externally applied voltage (E in Fig. 15-1) that produced this effect. Back emf is proportional to speed of the motor and may be expressed as

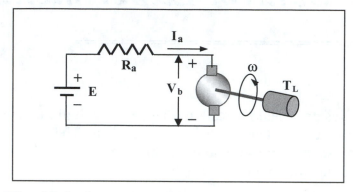

Fig. 15-1 Armature Circuit of a DC Motor

$$V_b = k_b \omega$$

where k_b is the back emf constant. Back emf limits armature current, which may be expressed as follows:

$$I_a = (E - V_b)/R_a$$

It is clear from this equation why a motor sometimes burns out. By pressing an electric drill too hard, for example, and forcing a drastic reduction in its speed, the back emf drops and armature current shoots up, possibly above the rated value as predicted by the above equation.

As mentioned earlier, a motor is an electromechanical transducer converting electrical energy to mechanical energy, which is usually expressed in terms of horsepower (hp) in the English system of units as follows:

$$P_m = (T_L)(\omega)/5252$$

where P_m is the motor's output power in hp, T_L is the load torque in ft-lb, and ω is the motor's speed in RPM. The input power to the motor is

$$P_i = EI_a$$

And the motor's efficiency may be expressed as

$$\eta = 100(P_m/P_i)$$

and power lost as heat is the difference $P_i - P_m$. Dimensional homogeneity must be observed in performing these mathematical operations, so one quantity or the other must be converted. The link conversion equation is 1 hp = 746 W.

Many types of motors are available. The DC motors may use a permanent magnet (PM DC motor), or they may use an electromagnet (field winding) resulting in several configurations including series wound, shunt wound, or compound wound depending on the way the armature and the field winding are connected (series, shunt, or a combination of the two). The PM DC motor is used generally in low-power applications, while the series, compound, and shunt motors are used in high-power industrial applications. In Chapter 12 shunt and series configurations using field coils were illustrated in experiments.

There are also the AC motors used frequently in household or consumer applications. Most of these are brushless because they don't use the commutator. And finally there are the stepper motors known for great accuracy in motion. We will use the DC and the stepper motors in experiments in this chapter.

Experiment 1: DC Motor Motion Control

Objective
In this experiment we will use LabVIEW to operate a small DC motor. LabVIEW software will be provided and you will use the Front Panel controls to vary the speed of the motor and observe the motor speed on the digital indicators. The software will generate control signals for three different motion profiles.

Parts
DAQ board (such as MIO-16E-10 or LabPC$^+$)
Extender board (associated with DAQ board)
Photogate (PASCO ME-9204B)
Small DC motor
LabVIEW (version 5.1 or higher)

Test Setup
Fig. 15-2 shows the mechanical setup that is used in this experiment. The DC motor is clamped in a vice next to the photogate, which is also positioned by being clamped to the vertical rod. Tab A is attached to a wire, which in turn is attached to the motor shaft. As the motor shaft spins, the tab breaks the photogate beam once every revolution.

The power supply provides the bias voltage for the photogate device. Both the DC motor and the photogate are connected to the DAQ extender board.

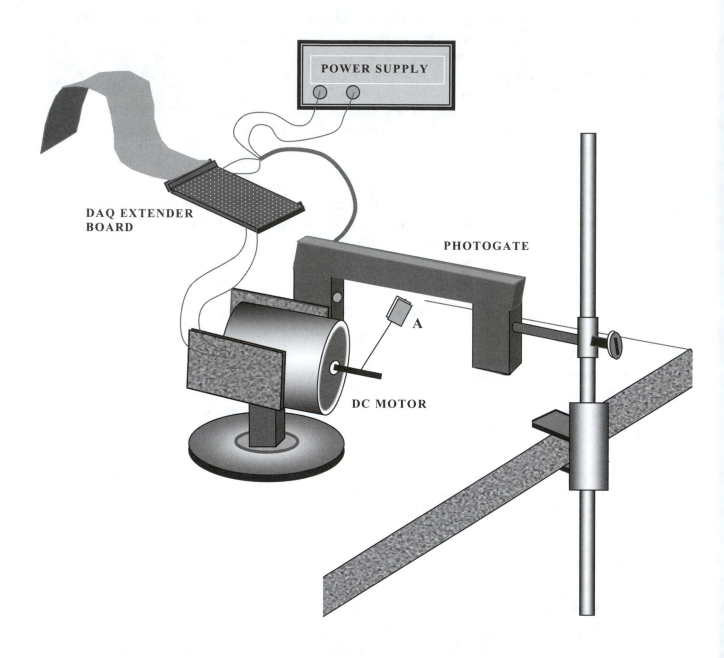

POWER SUPPLY

DAQ EXTENDER BOARD

PHOTOGATE

A

DC MOTOR

Fig. 15-2 The Mechanical Setup for the DC Motor Experiment

Procedure:

1. This experiment uses one photogate device. The photogate unit uses the stereo plug shown below.

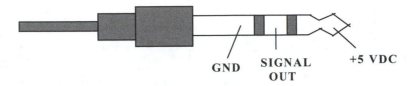

GND   SIGNAL OUT   +5 VDC

Connect GND, +5VDC to the photogate device. Connect Signal Out of the Photogate device to AI Ch. 0 of the DAQ extender board.

2. In this experiment we will use the DAC output on the extender board to control the speed of the DC motor. Connect the two input leads of the DC motor to the DAC output Ch. 0. A word of caution: If the DC motor that you are using is relatively large, requiring more than 50 mA of current, the DAQ extender board may not be able to drive the motor. In that case you should build an amplifier (possibly a Darlington) to drive the motor.

3. At this time LabVIEW software (version 5 or higher) should be installed on your computer and the data acquisition board (DAQ board such as Lab PC[+] or MIO-16E-10) should be plugged into the expansion slot inside the computer. Also, the extender board should be connected to the DAQ board, whose connector is visible in the back of the computer. The data acquisition driver software (NI-DAQ) should also have been loaded. Finally, the WDAQCONF.EXE utility should have been executed to configure your data acquisition board. All this is necessary before you run the data acquisition and analysis VI.

4. In this step you will be introduced to the Front Panel controls and indicators of DC Motor.vi. Open this VI (Book VIs>Motor Control.LLB>DC Motor.vi). The Front Panel of this VI is shown in Fig. 15-3.

    The *Motor Speed Data* Waveform Graph displays the pulses from the photogate in the Continuous Mode of operation. The DC Motor.vi software converts these pulses to motor speed in RPM and displays it on the Meter digital indicator labeled as *Motor Speed* (RPM). In the last mode, the Speed Profile mode, the Waveform Graph is used to display the speed profile. The Waveform Graph is not used in the other modes of operation.

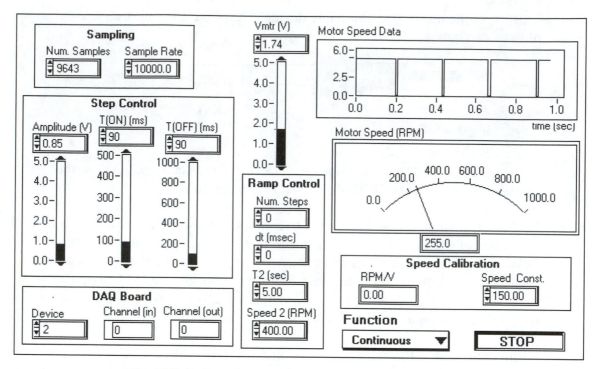

Fig. 15-3 The Front Panel of DC Motor.vi

The *Meter* digital indicator labeled as *Motor Speed (RPM)* has a needle display, and for better precision it also has a digital display. It is used to display motor speed in all modes except the Step mode.

The *DAQ Board* box contains one digital control, *Device,* and two string controls. The Device must include the number of your DAQ board. In the illustration of Fig. 15-3 this value is 2. The string controls *Channel (in)* and *Channel (out)* contain the data input channel and the DAC output channel. In Fig. 15-3 they are both 0.

The *Vmtr (V)* Vertical Fill Slide control is the source of voltage that is applied to the DC motor by way of the DAC channel. Use the Operating Tool to move the slide and thus select the voltage that you want to apply to the DC motor.

The **Sampling** box contains two digital controls, one for the number of samples and the other for the sampling rate. These values are used in the first mode of operation, the Continuous Mode.

The **Step Control** contains the Vertical Fill Slide controls used to affect the operation of the DC motor in the *Step* mode. Use the Operating Tool to change values on the slide.

The ***Ramp Control*** box contains four digital controls to shape the ramp speed response of the DC motor. These controls are used in the *Ramp* and the *Speed Profile* modes.

The ***Speed Calibration*** box contains one digital indicator that displays the instantaneous motor speed constant (RPM/V) in the Continuous mode of operation. The second object in this box is the digital control that stores the value determined by you in the calibration step. This value is used by the VI for display purposes in the last two modes of operation.

The *Menu Ring* labeled as ***Function*** allows you to select one of four modes of operation. Use the *Operating Tool* to make the selections.

The *Rectangular Stop Button* labeled as ***STOP*** is used to stop the execution of the VI. When you want to run the VI in the mode that you selected, first click on the *Start* button with the Operating Tool and then click on the *Run* button located in the upper left-hand side of the window. Click on the Stop button if you want to terminate the VI execution. Note that the face of the Stop button is white when it is ON, and turns to red when it is OFF.

5.   This VI has four modes of operation

> **Continuous**
> **Step**
> **Ramp**
> **Speed Profile**

To select one of these modes, click with the Positioning Tool in the *Function* Menu Ring.

Continuous Mode. This is the first mode of operation and probably most natural for the DC motor. You can control the speed of the motor by adjusting the Vertical Fill Slide digital control labeled *Vmtr (V)* with the Positioning Tool. The voltage that you select here will drive the DC motor from the DAC output channel. In this mode, the acquired data pulses will be displayed on the Waveform Graph and the computed speed will be continuously displayed on the *Motor Speed* digital indicator.

Step Mode. This mode of operation is very unnatural for the DC motor. A stepper motor can easily do this type of stepping motion. However, this mode has been included here as an exercise where the DC motor is forced to step. You will notice that the steps are clumsy and unequal. To operate in this mode, adjust the parameters in the Step Control box on the Front Panel. In this mode the data and the speed are not displayed. You merely observe the motion.

Ramp Mode. In this mode the motor will accelerate linearly toward its final speed and then decelerate to zero speed. The operation will be repeated indefinitely until you click on the *Stop* switch. To operate in this mode, adjust the parameter values in the *Ramp Control* box on the Front Panel. You have to enter the values for *Num. Steps* and *dt*. In this mode the motor speed in RPM will be displayed on the Meter digital indicator labeled as *Motor Speed (RPM)*.

Speed Profile Mode. In this mode the motor executes a specific speed sequence. For example, in the illustration shown below

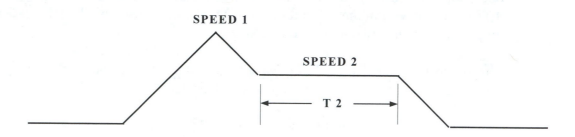

the motor ramps to Speed 1, and then it decelerates to Speed 2. It remains at Speed 2 for the time T2. Finally it decelerates to zero speed. The parameter values that affect the Speed Profile are in the *Ramp Control* box in the Front Panel.

6.   Enter the following settings:
In the **DAQ Board** box enter the values that are appropriate for your DAQ board.

In the **Sampling** box enter 10,000 for Sample Rate, and 15,000 for Num. Samples. With these values the VI will require approximately 1.5 seconds (15,000/10,000) to acquire and display the data. If you choose 10,000 for the Num. Samples value, the update rate will be once every second. If the Num. Samples value is too low, and the motor is running at a very low speed, the data acquired may include only one photogate pulse. This will cause the software to fail because it needs the time interval between two pulses in order to calculate the speed in RPM.

7.   In this step you will perform the **speed calibration**, a procedure that involves the determination of the speed constant. The last two modes will not run properly without the value for the speed constant.

To begin the calibration, select **Continuous** from the Function menu and click on the **Stop** button, using the *Positioning Tool*. In order for the VI to run, the *Stop* button must appear depressed.

Click on the **Run** button and execute the VI. Adjust V_{mtr}, the Vertical Fill Slide digital control, to vary the motor speed. Adjust V_{mtr} to 1 V and read the motor speed. Repeat this for 2 V, 3 V, 4 V, and 5 V up to maximum voltage. Record the voltages and the corresponding speeds in RPM. Determine the best straight line fit to this data and the value of the slope of this straight line. The slope represents the motor speed constant in RPM/V. The *Speed Calibration* box in the Front Panel includes the *RPM/V* digital indicator. This can be used as a check as it gives the value of the slope at different voltages.

Enter the value of the slope into the *Speed Const.* digital control in the *Speed Calibration* box of the Front Panel. In the illustration of Fig. 15-3, this value is 150 RPM/V. To ensure that this value comes up the next time you open this VI, make this value a default. To do so, pop up on the Speed Const. digital control and choose *Data Operation>Make Current Value Default* from the pop up menu.

While you are in the *Continuous* mode you are free to experiment with other values of *Sample Rate* and *Num. Samples* to see what effect they have on data acquisition and display. After you are done, click on the *Stop* button to termination mode operation.

8.   Choose ***Step*** from the *Function* menu, click on the *Stop* button and then run this mode by clicking on the *Run* button. You will have to play with the controls in the *Step Control* box on the Front Panel to force the motor to step. The Amplitude controls the voltage applied to the motor. The T(ON) controls the time when voltage is applied, and T(OFF) controls the time when no voltage is applied to the motor. After finishing, click on the *Stop* button to terminate this mode.

9.   Choose ***Ramp*** from the *Function* menu. In this mode you must provide the ramp control parameters in the *Ramp Control* box of Front Panel. Two values are needed in this mode: one for *Num. Steps* and the other for dt. Enter 20 into the *Num. Steps* digital control (this will divide the 5-volt range into 0.25 V steps) and 100 for dt (making each step 100 msec or 0.1 sec long). When you run this mode, the motor will take 2 seconds to reach its maximum speed and another 2 seconds to decelerate to zero speed.

Click on the *Stop* button (its face should be white when ON) and then run this mode by clicking on the *Run* button. The Motor Speed indicator will display continuously the motor speed. The Waveform Graph is not used in this mode. The ramping will be repeated again and again until you click on the Stop button. You may wish to experiment with other values of Num. Steps and dt. After finishing, click on the *Stop* button (its face color turns to red) to terminate this mode.

10.   Choose ***Speed Profile*** from the *Function* menu. In this mode you must provide the *Speed Profile* control parameters in the *Ramp Control* box of the Front Panel. All four values in that box, as well as the value of the *Speed Const.* are used in this mode. Enter the following values.

> Num. Steps = 20
> dt = 100
> T2 = 5 sec.
> Speed 2 = 400 RPM

Click on the *Stop* button and then run this mode by clicking on the *Run* button. In this mode, the *Motor Speed* indicator will provide the continuous display of motor speed and the *Motor Speed Data* Waveform Graph will display the speed profile at the end of motion. The speed profile cycle will be executed only once. To run it again, click on the *Stop* and the *Run* buttons. Feel free to experiment with other values.

Analysis

1. How did you determine the speed constant in step 7 of the Procedure?

2. Suppose the pulses shown below are those from the photogate when the motor operates in the Continuous Mode. Use this data to calculate the motor's speed in RPM.

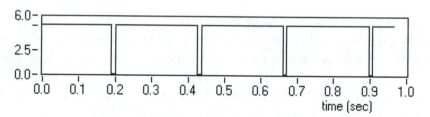

3. Suppose 4 VDC is applied to some motor whose speed constant is 180. What would be the expected speed of that motor in RPM?

4. The required speed profile for the motor that you used in this experiment is shown below.
    Determine the parameter values in the Ramp Control box in the Front Panel. Run the VI to verify the results.

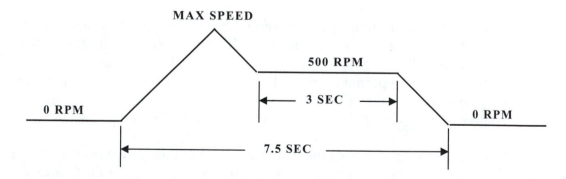

5. Suppose the motor is running in the Continuous Mode. The Sample Rate is set to 10,000 and the Num. Samples value is set to 2,000. Calculate the lowest motor speed in RPM that cannot be detected by this VI.

Stepper Motors

Theory

Stepper motors are electromechanical devices that move in discrete steps, unlike DC or AC motors, whose motion is continuous. In some ways stepper motors resemble digital devices because their motion is discrete and they are driven by pulses. To understand the operation of a typical stepper motor, consider a simple illustration in Fig. 15-4.

In this illustration a permanent magnet rotor is placed on a shaft so that it is free to rotate. The stationary member of the motor, called the stator, has four poles, and each pole has a winding. When the switch closes, current flows through the winding, thus magnetizing the pole. The pole is an electromagnet, which retains its magnetic properties as long as current flows through its winding, and loses its magnetism when the current is switched OFF. The switching of the current is accomplished by switches S1 through S4.

The polarity of the induced magnetism can be determined by applying the right-hand rule. In the illustration of Fig. 15-4, switch S1 is closed, and the current I through the phase winding magnetizes one end of the pole as the North pole.

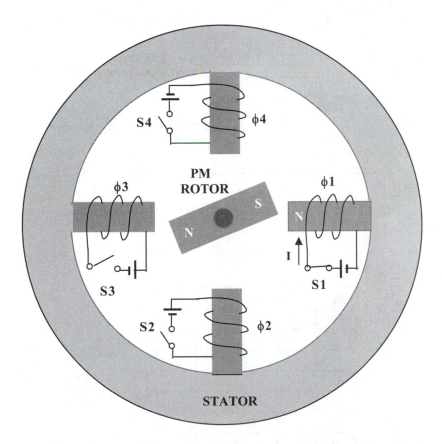

Fig. 15-4 A Simplified Diagram of a Stepper Motor

The rotor's South pole is close to the induced North pole, and is attracted to it since like poles attract and unlike poles repel.

If switch S1 is now opened and switch 2 closed, the next pole is magnetized and the rotor is forced to move 90°. Opening switch S2 and closing switch S3 causes the rotor to rotate another 90°. Repeating the 1-2-3-4 sequence of opening and closing switches S1 to S4 results in the rotor's clockwise rotation of 90° per step. The direction of rotation is reversed when the switching sequence is reversed. In the illustration, a 4-3-2-1 sequence will result in counterclockwise rotation.

Each winding, called a phase, is energized by applying current to it, and when the current is switched OFF, it is de-energized.

The stepper motor switching diagram shown in Fig. 15-5 uses the symbol for the stepper motor. The windings, or phases φ1 to φ4, are switched by switches S1 to S4. Closing switch S1 energizes φ1. Other phases are energized and de-energized in a similar

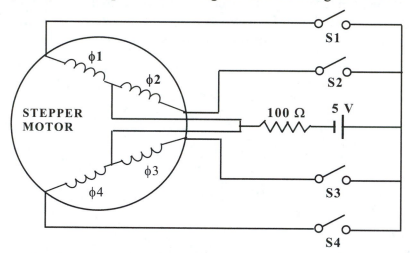

Fig. 15-5 Stepper Motor Switching Diagram

manner. The phase switching for a motor with a 4° step angle is illustrated in the table shown in Fig. 15-6.

θ_S	φ1	φ2	φ3	φ4
0°	ON			
4°		ON		
8°			ON	
12°				ON
16°	ON			
20°		ON		
24°			ON	
28°				ON

Fig. 15-6 Full Step Sequence for a Stepper Motor with a 4° Step Angle

Toggle switches have been used in previous explanations of stepper motor operation to provide the switching of phases. In practice, however, toggle switches are too slow. Instead, pulses are used to drive the phase windings. Running a stepper motor at 1000 steps/sec is not uncommon. To an observer such a fast speed makes the stepper motor motion seem to be continuous and not discrete. In reality, the stepper motor must momentarily stop after each step.

Step Angle, Stepping Rate, and Motor Speed

An expression for the step angle θ_S may be deduced by considering the stepping rate, R_S, the number of steps that the motor rotates in one revolution. Hence

$$R_S(steps/rev) \times \theta_S(degrees/step) = 360°$$

because one revolution is equal to 360°. The equation for the step angle becomes

$$\theta_S = 360/R_S \quad (degrees) \tag{15-1}$$

Motor speed, S, is often expressed in steps/sec. This can be converted to RPM as follows
$$\omega = S(steps/sec) \times \theta_S(deg/step) \times (1\ rev/360\ deg) \times (60\ sec/min)$$

Hence
$$\omega = (S)(\theta_S)/6 = 60S/R_S \quad (RPM) \tag{15-2}$$

where $60S/R_S$ is obtained through the use of Eq. 15-1. A common stepper motor has an $R_S = 200$ steps/rev. The stepping angle can be calculated as 1.8°.

Half Stepping

In earlier discussion it was assumed that the step angle is a full step angle. It is possible to make the stepper motor move in half steps by energizing more than one phase at a time. Suppose that in Fig. 15-4 the rotor is aligned with the phase 1 pole and phase 1 is energized. Next, phase 1 and phase 2 are energized together, causing the rotor to move halfway between poles 1 and 2. When phase 1 is de-energized, the rotor will align with pole 2. As this type of sequence continues where two phases are energized and then one of them is de-energized, the stepper motor will make half steps. This is illustrated in the table of Fig. 15-7 for a stepper motor with a full step angle of 4°.

θ_S	$\phi1$	$\phi2$	$\phi3$	$\phi4$
0°	ON			
2°	ON	ON		
4°		ON		
6°		ON	ON	
8°			ON	
10°			ON	ON
12°				ON
14°	ON			ON
16°	ON			

Fig. 15-7 Half Step Sequence for a Stepper Motor with a 4° Step Angle

Microstepping

There are applications that require the stepper motor to make extremely fine steps. In some industrial operations a stepper motor moves a load, making full step angle, and then at some predetermined point it has to make fine steps in order to properly position the load. In initiating the microstepping process we must consider two quantities: the microstep angle θ_{MS} and the microstep voltage V_{MS}. They are defined as follows

$$\theta_{MS} = \theta_S/N \qquad (15\text{-}3)$$
$$V_{MS} = V_{FS}/N \qquad (15\text{-}4)$$

where θ_S is the full step angle defined by Eq. 15-1, V_{FS} is the full step voltage that must be applied to the phase winding for a full step angle, and N is the microstep scale factor. For example, if a stepper motor's $\theta_S = 40^\circ$ and the $V_{FS} = 12$ V, then in order for the motor to microstep with 4° steps, $N = 10$ must be selected, resulting in a microstep voltage of 1.2 V.

Two phases are involved in the microstep process; one phase is driven by the ascending staircase of steps, while the other phaseis driven by the descending staircase of steps. The following example will illustrate a microstepping process.

Example 1

Consider a microstepping application in an industrial process where a bottle moved on a conveyor belt driven by a stepper motor must be positioned precisely under the capping machine, which will place a cap on top of the bottle and seal it, as shown in Fig. 15-8.

In order to cap the required of number of bottles per hour, the step motor must move the bottle quickly along the conveyor belt; and when the bottle is close to the capping machine, it must initiate the microstep sequence in order to properly position the bottle.

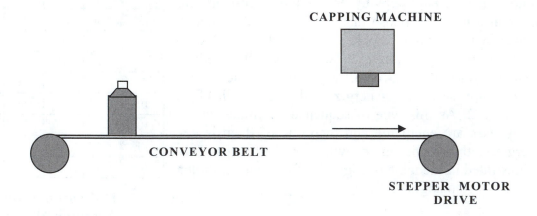

Fig. 15-8 Industrial Microstepping Application

The following parameters are known: R_S=75 steps/rev and V_{FS} =10 V. It was determined that from the starting position the motor must make 100 revolutions and must initiate the microstep sequence on the 101st revolution between 16° and 20°, microstepping with 0.8° steps.

Solution

From Eq. 15-1 the full step angle is

θ_S = 360/R_S = 360/75 = 4°.
Because the microstep angle is 0.8°, the value of N must be 5 (4/0.8).

From Eq. 15-4 the microstep voltage
V_{MS} = V_{FS}/N = 1/5 = 2 V

θ_S	$\phi1$	$\phi2$	$\phi3$	$\phi4$
0°	ON			
4°		ON		
8°			ON	
12°				ON
16°	ON			
20°		ON		
24°			ON	
28°				ON

For convenience the full step sequence for a 4-phase stepper motor with a 4° full step angle from Fig. 15-6is repeated here for convenience.

From this table it can be seen that the microstep must start at 16° when V_{FS} is applied to the phase 1 ($\phi1$) winding and no voltage is applied to $\phi2$. The microstepping sequence must stop at 20° when V_{FS} is applied to $\phi2$ and no voltage is applied to $\phi1$. Between these extremes both phases are energized by the microstep voltages. The microstep incremental values are shown in tabular and in graphical forms in Fig. 15-9.

θ_{MS}	$\phi1$	$\phi2$	$V_{\phi1}$	$V_{\phi2}$
16.0°	ON		10v	0v
16.8°	ON	ON	8v	2v
17.6°	ON	ON	6v	4v
18.4°	ON	ON	4v	6v
19.2°	ON	ON	2v	8v
20.0°		ON	0v	10v

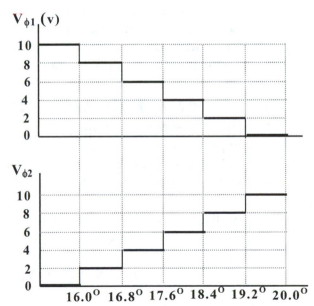

Fig. 15-8 Microstepping Sequence Table and Phase Excitation Waveforms

The actual implementation of the microstep sequence ramping waveform is done by the indexer circuit, or the translator circuit as it sometimes called. The drive detail is shown in Fig. 15-10.

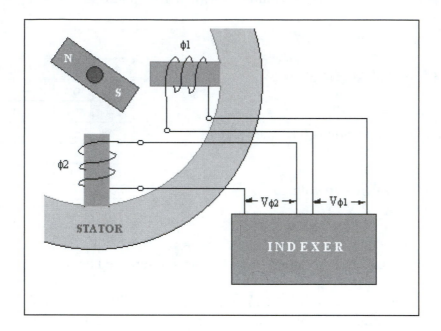

Fig. 15-10. The Indexer Drives Phases 1 and 2 in a Microstepping Sequence

Microstepping technique where a full step angle is subdivided into N small angles provides us with extremely precise positioning. In the above example the value of N was set to 5 but in general practical indexers provide N values of 10, 25, 50, and even 100. A read/write head must be precisely positioned over the desired track in a floppy drive, a perfect application for a stepper motor.

Pulse Width Modulation (PWM)

The ramping waveforms shown in Fig. 15-9 that accomplish the required drive in a microstepping sequence are cumbersome because they analog and unsatisfactory for a microprocessor or a computer control. Digital devices work with pulses, and they are more likely to be used by an indexer board in a microstepping control.

We know from pulse theory that the average value of the pulse waveform shown below can be expressed as follows

$$V_{avg} = (E)(d) \ V_{dc} \qquad (15\text{-}5)$$

where the duty cycle is defined as follows

$$d = T_1/(T_1 + T_2)$$
$$= (T_1)(f) \qquad (15\text{-}6)$$

Because $T_1 + T_2$ is the period of the square wave, its reciprocal is the frequency f of the pulses.

Combining Eq. 15-5 and 15-6 and solving for the pulse width T_1 we get

$$T_1 = V_{avg} / [(E)(f)] \qquad (15\text{-}7)$$

From Eq. 15-7 it is clear that if the frequency of the pulses is kept constant, then for each V_{avg} there is a unique value of T_1. Thus, by adjusting the width of the pulse we adjust the equivalent DC voltage, hence the pulse width modulation (PWM).

For example, consider the microstepping table in Fig. 15-9. In order for the rotor to advance to $\theta_{MS} = 19.2°$, the phase 1 winding must be supplied 2 V_{dc}, and the phase 2 winding must be supplied with 8 V_{dc}. The 1 kHz, 12 V_{pk} square wave shown below will accomplish the same thing. Check the timing.

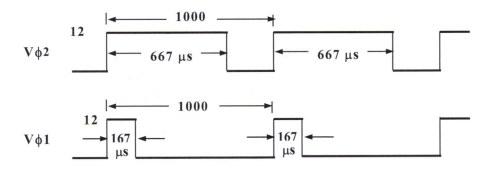

Speed Torque Characteristics

The torque-producing ability of the stepper motor as a function of the operating speed is usually provided by the manufacturer as a curve, such as the speed torque curve shown in Fig. 15-11. This curve shows two operational ranges: the start-stop range and the slew range (gray area).

When the motor operates in the start-stop range, it can instantly start, stop its motion, or reverse direction without losing steps. Starting or stopping instantly may seem odd especially when one considers a DC motor, which has rotor inertia and, therefore, because of its mechanical time constant, cannot suddenly stop. The stepper motor also has inertia, but we must remember that the stepper motor executes discrete motion starting to move at the beginning of each step and stopping at the end of each step regardless of how fast it moves. As illustrated in the diagram below, the motor must stop before starting the next step.

The solid curve labeled *Max Running Torque, Start-Stop (Bidirectional)* in Fig. 15-10 represents maximum torque that the motor can develop at a given speed. For example, at

400 steps/s the maximum running torque is 78 oz-in. If the motor is driving a 40 oz-in load at 400 steps/s then the additional available torque is 38 oz-in. At this speed the motor can start, stop, or reverse the direction of rotation.

The solid curve labeled *Max Running Torque, Slew (Unidirectional)* also represents the maximum running torque that the motor can develop. However, as indicated later, the motor cannot start, stop, or change the direction of rotation without losing steps inside the slew range.

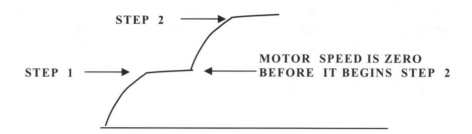

The speed torque curve in Fig. 15-11 shows a maximum torque at zero speed. As the motor speed increases, the torque that the motor can drive is reduced. For example, if the load torque is 40 oz-in, then according to Fig. 15-11 the motor must not exceed 700 steps/sec. Of course, the motor may operate anywhere along the horizontal 40 oz-in line. At 820 steps/s, the maximum response point, no load is allowed.

Example 2

Using the speed torque characteristics in Fig. 15-11, determine the least time that a motor driving a 50 oz-in frictional load would require to advance the load through a 240° angle. Neglect inertial effects and assume that the motor's step angle is 3°.

Solution

To determine the least time that the motor requires we must know the maximum speed. From Fig. 15-11, the maximum speed is 620 steps/sec for a load of 50 oz-in. The motor must make 80 steps (240/3) to move 240° at 3° per step. The required minimum time must be 129 ms (80/620).

The slew range, the gray area in Fig. 15-11 is the region where the motor cannot start or stop suddenly or change direction of rotation without losing steps. For example if the motor in Fig. 15-11 is to drive a 50 oz-in load at 800 rpm, its operating point will be located inside the slew range. If the drive pulses are suddenly applied, the motor will start slipping and lose some steps.

Losing steps is a severe failure in a positioning application because the system will apply a predetermined number of pulses to the motor with the expectation that the motor will execute a predetermined number of steps to reach the destination point. Take a floppy drive as an example, where the stepper motor positions the read-write head. Suppose that it was determined that 200 pulses are required to position the read-write head over track number 50 and some steps were lost because the motor was operating in the slew range. The read-write head may be wrongly positioned over track 48.

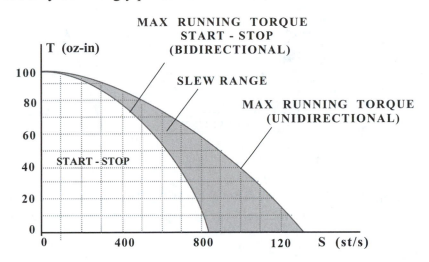

Fig. 15-10 Stepper Motor Speed Torque Characteristics

If operation inside the slew range is unavoidable, ramping must be used. Ramping requires a gradual acceleration from the start-stop range.

Suppose that the desired operation, T_L at S_2, is in the slew range as shown in Fig. 15-12a. We must first establish an operating point in the start-stop range, shown in Fig. 15-12a as T_L at S_1, and then accelerate or ramp to the (S_2, T_L) point. The ramping operation is shown in Fig. 15-12b. The value of acceleration may be expressed as

$$A = (S_2 - S_1)/\Delta t \quad (\text{steps/sec}^2) \qquad (15\text{-}8)$$
$$= (2\pi)(S_2 - S_1)/(R_S \Delta t) \quad (\text{rad/sec}^2)$$

The motor must make the transition from the start-stop range to the slew range in time Δt. In order to make this transition the motor must develop sufficient torque to drive T_L and additional torque to accelerate total inertia due to the load and the rotor. The total inertial torque is

$$T_J = J_T A \qquad (15\text{-}9)$$

Example 3

The stepper motor with R_s = 120 steps/rev drives a frictional load T_L = 20 oz-in at 500 steps/s. The combined inertia of the load and the rotor is 0.15 oz-in-s^2. Determine the shortest time required to accelerate the load to 1000 steps/s in the slew range. Refer to the speed torque curve in Fig. 15-11.

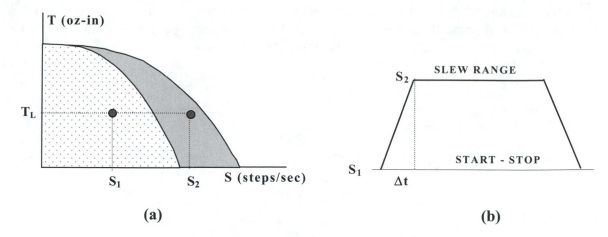

(a)   **(b)**

Fig. 15-12 (a) Driving a Load in the Slew Range, (b) Ramping from the Start Stop Range into the Slew Range

Solution

The total inertial torque is calculated as follows using Eq. 15-8 and 15-9:

$$T_J = J_T(2\pi)(S_2 - S_1)/(R_S\Delta t)$$
$$= (0.15)(2\pi)(800 - 400)/(120\Delta t)$$

From Fig. 15-10 we see that the motor develops maximum torque of 40 oz-in at 1000 steps/s and 74 oz-in at 500 steps/s. Since the actual net torque available to accelerate the inertial load into the slew range is between $(40 - 20)$ and $(74 - 20)$, it is common practice to use the average of the two resulting in $T_{avai} = [(40 - 20) + (74 - 20)]/2 = 37$ oz-in. Using this value in the above equation

$$37 = (0.15)(2\pi)(800 - 400)/(120\Delta t)$$

and solving for Δt, we get

$$\Delta t = (0.15)(2\pi)(800 - 400)/(120 \times 37) = 84.9 \text{ ms}$$

Experiment 15-2: Stepper Motor Motion Control 1

Objective

In this experiment we will use LabVIEW to operate a stepper motor. LabVIEW software will be provided, and you will use the Front Panel controls to vary the speed of the motor and observe the motor speed on the digital indicators. The software will generate signals to control the stepper motion speed.

Parts

DAQ board (such as MIO-16E-10 or LabPC$^+$)
Extender board (associated with DAQ board)
Four-phase stepper motor
LabVIEW (version 5.1 or higher)
Stepper Motor 1.vi

Test Setup

The test setup for the stepper motor experiment is shown in Fig. 15-13. Depending on the size of the stepper motor, its windings require a substantial drive current perhaps in excess of 100 or 200 mA. Because the DAQ board is unable to source this current, we need a current amplifier. A current amplifier that uses Darlington transistors is shown in Fig. 15-14. Darlington devices are available commercially and are inexpensive. The NTE 48 is rated at 3 A collector current.

Darlingtons provide current switching to each of the four stepper motor phase windings. Each of the phase windings is in the collector circuit. The switching is done by the pulses from port C of the 8255 PIA on the DAQ board. A pulse applied to the base of the Darlington switches it ON and applies current to the phase winding during the LOW to HIGH transition, and switches it OFF during the HIGH to LOW transition. The 100 Ω resistor and diode combination across each phase winding eliminates inductive kick by providing a discharge path for the phase current when the Darlington switches OFF.

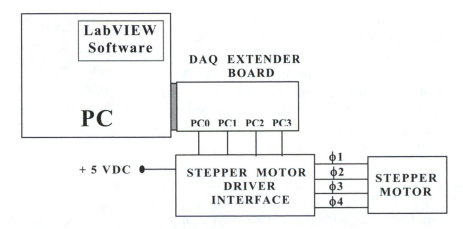

Fig. 15-13 Test Setup for the Stepper Motor Experiment

Inductive kick occurs when the current through an inductor is suddenly cut off. At this time the voltage across the coil increases without limit. If it wasn't for the discharge path provided by the resistor in series with the diode, the Darlington would be destroyed. When the Darlington is ON, the diode is reverse biased, thus disabling the discharge path.

Pulses from port C energize the phase windings sequentially, causing the stepper motor to run. The speed is controlled by adjusting the frequency of the pulses in the software.

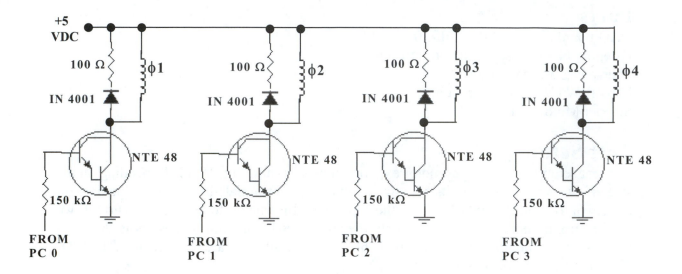

Fig. 15-14 Stepper Motor Drive Interface

Software Description

The Front Panel of the Stepper Motor 1.vi program used in this experiment is shown in Fig. 15-15. It includes five recessed boxes, four of which include objects that pertain to the selected test. The fifth recessed box includes the waveform chart that is shared for display purposes by several tests.

The Test Menu, a Menu Ring control, provides the user with six options as shown in the illustration here. The first three options are experiment options, meaning that the test setup must correspond to that shown in Fig. 15-13 because software-generated signals are used to control the operation of the stepper motor. The last three options are simulation options that provide the user with the ability to simulate stepper motor operation.

| Exp: Speed |
| Exp: Ramp Up |
| Exp: Ramp Down |
| Sim: Speed |
| ✓ Sim: Ramp Up |
| Sim: Ramp Down |

The Exp: Speed option allows the user to control the speed and direction of rotation. The controls and indicators are in the recessed box labeled *Exp: Speed*. The Vertical Slide control labeled *Motor Speed* is a control for adjusting speed from 1 to 1000 steps/s. The Boolean control labeled *Direction* provides the direction of rotation. Change the state of the direction control with the Operating Tool. Enter the Device number that corresponds to your DAQ board. The four LEDs provide visual indication of the currently energized phase. The *Push to Start/Stop* Boolean control is used to terminate VI execution. The test setup must correspond to that shown in Fig. 15-13 because software-generated signals are used to control the operation of the stepper motor.

The **Exp: Ramp Up** option accelerates the stepper motor from 1 steps/s to the value set on the Vertical Slide control labeled *Slope Control*. Test setup must correspond to that shown in Fig. 15-12 because software-generated signals are used to control the operation of the stepper motor. This digital control sets the terminal ramp speed, thus controlling the ramp slope. The LEDs display the currently energized phase. The *Push to Start/Stop* Boolean control is used to start or stop the operation. It must be pushed with the Operating Tool to start and then pushed again to stop.

The ramping waveform will be displayed on the Waveform Chart. The string indicator above the chart will display the name of the current ramping test.

The **Exp: Ramp Down** option decelerates the stepper motor from the value set on the Vertical Slide control labeled *Slope Control* to 1 steps/s. This digital control sets the terminal ramp speed, thus controlling the ramp slope. The test setup must correspond to that shown in Fig. 15-13 because software-generated signals are used to control the operation of the stepper motor.

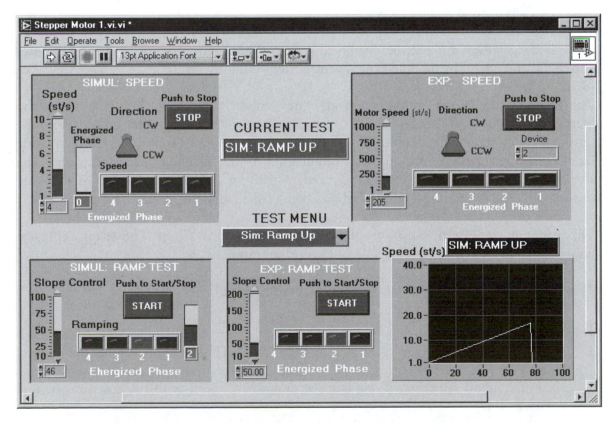

Fig. 15-15 Front Panel of Stepper Motor 1.vi

The LEDs display the currently energized phase. The *Push to Start/Stop* Boolean control is used to start or stop the operation. It must be pushed with the Operating Tool to start and then pushed again to stop.

The ramping waveform will be displayed on the Waveform Chart. The string indicator above the chart will display the name of the current ramping test.

The **Sim: Speed** option allows the user to control the speed and direction of rotation. This is a simulation test and not the experiment. The controls and indicators are in the recessed box labeled *Exp: Speed*. The Vertical Slide control labeled *Motor Speed* is a control for adjusting speed from 1 to 1000 st/s. The Boolean control labeled *Direction* provides the direction of rotation control. Change the state of the direction control with the Operating Tool. The four LEDs and the Vertical Slide labeled *Energized Phase* provide visual indication of the currently energized phase. The *Push to Stop* Boolean control is used to terminate VI execution.

The **Sim: Ramp Up** option accelerates the stepper motor from 1 steps/s to the value set on the Vertical Slide control labeled *Slope*. This digital control sets the terminal ramp speed, thus controlling the ramp slope. The LEDs and the Vertical Slide labeled *Energized Phase* display the currently energized phase. The *Push to Start/Stop* Boolean control is used to start or stop the operation. It must be pushed with the Operating Tool to start and then pushed again to stop.

The ramping waveform will be displayed on the Waveform Chart. The string indicator above the chart will display the name of the current ramping test.

The **Exp: Ramp Down** option decelerates the stepper motor from the value set on the Vertical Slide control labeled *Slope Control* to 1 steps/s. This digital control sets the terminal ramp speed, thus controlling the ramp slope. The LEDs and the Vertical Slide labeled *Energized Phase* provide visual indication of the currently energized phase. The *Push to Start/Stop* Boolean control is used to start or stop the operation. It must be pushed with the Operating Tool to start and then pushed again to stop.

The ramping waveform will be displayed on the Waveform Chart. The string indicator above the chart will display the name of the current ramping test.

The **Current Test** string indicator displays the test currently performed.

Procedure

1. Configure the test setup as shown in Fig. 15-13. The user must build the stepper motor drive circuit as shown in Fig. 15-14. The type of stepper motor that you use is not critical. Any four-phase stepper motor will work with the software of this experiment.

LabVIEW full development software version 5.1 or higher must be installed and the DAQ board configured. The device or the board number will be assigned during the configuration.

2.   Open the Stepper Motor 1.vi program, switch to the Front Panel, and become familiar with different tests, controls, and indicators. The Software Description section provides all the necessary information.

3.   Run simulation tests, as they will prepare you for the experimental tests.

4.   Run the *Speed* experiment.

Select a low speed such as 1 step/s and count the number of steps that the motor makes in one half revolution. Record this value.

Increase the speed of the motor gradually until you notice the motor slipping. Record this speed.

Set the motor speed to 10 steps/s and measure the time that the motor requires to make 5 revolutions.

Change the direction of rotation (any speed) and note whether the direction of rotation is reversed. If the direction of rotation as set on the Front Panel (CW or CCW) does not correspond to the actual direction of rotation, reconnect the phase windings of the motor until they are the same.

5.   Run the Ramp Up experiment. Set *Slope Control* first to 25 and run the VI. Then repeat for 50 and 100 settings of *Slope Control*. Record the final speed of the motor in all cases.

6.   Run the Ramp Down experiment. Set *Slope Control* first to 25 and run the VI. Then repeat for 50 and 100 settings of *Slope Control*. Record the initial speed of the motor in all cases.

Analysis

1.   Using the acquired data, calculate the stepping rate R_s, and the full step angle θ_s.

2.   With no applied load, what is the maximum speed in the Start-Stop range?

3.   Based on the measured time required for 5 revolutions, calculate motor speed in RPM.

Calculate the speed once again using the 10 steps/s value. Within 10%, are the two values the same? Explain if they are different.

4. Did the direction of rotation correspond to the CW and CCW settings on the Front Panel? If not, explain what you did to fix the problem. Be specific.

5. Inspect the Block Diagram and determine the equation for Δt that the software is using to set the time between steps in the Ramp Up test. According to this equation, what is the final value of Δt? How many iterations does LabVIEW use to reach the final speed?

6. Inspect the Block Diagram and determine the equation for Δt that the software is using to set the time between steps in the Ramp Down test. According to this equation, what is the final value of Δt? How many iterations does LabVIEW use to reach the final speed?

7. Calculate the *Slope Control* value for a motor speed of 32 steps/s in the Ramp Up operation using one of the equations that you obtained.

Chapter 16
Control System Experiments

Introduction

Second-order systems occur frequently in electronics and in physics and for that reason they deserve special attention. There are two models that can be used for analysis. The first model is a second-order differential equation

$$\tau \frac{d^2(y)}{dt^2} + k_1 \frac{dy}{dt} + k_2 \, y(t) = k \, x(t) \tag{16-1}$$

and the second model is a second-order transfer function expressed as follows:

$$G(s) = \frac{C(s)}{R(s)} = \frac{\omega_n^2}{s^2 + 2\zeta\omega_n^2 s + \omega_n^2} \tag{16-2}$$

This is often referred to as the universal second-order transfer function where

ζ = damping ratio, a dimensionless quantity that depends on system losses

ω_n = natural resonant frequency

The differential equation allows for the initial conditions and the transfer function model does not. If the transfer function model is used for analysis, the system must be at rest, i.e., no initial conditions.

Fig. 16-1 illustrates several systems that include mechanical translational and rotational systems, a position control servo and a Butterworth filter. They all fall into the same category generally referred to as the second-order systems. Despite the fact that these systems are very different, their response to a step input, for example, is identical. It is therefore desirable to have general theory that can be applied to any second-order system

This chapter considers the behavior of second-order systems in the Time Domain (TD) and in the Frequency Domain (FD). In the time domain only the step response is considered. We begin with the step response of the second-order systems in the time domain.

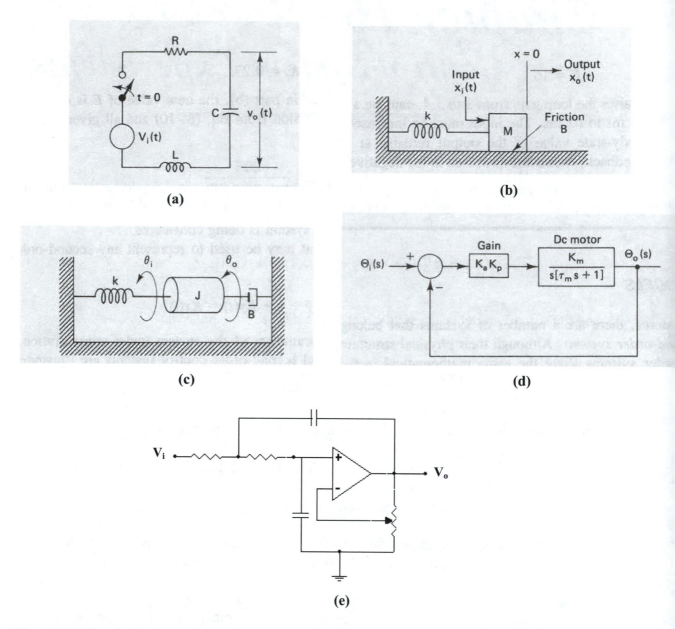

Fig. 16-1 Typical Second-order Systems: (a) RLC Circuit; (b) Mechanical
Translational System; (c) Mechanical Rotational System; (d) DC
Motor Position Control Servo; (e) Butterworth Low Pass Filter

Time Domain Step Response

The second-order system transfer is used here for analysis. The input $r(t) = Eu(t)$, whose LaPlace transform $R(s) = E/s$, is substituted in Eq. 16-2, resulting in the following

$$C(s) = \frac{E\omega}{s(s^2 + 2\zeta\omega_n s + \omega_n^2)} \qquad (16\text{-}3)$$

The poles or the roots of the denominator are

$$s_1 = -\zeta\omega_n + \omega_n(s^2 - 1)^{\frac{1}{2}} \qquad (16\text{-}4)$$
$$s_2 = -\zeta\omega_n - \omega_n(s^2 - 1)^{\frac{1}{2}}$$

The character of the roots s_1 and s_2 and the value of ζ determine the shape of the time domain response. Depending on the value of ζ, the TD responses are characterized as follows:

$$\zeta = 0 \quad \text{undamped response}$$
$$\zeta < 1 \quad \text{underdamped response}$$
$$\zeta = 1 \quad \text{critically damped response}$$
$$\zeta > 1 \quad \text{overdamped response}$$

Undamped Response
In this case, $\zeta = 0$ and Eq. 16-3 reduces to

$$C(s) = \frac{E\omega_n}{s(s^2 + \omega_n^2)}$$

The inverse LaPlace transform $L^{-1}C(s)$ yields

$$c(t) = E(1 - \cos(\omega_n t)) \qquad (16\text{-}5)$$

Fig. 16-2 illustrates an undamped step response of a second-order system with $E = 10$ and $\omega_n = 100$. As can be seen from the illustration, the sinusoidal response is shifted upward by the value of the step input, as predicted by Eq. 16-5, and its period is 62.8 ms ($2\pi/100$).

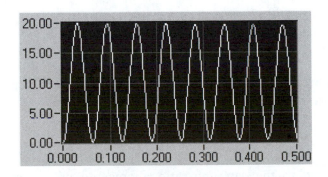

Fig. 16-2 Step Response of an Undamped Second-order System

Underdamped Response

The complex conjugate roots of Eq. 16-4 for $\zeta < 1$ may be expressed as follows:

$$s_1 = -\zeta\omega_n + j\omega_n\beta$$
$$s_2 = -\zeta\omega_n - j\omega_n\beta$$

where $\beta = \sqrt{(1 - \zeta^2)}$ with ζ and β related by the right triangle shown below. From the triangle it follows that the angle $\phi = \cos^{-1}(\zeta)$.

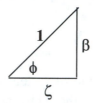

After taking the inverse LaPlace transform of

$$C(s) = \frac{E\omega_n}{s(s - s_1)(s - s_2)}$$

we obtain

$$c(t) = E[1 - e^{-\zeta\omega_n t}\sin(\omega_n\beta t + \phi)] / \beta \qquad (16\text{-}6)$$

where β and ϕ have been defined previously. Another quantity that appears in Eq. 16-6 is the damped frequency of oscillation defined as follows:

$$\omega_d = \omega_n\beta \qquad (16\text{-}7)$$

A typical step response of an underdamped second-order system is shown in Fig. 16-3. C_1 and T_1 identify the amplitude and time, respectively, of the first peak. In general the kth peak amplitude and time may be calculated using the following equations:

$$T_k = k\pi/(\omega_n\beta) \qquad (16\text{-}8)$$
$$C_k = E[1 + (-1)^{k+1}e^{-k\pi\zeta/\beta}]$$

Percent overshoot, representing the amount by which the first peak of the response exceeds the input, expressed as percent of input, is

$$POT = 100(C_1 - E)/E = 100e^{-\pi\zeta/\beta} \qquad (16\text{-}9)$$

POT is inversely related to the damping ratio, increasing with decreasing ζ and vice versa.

Amplitude (V) vs. Time (sec)

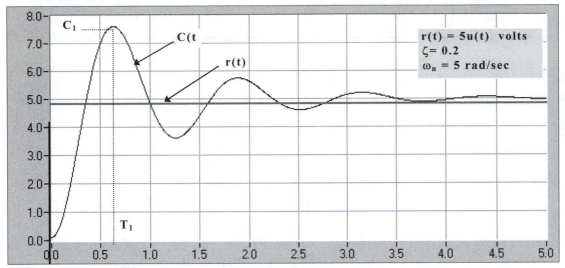

Fig. 16-3 Step Response of an Underdamped/Second-Order System

Critically Damped Response

When $\zeta = 1$, both roots in Eq. 16-2 are equal to ω_n, and C(s) in Eq. 16-2 reduces to

$$C(s) = \frac{E\omega_n}{s(s - s_1)(s - s_2)}$$

The inverse LaPlace transform of the above C(s) yields

$$c(t) = E[1 - e^{-\omega_n t}(\omega_n t + 1)] \qquad (16\text{-}10)$$

The graph of this response for $\omega_n = 5$ rad/sec and $E = 5$ V is shown in Fig. 16-4.

The 10% to 90% rise time of this response may be expressed as

$$T_r = 3.3\tau \quad \text{where} \quad \tau = 1/\omega_n$$

In the above response, the rise time is 0.66 s.

Overdamped Response

In the case of the overdamped system ($\zeta > 1$), the roots of the characteristic equation in Eq. 16-4 are real and non-repeated:

$$s_1 = -\zeta\omega_n + \omega_n\alpha$$
$$s_2 = -\zeta\omega_n - \omega_n\alpha$$

where $\alpha = (s^2 - 1)^{1/2}$

Amplitude (V) vs. Time (sec)

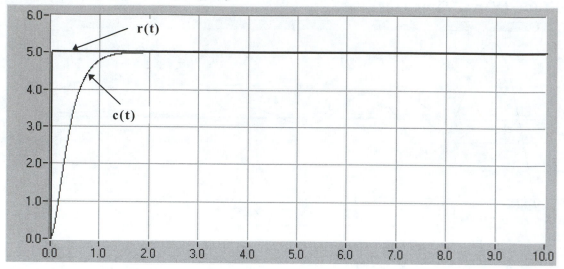

Fig. 16-4 Step Response Of A Second-order Critically Damped System

The inverse LaPlace transform of

$$C(s) = \frac{E\omega_n}{s(s - s_1)(s - s_2)}$$

yields

$$c(t) = E[1 - C_1 e^{t/\tau 1} + C_2 e^{t/\tau 2}] \qquad (16\text{-}11)$$

where

$$C_1 = (\zeta + \alpha)/(2\alpha) \qquad \tau_1 = [\omega_n(\zeta - \alpha)]^{-1}$$
$$C_2 = (\zeta - \alpha)/(2\alpha) \qquad \tau_2 \quad = \quad [\omega_n(\zeta \quad + \quad \alpha)]^{-1}$$

C_1 and C_2 represent the amplitudes of the transient terms, and τ_1 and τ_2 are their time constants.

A typical step response of an overdamped system with $\omega_n = 100$ and $\zeta = 5$ is illustrated in Fig. 16-5. As expected, the system shows a slow response due to excessive energy losses.

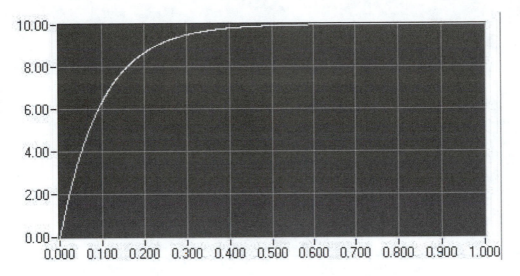

Fig. 16-5 Step Response of a Second-order Overdamped System

Frequency Domain Response

Underdamped Response

The roots for the underdamped ($\zeta < 1$) system as expressed by Eq. 16-4 are as follows

$$s_1 = -\zeta\omega_n + \omega_n\beta$$
$$s_2 = -\zeta\omega_n - \omega_n\beta$$

After the roots are substituted in the transfer function in Eq. 16-2, we get

$$G(s) \ = \frac{C(s)}{R(s} \ = \ \frac{\omega_n^2}{(s - s_1)(s - s_2)} \qquad (16\text{-}12)$$

After substituting $s = j\omega$ in Eq. 16-8 and solving for the magnitude and phase of $G(j\omega)$, we get the following:

$$M(\omega) = |G(j\omega)| = [(1 - (\omega/\omega_n)^2)^2 + 4\zeta^2(\omega/\omega_n)^2]^{-0.5} \qquad (16\text{-}13)$$

$$\phi(\omega) = \{\tan^{-1}[\ ((\omega/\omega_n) - \beta)/\zeta\ ] + \tan^{-1}[\ ((\omega/\omega_n) + \beta)/\zeta] \qquad (16\text{-}14)$$

A second-order underdamped system's Bode plots for three values of ζ are shown in Fig. 16-6. The magnitude plots exhibit a resonant peak that is inversely related to ζ.

M(ω) db

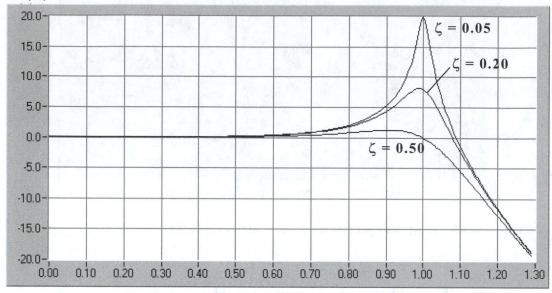

y = ω/ω_n

φ(ω) deg

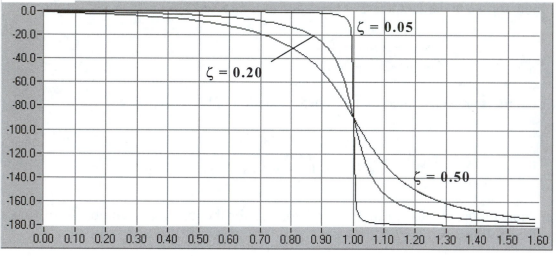

y = ω/ω_n

**Fig. 16-6 Bode Plots for an Underdamped Second-Order System:
(a) Magnitude Response; (b) Phase Response**

Critically Damped Response

The roots s_1 and s_2 are both equal to ω_n for $\zeta = 1$. The magnitude and phase are as follows:

$$M(\omega) = \omega_n^2 / (\omega^2 + \omega_n^2) \tag{16-15}$$
$$\phi(\omega) = -2\tan^{-1}(\omega/\omega_n) \tag{16-16}$$

Typical Bode plots for a critically damped second-order system are shown in Fig. 16-7.

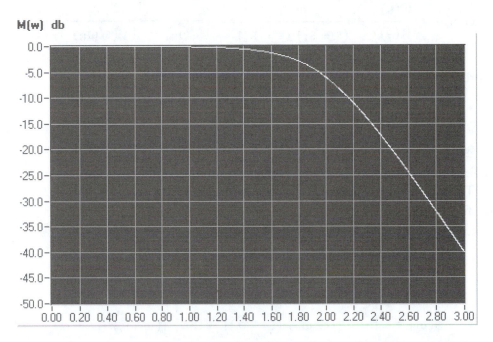

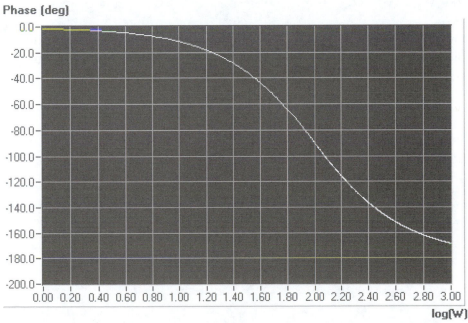

Fig. 16-7 Bode Plots for a Second-Order Critically Damped System

Overdamped Response

When the system is overdamped with $\zeta > 1$, its roots are real and unequal:

$$s_1 = -\zeta\omega_n + \omega_n\alpha$$
$$s_2 = -\zeta\omega_n - \omega_n\alpha \qquad \text{where } \alpha = (s^2 - 1)^{\frac{1}{2}}$$

The second-order system transfer function may be expressed as follows in terms of these roots:

$$G(s) = \frac{C(s)}{R(s)} = \frac{\omega_n^2}{(s - s_1)(s - s_2)} = \frac{1}{(s/\omega_{B1} + 1)(s/\omega_{B2} + 1)} \qquad (16\text{-}17)$$

with the break frequencies that are expressed as follows:

$$\omega_{B1} = \omega_n(\zeta - \alpha)$$
$$\omega_{B2} = \omega_n(\zeta + \alpha)$$

The transfer function in Eq. 16-17 is in the form that can be used to plot the asymptotic response. To get the expression that can be used to plot the exact response, we must first substitute for s_1 and s_2, let $s = j\omega$ in Eq. 16-17, and simplify. The following magnitude and phase expressions are extracted from $G(j\omega)$:

$$M(\omega) = \frac{1}{\sqrt{[(\omega/\omega_n)^2 + (\zeta - \alpha)^2][(\omega/\omega_n)^2 + (\zeta + \alpha)^2]}} \qquad (16\text{-}18)$$

$$\phi(\omega) = -\tan^{-1}[(\omega/\omega_n)/(\zeta - \alpha)] - \tan^{-1}[(\omega/\omega_n)/(\zeta + \alpha)] \qquad (16\text{-}19)$$

The frequency response of a typical overdamped second-order system is shown in Fig. 16-8. The plots use $\zeta = 5$ and $\omega_n = 100$. Both responses show unevenness due to the two break points.

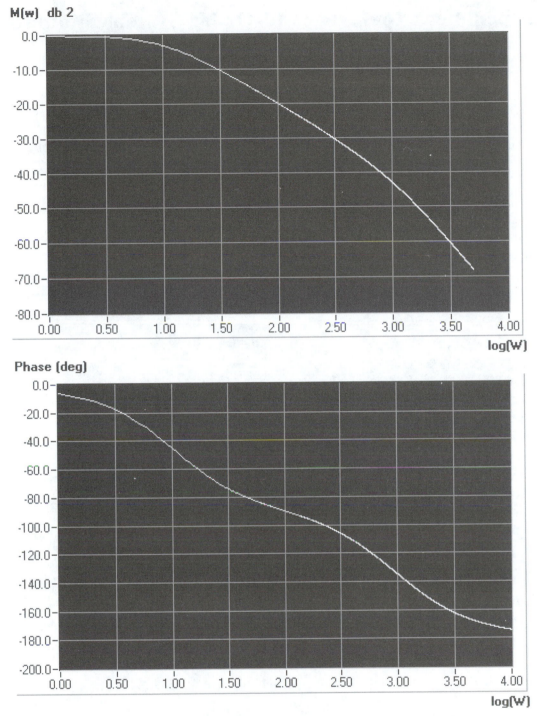

Fig. 16-8 Bode Plots for a Second-Order Overdamped System

Experiment 16-1: Second-Order System Simulation

Introduction

This exercise uses LabVIEW software to simulate the responses of a second-order system in the time domain and in the frequency domain. The user will adjust the second-order system parameter values on the Front Panel and examine and assess system responses. Menu options provide underdamped, critically damped and overdamped system configurations.

Software

Open the Second Order System.vi that is included on the CD (Book VIs>Second-Order System) and examine the Front Panel.

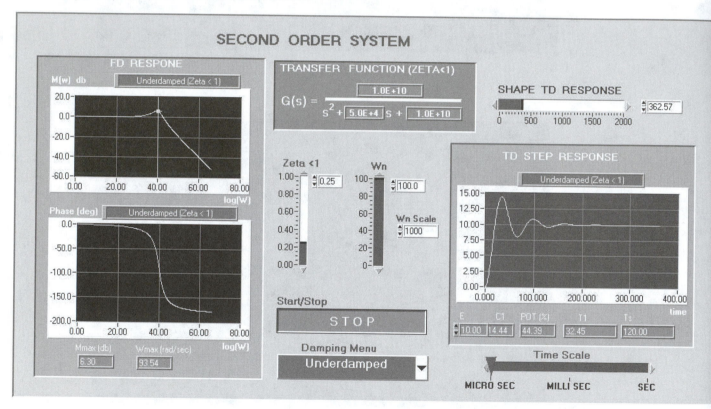

Fig. 16-9 Front Panel of Second-order System.vi

The **Damping Menu** menu ring control includes Underdamped, Critically Damped, and Overdamped response options. Use the Operating Tool to select the desired damping.

The **Start/Stop** Boolean control initiates and terminates VI execution. To run the VI, the switch must be in the RUN position.

The **FD Response** recessed box includes two waveform graphs that display magnitude and phase responses commonly called the Bode plots. The two digital indicators display the value of the peak in dB and the radian frequency of the peak. These indictors disappear if the user selects the critically damped or the overdamped response. The string indictor above the graph displays the type of response in progress.

The **TD Step Response** recessed box includes a waveform graph that displays the time domain step response of the system for the selected ζ and ω_n values. The digital control E is the height of the step entered by the user. The remaining digital indicators display time domain response waveform selected values such as the percent overshoot, the first peak amplitude and time, and the settling time. These digital indicators disappear if the user chooses the overdamped or the critically damped response. A rise time digital indicator appears for the critically damped and overdamped responses. The string indicator above the graph displays the type of response test being conducted.

The **Transfer Function** (for $\zeta < 1$ only) recessed box displays the system transfer function for the underdamped system in the universal format for the values of ζ and ω_n selected by the user.

The ω_n vertical slide digital control passes values of ω_n entered by the user to the Block Diagram. This slide is active for all cases of damping.

The ζ vertical slide digital control passes values of the damping ratio set by the user to Block Diagram. Depending on the user's selection from the Damping Menu, the appropriate vertical slide will appear, that is $\zeta < 1$ or $\zeta > 1$.

The **Time Scale** horizontal slide control sets the time scale for a suitable time domain display. It is active for all cases of damping.

The ω_n **Scale** digital control provides scaling of the natural resonant frequency. For example, ω_n may have a value of 100,000 for an RLC circuit. This value may be implemented by setting ω_n to 100 and the ω_n Scale control to 1000; or a 50, 2000 combination will also work.

The **Shape TD Response** horizontal slide control allows the user to make adjustments until the TD response has a satisfactory appearance.

Procedure
Open Second-order System.vi from the Book VIs folder.

1. Set the following parameter values for the underdamped system: 20 volt step, 0.5 damping ratio, and 150 natural resonant frequency, and run the VI. Record T_1, C_1, POT, T_s, M_{max}, ω_{max}, and the transfer function.

2. Set the following parameter values for the critically damped system, 12-volt step input and 200 rad/s natural resonant frequency, and run the VI. Record T_r in the TD response recessed box. Use the cursor in the FD Response box to record the values

of M(200) and M(400). Remember that the ω scale is in dB. Also, record the transfer function.

3. Repeat step 2, changing ω_n to 400.

4. Set the following parameter values for the overdamped system, 12-volt step input and 200 rad/s natural resonant frequency, and run the VI. Record T_r, ω_{b1} and ω_{b2}, and the transfer function.

Analysis

1. Use appropriate theory to verify the recorded values in steps 1–4 of the Procedure.

2. Given the time domain step response of the second-order system shown in Fig. 16-10, where the time scale is in seconds, determine the Bode plots for this system.

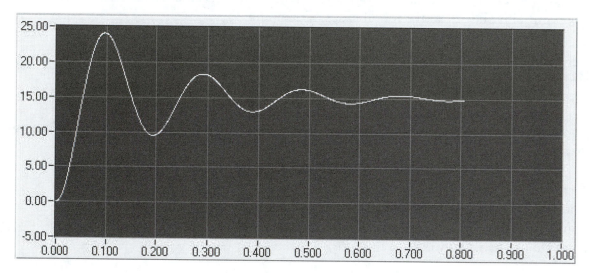

Fig. 16-10 Time Domain Response for Problem 2

3. Given the Bode plots for a second-order system, determine the corresponding time domain response. Show all important values.

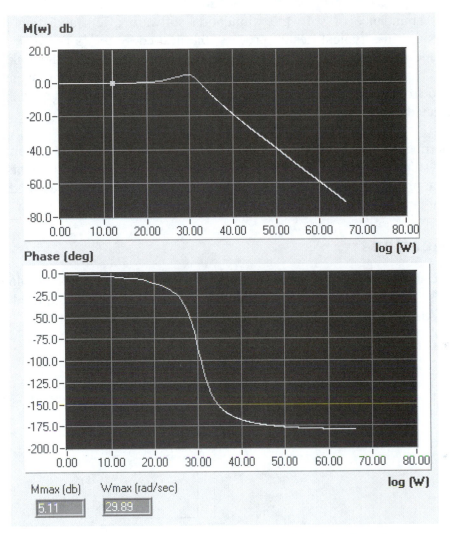

Fig. 16-11 Frequency Domain Response for Problem 3

Experiment 16-2: Design Project

In today's abundance of software, simulation software offers a distinct advantage in system design. In a typical design process, the desired system's response is attained in software and the system's parameter values are extracted. After the link between the parameters and system components is established, system component values are determined from the parameter values obtained in simulation. At this time the system can be constructed and tested.

This experiment focuses on the simulation as an important tool in system design. The selected system is a second-order filter. The simulation data, measurements, and theory-based data are compared and significant differences resolved. Featured also in this experiment is the computerized data acquisition using LabVIEW.

Specifications

Fig. 16-12 shows the design specification. The circuit is a second-order low-pass filter. The values of R_1, R_2, R_3, and C are to be determined from the required time domain response. Also, the associated Bode plots (magnitude) must be determined.

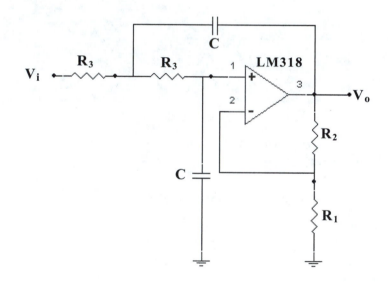

Fig. 16-12 System Specification: The Circuit and the Time Domain Response

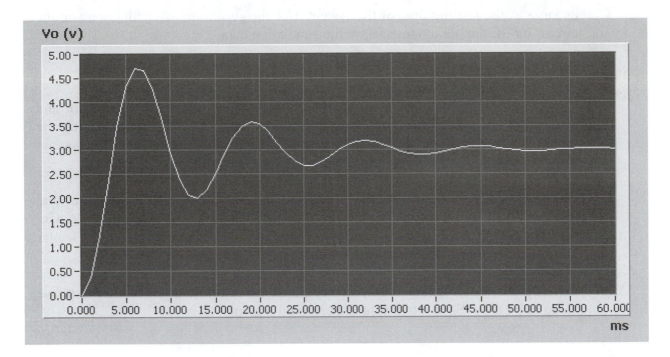

Fig. 16-12 System Specification: The Circuit and the Time Domain Response (continued)

Design Procedure

1. Determine the transfer function for the circuit in Fig. 16-12.

2. Determine the design equations by comparing the universal transfer function to the system transfer obtained in step 1. The design equations relate the universal second-order system parameters ζ and ω_n to the system parameters R, L, and C.

3. Determine the values of ζ and ω_n from the required time domain response given in Fig. 16-12.

4. Use the values obtained in step 3 in Second-order System.vi. Compare the response obtained with the response in Fig. 16-2. This is just a check to ensure that the values obtained in step 3 are correct.

5. Determine the values of R, L, and C from the design equations. As you will see, there are more parameters than equations. This means that you have to pick one value arbitrarily and then calculate the others.

6. Simulate the system's time domain performance on Electronics Workbench, Multisim, or other available software, using the R, L, and C values in the circuit. The input is a step whose value is inferred from the response in Fig. 16-12. A square wave may be used in place of a step; however, its half period must be at least

as large as the settling time of the system. Use the system transfer function and the circuit for the simulation tests. Save the simulation data.

7.   Construct the circuit shown in Fig. 16-12.

8.   Test the system performance in the time domain. Use LabVIEW to acquire and display on one waveform graph two channels of data, channel 0 representing the input (square wave) and channel 1 representing the response. Save the time domain response data to a spreadsheet or on the graph for future reference.

9.   Obtain system performance in the frequency domain using the GPIB interface and display the magnitude plot on a waveform graph. Save the frequency domain response data to a spreadsheet or on the graph for future reference. Refer to Chapter 18 for details on the Bode plot data acquisition.

10. Develop the system's (Fig. 16-2) step response based on theory. Plot this response using LabVIEW. Save it to a spreadsheet or on a graph.

Formal Report

Prepare this report in accordance with the formal report guidelines. Note that the ultimate goal of a design project or any meaningful lab experiment for that matter, is to achieve an agreement between *simulation, theory,* and the *real data* within an agreed-upon or otherwise specified margin of error. The error may be due to component tolerances, signal inaccuracy, or fluctuations. The computerized testing virtually excludes errors caused by humans. Differences between simulation data, theory, and real data above the specified amount must be reconciled by retesting or by theory-based explanation.

Experiment 16-3: DC Motor Characteristics

This experiment has a dual purpose. The first is to familiarize the user with an analog servo trainer. The second objective is to measure the DC motor parameters. This includes the motor's speed constant k_m, and the mechanical time constant τ_m. LabVIEW is used to collect and process the acquired data.

Equipment

Analogue Unit 33-110 (Feedback Corp.)
Mechanical Unit 33-100 (Feedback Corp.)
Power supply (Feedback Corp.)
LabVIEW (version 5.1 or higher)
DAQ board
Extender board

Introduction

Fig. 16-13 shows the hardware and the software configuration for measuring the DC motor parameters.

The Mechanical Unit includes the DC motor and the generator, their shafts mechanically coupled. The output voltage of the generator (also called a tachogenerator) is proportional to motor speed. It can therefore be used to monitor the motor's speed. The motor speed is reduced 32 : 1. Thus, the digital meter on the Mechanical Unit indicates the reduced speed.

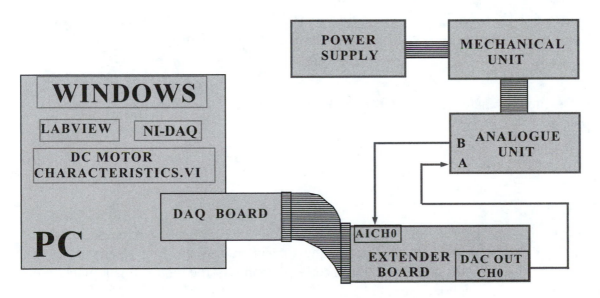

Fig. 16-13 DC Motor Characteristics Test Setup

The Analogue Unit is an I/O board for the Mechanical Unit. In addition, the board includes summing amplifiers, potentiometers, and other components that are configured with jumpers to form a closed-loop speed control or position control servo. PID and other controllers such as PI or PD can also be configured on the board.

The DAQ board that is almost completely inside the computer is the required interface for acquiring data. All necessary connections are made on the extender board that gives access to various I/O pins on the DAQ board.

The software in Fig. 16-13 includes the LabVIEW and the DC Motor Characteristics.vi program for acquiring data. NI-DAQ is the driver for the DAQ board.

In this experiment the motor is operated open loop, as shown in Fig. 16-14. The power amplifier (PA) with unity voltage gain supplies the required current to the motor.

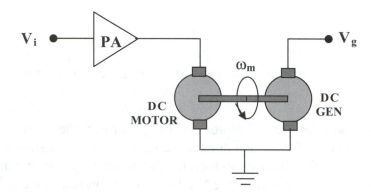

Fig. 16-14 Open Loop Operation of the DC Motor. The Shafts of the Motor and the Generator are Mechanically Coupled

The DC motor transfer function $\Omega(s)/V_i(s)$ is that of a first-order system as follows:

$$\frac{\Omega(s)}{V_i(s)} = \frac{k_m}{\tau_m s + 1} \qquad (16\text{-}20)$$

where k_m is the motor's speed constant that is usually expressed in krpm/v, although when its value is used in an equation, it must be in (rad/sec)/v. The mechanical time constant τ_m of the motor determines the speed of response of the motor. It is a function of inertia and viscous friction.

It can be shown that the motor speed constant depends on the load viscous friction and on other parameters as follows:

$$k_m = \frac{k_t}{B_T + k_t k_b} \tag{16-21}$$

where

$k_t = k_i/R_a$ (oz-in/v)

k_i (oz-in/amp) motor torque constant

R_a (Ω) armature winding resistance

$B_T = B_L + B_m$ (oz-in-sec/rad) total viscous friction

  B_L is the load viscous friction and B_m is the friction
  due to windage, bearing friction, etc.

k_b (v-sec/rad) back emf constant often expressed in v/krpm

If $B_T \ll k_t k_b$, then k_m reduces to approximately $1/k_b$.

The motor's mechanical time constant depends on the inertia and the viscous friction as follows:

$$\tau_m = \frac{J_T}{B_T + k_t k_b} \tag{16-22}$$

where J_T combined inertia of the load and the armature.

Because the motor is a first-order system, its step response is that of any first-order system as described by the following equation:

$$\omega(t) = E k_m (1 - e^{-t/\tau_m}) \tag{16-23}$$

and a typical time domain response of the motor is shown in Fig. 16-15, the input is a 5 V step, and the motor's speed constant is 345 rpm/v.

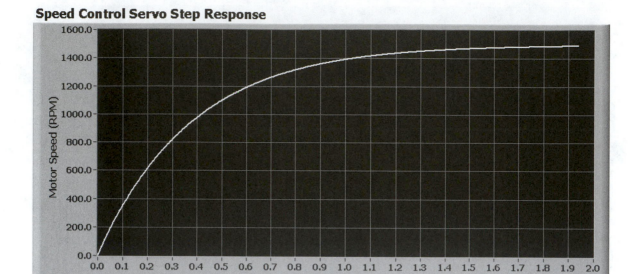

Speed Control Servo Step Response

Fig. 16-15 Step Response of the DC Motor

Software

DC Motor Characteristics.vi, which collects and processes the data, is available. It may be found in the Book VIs>Servo folder and its Front Panel is shown in Fig. 16-16. Open this VI and examine the Block Diagram and the Front Panel.

The Front Panel is subdivided into three sections: motor *Performance*, the *Settings,* and two X-Y graphs that display the motor's characteristics.

The *Performance* recessed box uses the gauge, a digital indicator, to display the motor's current speed. Included also is the *Time Remaining* indicator.

The *Settings* recessed box includes the parameter values that the user must input, and a toggle switch for selecting the test to be performed.

The two X-Y graphs display the acquired data. The upper graph displays the generator voltage versus the input voltage. The slope of this straight line, equal to $k_m k_g$, is the transfer function of the motor-generator combination. The lower graph displays the motor speed constant as a function of the input voltage. As expected, this is a horizontal line.

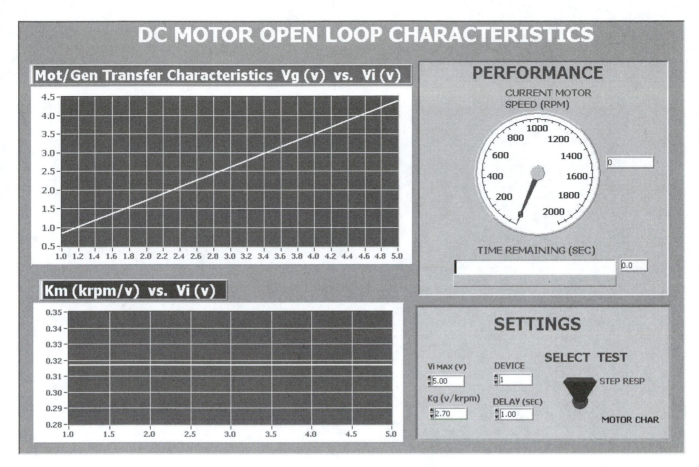

Fig. 16-16 The Front Panel of DC Motor Characteristics.vi

Procedure

1. Examine and become familiar with various inputs, outputs, and controls on the Analogue Board of the servo trainer.

2. In this step the values of the tachogenerator constant k_g and the motor speed constant k_m are measured. We will use k_g in the LabVIEW program shortly, and k_m is used as a reference constant later. Apply 4 Vdc from the external DC power supply to input A in Fig. 16-17. Read the speed on the Mechanical Unit and the generator voltage at point B in Fig. 16-17. Calculate k_m in krpm/v and k_g in v/krpm. The value of k_m is equal to the speed read on the Mechanical Unit divided by 4, and k_g is equal to the generator voltage divided by the motor speed in krpm. Save these values.

3. Wire the open-loop motor circuit as shown in Fig. 16-17 and in Fig. 16-13. Note the reference points A and B.

Control System Experiments   425

4. Open DC Motor Characteristics.vi from the Servo folder of Book VIs. Examine the Front Panel. Once the selected test is executed, many of the overlapping objects that are not required for the test will disappear.

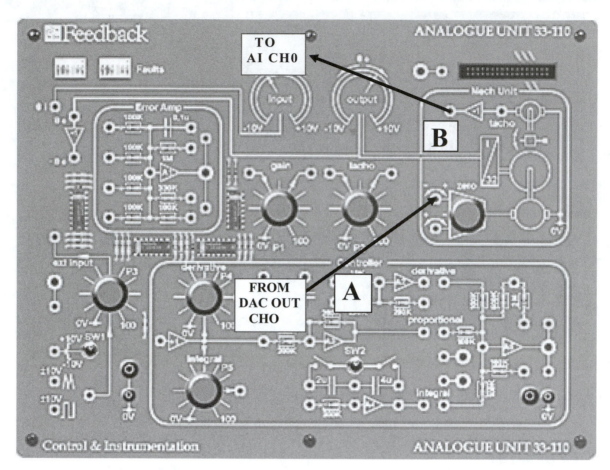

Fig. 16-17 The Wiring Diagram for DC Motor Characteristics.vi

Step Response Test

5. Because some Front Panel controls and indicators are hidden, enter a wrong device number, such as 100, and run the VI. A pop-up window informs you of device error. Click on Stop. At this time proper controls and indicators required by this test appear in the Performance recessed box. Enter the correct device number, the k_g value. Also enter Num Samples and Scan Rate (recommended 15,000 and 5000, respectively)

Run the VI. Upon completion of the test, motor's step response (ω_m (rpm) vs. time (sec)) is displayed on the upper X-Y graph. Also, the value of the mechanical time constant τ_m is displayed in the *Performance* recessed box.

DC Motor Characteristics Test

6. Because some Front Panel controls and indicators are hidden, enter a wrong device number, such as 100, and run the VI. A pop-up window informs you of device error. Click on Stop. At this time proper controls and indicators required by this test appear in the Performance recessed box. Enter the values of Vimax, device number, the k_g value, and the time delay.

The value of the Delay depends on the mechanical time constant of the motor. It should be equal to about 5 motor time constants ($5\tau_m$). The time constant value was measured in the Step Response Test.

The value of Vi max is the largest permissible analog input voltage to the DAQ board. For example, LabPC$^+$ has a dynamic range of 10 V. If this voltage is configured as $\pm$ 5 V, then the value of Vi max to be entered by the user is 5.

The time required to complete this test depends on the value of the Delay. Select the *DC Motor Characteristics* test from the *Settings* recessed box and run the *Time Remaining* and the *Motor Speed* digital indicators in the *Performance* recessed box in the Front Panel show the progress of the test. After the test is completed, the transfer characteristics of the motor-generator combination and the value of k_m versus Vi are displayed on the X-Y graphs. Save this data to a spreadsheet or inside the X-Y graph by clicking on it and selecting Data Operations>Make Current Value Default. To save data to spreadsheet, a slight modification is necessary in the Block Diagram using *Write to Spreadsheet File.vi* from the File I/O palette in the Functions floating palette. This completes the Motor Characteristics test.

Analysis

1. How does the value of k_m calculated in step 2 of the Procedure compare with the computer-generated X-Y graph in step 6? Should the curve be a horizontal line? Explain.

2. Use the value of the mechanical time constant from step 5, the value of the motor speed constant obtained in step 6, and the 4 V step input in the motor step response equation. Use theory to calculate the motor speed in rpm at 200 ms, 400 ms, and 600 ms. Compare the calculated values with data in step 6. Use cursor to read values from the X-Y graph.

3. Calculate the $k_m k_g$ product using the values from step 2 of the Procedure. Determine this value from the transfer characteristics test in step 6 of the Procedure. How do the two values compare? Justify significant differences (>15%).

Experiment 16-4: Speed Control Servo

In this experiment the performance of a closed loop speed control system in the time domain is investigated. System transient and steady state performance is examined. Parameters such as the system's speed constant and its time constant are measured and compared to those of the open loop. The steady state error, a system performance characteristic that is normally not associated with the open loop system, is also measured. A data acquisition interface and appropriate LabVIEW software are used to collect and process data.

Equipment

Analogue Unit 33-110 (Feedback Corp)
Mechanical Unit 33-100 (Feedback Corp)
Power supply (Feedback Corp)
LabVIEW (version 5.1 or higher)
DAQ board
Extender board

Introduction

Fig. 16-18 shows the block diagram of the speed control servo. The negative feedback is provided by the tachogenerator whose shaft is mechanically coupled to the shaft of the motor. The potentiometer P1 provides scaling of the generator signal.

$$\frac{\Omega(s)}{V_i(s)} = \frac{k_s}{\tau_s s + 1} \tag{16-24}$$

The transfer function for the closed loop speed control servo and that of the motor alone are identical in form. The subscript s in the transfer function above describes the closed loop system parameters. A through E are references for the corresponding points in the wiring diagram and the Block Diagram. As can be seen from the wiring in Fig. 16-22, the summing amplifier with gain of 10 in series with the potentiometer P2 provides an adjustable gain k_a. The potentiometer P1 scales the generator voltage.

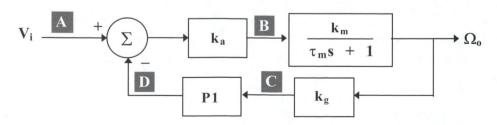

Fig. 16-18 Block Diagram for the Speed Control Servo

The speed constant k_s for the closed loop speed control servo is described as follows:

$$k_s = \frac{k_a k_m}{1 + k_a k_m k_g} \tag{16-25}$$

where k_g is the generator constant (v-s/rad or v/krpm) and k_m, the motor's speed constant, was discussed in the preceding experiment. If the loop gain $k_a k_m k_g$ is much greater than unity, then k_s reduces to $1/k_g$.

The time constant τ_s for the closed loop speed control system is described by

$$\tau_s = \frac{\tau_m}{1 + k_m k_g k_a} \tag{16-26}$$

Because the loop gain $k_a k_m k_g$ for the closed loop system is generally greater than unity, the closed loop system's time constant is smaller than the motor's time constant. This means that the step response of a motor is faster when it is operated closed loop than that of the open loop. The shape of the response is the same as that of any first-order system.

The steady state error is the difference between input and output as the time approaches infinity. In the s-domain, however, this difference is expressed as follows

$$E(s) = R(s) - C(s) = R(s) / [1 + G(s)] \tag{16-27}$$

Eq. 16-27 is based on a unity feedback system. The speed control system in this experiment does not have a unity feedback; however, it can be converted to a unity feedback through block reduction techniques. The error depends on the type of input $R(s)$ (step, ramp, or parabolic) and system *Type,* with system type being defined as the number of integrations (1/s terms) in the forward path. The speed control system is a Type 0 system. It has a finite steady state error for a step input and infinite error for ramp or parabolic inputs.

Note: The potentiometer P1 scales the generator voltage. Any calculations that use k_g must also include the P1 setting. For example, if P1 is set to 0.5, then the calculations msut 0.5 k_g.

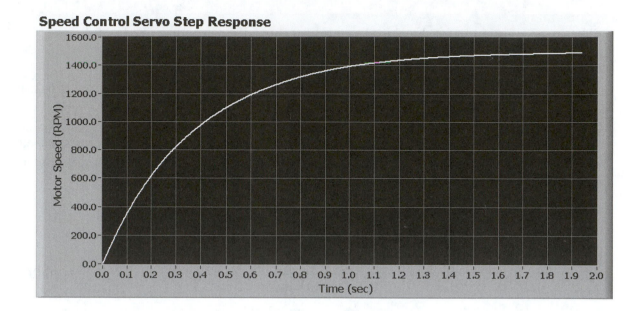

Fig. 16-19 Step Response of a Speed Control System

Steady state error is measured in the time domain. The above equation for E(s) can be translated to the time domain through the use of the Final Value Theorem as follows

$$e_{ss} = \lim_{t \to \infty} e(t) = \lim_{s \to 0} E(s) \tag{16-28}$$

The modified block diagram in Fig. 16-20 is equivalent to that shown in Fig. 16-18. In this diagram $G(s) = (k_a k_g k_m)/(\tau_m s + 1)$.

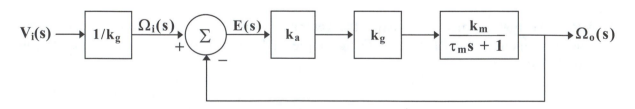

Fig. 16-20 The Modified Speed Control Servo Block Diagram

Combining Eq. 16-27 and 16-28 we obtain

$$e_{ss} = 1/[1 + K_p] \qquad (16\text{-}29)$$

where
$$K_p = \lim_{s \to 0} [\, G(s)]$$

The final form of the steady state error from Eq. 16-29 used to evaluate the system steady state error is

$$e_{ss} = \Omega_i / [1 + k_a k_m k_g] \qquad (16\text{-}30)$$

where $\Omega_i = V_i / k_g$.

Fig. 16-19 illustrates step response of a closed loop speed control system. The step input is 1500 rpm and the time constant is 0.4 seconds. At four time constants the response as shown by the cursor is 1472.5 rpm.

Software

The Front Panel of the Speed Control Servo is shown in Fig. 16-21. The recessed boxes group various inputs and response data. The *Settings* recessed box includes digital controls that pass numerical values to the Block Diagram. It also includes the *Choose Test* vertical slide. The *Operate Value* tool is used to select the desired test. There are four tests to choose from. The Step Response test produces the time domain speed response of the motor to the step input generated by the LabVIEW software. The Step Response Data recessed box shows data acquired by the software and the response waveform as shown in the illustration.

The Motor Speed Test recessed box includes the Input Speed Vertical Slide and the gauge that displays the actual speed of the motor. Because the gauge exhibits a small amount of jitter, the user can freeze the display by clicking on No in the Measure box. The steady state error indicator displays the difference between the input speed and the actual speed.

The remaining two tests evaluate the steady state error and the speed of response as loop gain is varied. The loop gain variation is accomplished by the gain change in the forward path. The user is prompted to change the gain to the value displayed in the Front Panel by a flashing indicator k_a. This procedure is repeated ten times as the gain is varied from 1 to 10 in unit increments. The appropriate graphs and digital controls will appear for a selected test and vanish in a different test because they are not required there.

Procedure

1. Wire the closed loop servo as shown in Fig. 16-18 and 16-22.

2. Open Speed Control Servo.vi from the Servo folder of Book VIs. Examine the Front Panel and become familiar with all controls and indicators.

3. Because some Front Panel controls and indicators are hidden, enter a wrong device number, such as 100, and run the VI. A pop-up window informs you of device error. Click on Stop. At this time proper controls and indicators required by this test appear in the Performance recessed box.

4. **Step Response Test**

   Enter the correct device number, the k_g value, the Num Samples and Scan Rate (15,000 and 5000 recommended, respectively; for best results, higher values may be used), and the value of $V_i = 2v$ in the *Settings* recessed box. V_i is the input voltage that represents the desired speed of the motor. Set the gain k_a to 5 by adjusting P2 on the Analogue Unit. The value of k_g has been determined in the previous experiment. Set P1 to 0.4; the value of P1 may be readjusted for best results.

   Set the *Choose Test* switch to *Step Response* and run the VI. Save data generated by the VI.

5. **Motor Speed Test**

   Repeat step 3. Set the gain k_a to 5 by adjusting P2 on the analogue unit. Enter the value of k_g in the Settings box. Set the *Choose Test* switch to *Motor Speed* and run the VI. Vary the *Input Speed* Vertical Slide control in the Motor Speed recessed box and observe the *Motor Speed* on the gauge indicator and the steady state error. Record the steady state error at the motor speeds of 500, 750, 1000, and 1500 rpm. Also record the corresponding input speeds.

6. **Steady State Error Test**

   Repeat step 3. In this test, the VI will plot the steady state error as a function of loop gain. Enter the Num. Sampl., the Sampl. Rate (use values from step 4), $V_i = 2V$, and the values of k_m and k_g. The delay allows the motor to reach the steady state speed before the next input is applied. Use 3 seconds for the delay value. Use the value of P1 from step 4.

   Run the VI and save all VI generated data.

7. System Time Constant Test

Repeat step 3. Enter parameter values in the *Settings* recessed box. They are same as in the previous test. Choose the *Time Const.* test and run the VI. The prompt will appear instructing you to set the gain on the analogue unit to the value indicated by the blinking digital indicator k_a. Adjust the gain and click on OK inside the prompt box. This procedure is repeated 10 times. Upon completion of the test, the System Time Constant vs. Loop Gain curve will be displayed on the X-Y graph. Save this data.

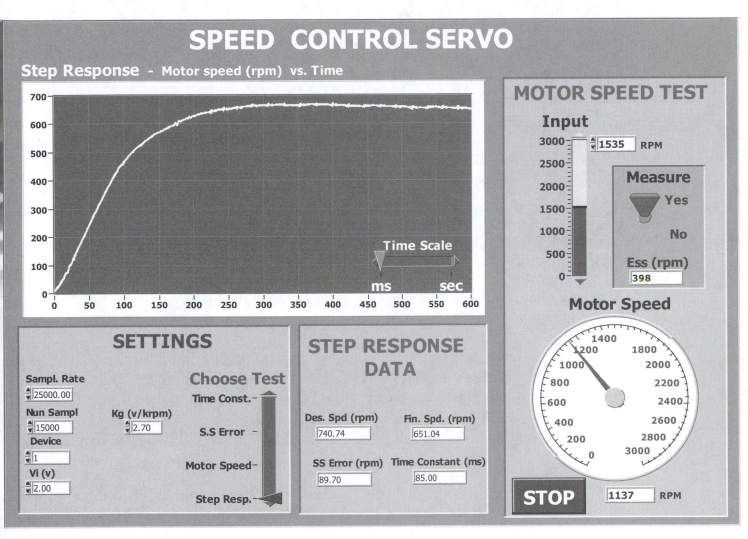

Fig. 16-21 Front Panel of Speed Control Servo.vi

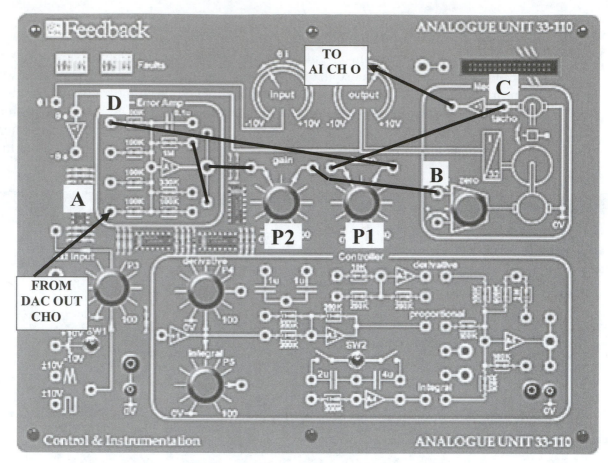

Fig. 16-22 Wiring Diagram for the Speed Control Servo

Analysis

1. The step response test measures the closed loop time constant of the speed control servo. Verify this value using theory.

2. Verify the steady state error in steps 5 and 6 of the Procedure using appropriate theory.

3. In step 6 of the Procedure, LabVIEW displays the Steady State Error vs. Loop Gain curve on the X-Y graph. This curve is based on the acquired motor speed data. Modify the VI to display on the same graph the Steady State Error curve based on theory. Run the VI and save data.

4. In step 7 of the Procedure, LabVIEW displays the System Time Constant vs. Loop Gain curve on the X-Y graph. This curve is based on the acquired motor speed data. Modify the VI to display on the same graph the System Time Constant vs. Loop Gain curve based on theory. Run the VI and save data.

5. Provide justification for significant deviation (>15%) between the theory and the measurements in steps 1 and 2 of the Analysis.

Experiment 16-5: Position Control Servo

This experiment investigates the performance of a closed-loop position control system. The system's step response is examined and its performance evaluated. LabVIEW is used to collect and process measured data. There are three parts to this experiment: simulation, theory, and measurements. The data from the three parts must be within an acceptable margin of error. The LabVIEW software collects and processes the acquired data. The processed data provides the user with experimental information to be used for system performance interpretation and analysis.

Equipment

Analogue Unit 33-110
Mechanical Unit 33-100
Power supply
LabVIEW (version 5.1 or higher)
DAQ board
Extender board

Introduction

The block diagram of the position control system is shown in Fig. 16-23. It may appear similar to the speed control system in Fig. 6-18; however, upon a closer examination it is clear that the two are very different. The controlled variable is Θ_o and there is an s term in the denominator of the motor's transfer function, indicating that an integration takes place (1/s). Since the output of the speed control system is $\omega = d\theta/dt$ and the LaPlace transform yields $\Omega(s) = s\Theta_o(s)$, then substitution for $\Omega(s)$ in Eq. 16-20 yields the motor transfer function as expressed by the following equation and also shown in Fig. 16-23.

$$\frac{\Theta_o(s)}{V_i(s)} = \frac{k_m}{s(\tau_m s + 1)} \qquad (16-31)$$

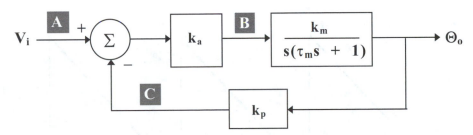

Fig. 16-23 Block Diagram for the Position Control Servo

In the speed control system, velocity feedback is provided by a generator. In the position control system, however, the angular position of the motor's shaft must be sensed. The transducer in this case is a potentiometer whose shaft is mechanically coupled to the shaft of the motor.

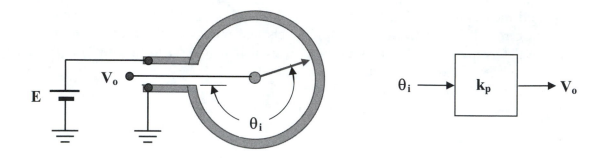

Fig. 16-24 The Potentiometer and Its Block Diagram Representation

The potentiometer illustration in Fig. 16-24 also shows its block diagram representation. Its transfer function $k = E/(2\pi)$ volts/radian (or volts/degree if 360 is used instead of 2π). Actually as the wiper arm rotates together with the shaft of the motor, it must jump over the dead zone, which is usually a few degrees, so the 2π approximation used above is acceptable. The servo potentiometer is not cheap because it is designed to withstand the wear and tear of repetitive operation; its shaft is mounted on ball bearings in order to minimize friction that causes wear.

Another approach to position sensing is to use an absolute Gray code encoder. This is a digital approach to position encoding. A four-bit encoder has a resolution of $360/2^4 = 22.5°$, which may not be acceptable in some applications. An encoder wheel that has 8 encoding tracks produces $2^8 = 256$ unique codes with associated resolution of $360/256 = 1.4°$ that is much better. The potentiometer used in this experiment produces a continuous output as a function of the input angle. If the potentiometer was repeatedly rotated, its output viewed on the oscilloscope is a ramp as shown below.

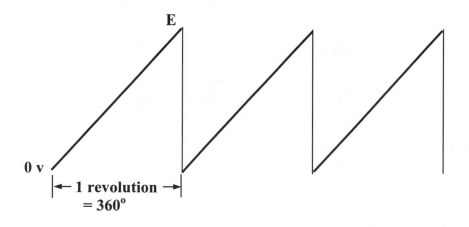

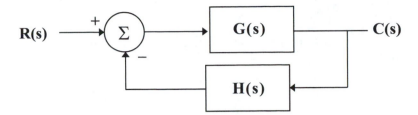

Fig. 16-25 General Closed Loop Negative Feedback System

A general closed loop system using negative feedback, shown in Fig. 16-22, has a the transfer function

$$\frac{C(s)}{R(s)} = \frac{G(s)}{1 + GH(s)} \tag{16-32}$$

In comparing the position control system in Fig. 16-23 with the general system in Fig. 16-25, it is clear that

$$G(s) = \frac{k_a k_m}{s(\tau_m s + 1)} \qquad H(s) = k_p$$

The following transfer function is obtained after the above expressions for G(s) and H(s) are substituted in Eq. 16-32:

$$\frac{\Theta_o(s)}{\Theta_i(s)} = \frac{k_a k_m k_p / \tau_m}{s^2 + (1/\tau_m)s + k_a k_m k_p / \tau_m} \tag{16-33}$$

In Eq. 16-33, $\Theta_i(s) = V_i/k_p$, which is the equivalent input angle that represents the desired output angle $\Theta_o(s)$. In comparing the transfer function for the position control system in Eq. 16–33 with that of the universal second-order system expressed by Eq. 16-3, the relationship or the link is established between the universal system parameters ζ and ω_n and those of the position control system as follows:

$$\omega_n = \sqrt{k_a k_m k_p / \tau_m} \tag{16-34}$$

$$\zeta = \frac{1}{2\sqrt{k_a k_m k_p \tau_m}} \tag{16-35}$$

The position control system is underdamped as it often is. Consequently, Eq. 16-6 through 16-9 may be used in evaluating the step response of the position control system. The shape of the step response is identical to that shown in Fig. 16-3.

Software

Open Position Control Servo.vi from the Servo folder in the Book VIs folder. Examine the Front Panel shown in Fig. 16-26. The *Settings* recessed box includes digital controls that pass values set by the user to the Block Diagram. The VI uses these values to compute various system parameters shown in the *Step Response Data* recessed box, including the settling time (T_s), percent overshoot (POT), damping coefficient (ζ), system time constant (Taus), and the damped (ω_d) and natural (ω_n) resonant frequencies. The graph displays the input angle step and the output angle response.

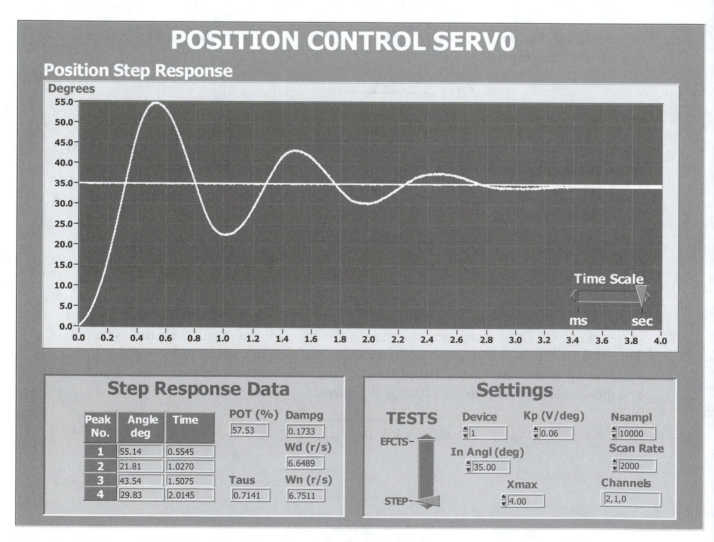

Fig. 16-26 Front Panel of Position Servo.vi, the Step Test

Also computed and displayed in an array fashion are the first four response peaks. The Waveform Graph displays the input angle step and the output angle response.

There are two tests that are selected by the *Tests* vertical slide control. The *Step* test applies a step to the position control servo, and displays its response and the input angle selected by the user on the same waveform. LabVIEW also computes other response characteristics and displays their values on digital indicators.

The second test called *Efcts* investigates the effects of viscous friction on the system's damping ratio (ζ) and the overshoot. It is a known fact that viscous friction increases with speed. Should system damping remain constant as the input angle is increased? Interesting results emerge from this test, and the objective is to reach satisfactory conclusions that justify the acquired data.

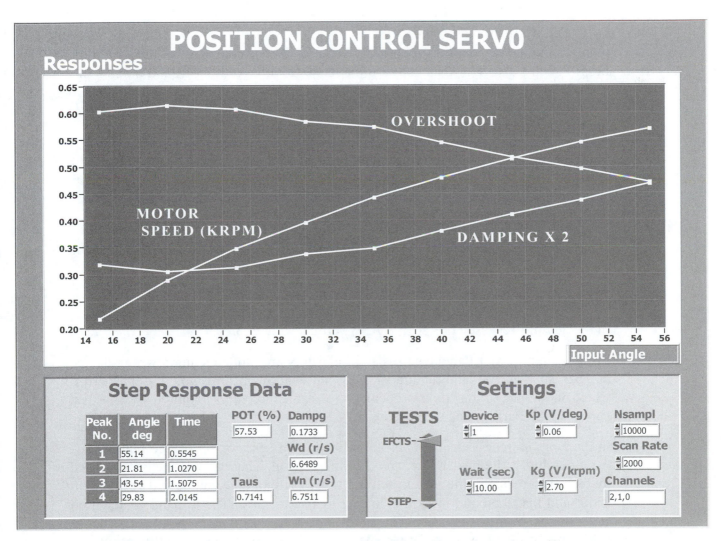

Fig. 16-26 Front Panel of Position Servo.vi, the Efcts Test (continued)

Procedure

1. Wire the position control servo as shown in Figures 16-23 and 16-27. The low pass filter shown below is used to filter potentiometer wiper arm noise. This noise, if left unfiltered, may interfere with the measurement of settling time where the position response voltage level and the signal-to-noise ratio are both low. With the exception of the filter, Fig. 16-27 closely resembles Fig. 16-23.

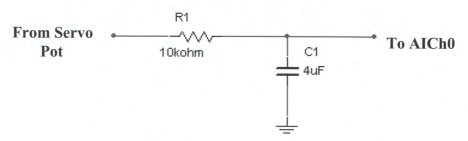

DAC Out Ch0, the D/A converter on the DAQ board, provides the input step representing the desired output position. The user chooses the value of the step in the Front Panel. LabVIEW then converts the angle to voltage using k_p and outputs this voltage step at DAC Out Ch0. As shown in Fig. 16-27, position response is wired to AICh0, the input angle to AICh1, and the tachogenerator is wired to AICH2. (*Note:* Step 1 may be omitted if noise is not a problem.)

2. Measure the value of the potentiometer constant k_p. This is done by measuring the voltage change at the θ_o output on the Analogue Unit after rotating the potentiometer calibrated dial on the Mechanical unit by 90°. The measured voltage is then divided by 90 for k_p in v/deg or by $\pi/2$ for v/rad respectively.

3. Open Position Control Servo.vi from the Servo folder of Book VIs. Examine the Front Panel and become familiar with all controls and indicators.

Step Response Test

4. In the *Settings* recessed box set the *Tests* vertical slide to *Step*.

5. Because some Front Panel controls and indicators are hidden, enter a wrong device number, such as 100, and run the VI. A popup window informs you of device error. Click on Stop. At this time proper controls and indicators required by this test appear in the Performance recessed box.

6. In the *Settings* recessed box input the following values: correct device number, k_p, the input angle, the band value, and the Xmax value. The value of the band is used by the VI to measure the system settling time. For example, suppose that the input is 40° and that the band is set by the user to $10\% = \pm 5\%$. The 4° band centered on 40° extends from 38° to 42°. The point where the oscillatory position response enters this band and is henceforth contained by the band, marks the settling time.

Note that the band control and the settling time indicator is not included in the Front Panel. The software to measure settling time is to be designed by the user. See step 1 of the Analysis.

The X_{max} input determines the maximum time along the X-axis. In order to use the X_{max} control, the AutoScaleX function of the graph must be disabled. Also, input the number of samples Nsampl, Scan Rate, and the channels to be acquired. The channels acquired must be in descending order and the "0" channel must be last. In this test the position response is wired to AICh0 and the input angle to AICh1. Consequently, enter 1,0 in the Channels string control. The time required to acquire 10,000 scans at the scan rate of 2,000 scans/s is 5 seconds (10000/2000).

See step 1 of Analysis. Settling time software must be designed by you.

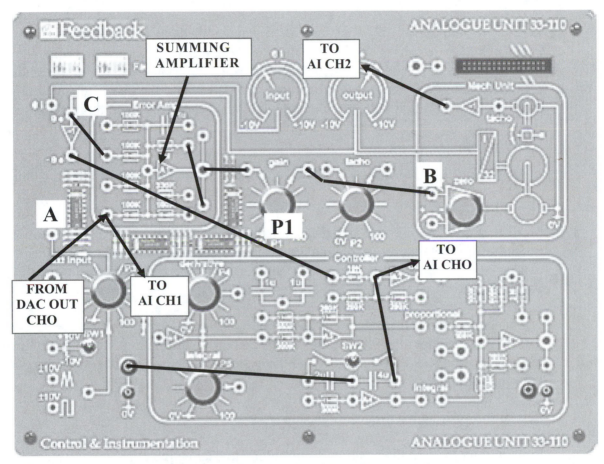

Fig. 16-27 Wiring Diagram for the Position Control Servo

7. Set the gain k_a to 5. This is done by setting the potentiometer P1 halfway. The summing amplifier gain is set to 10. The series potentiometer P1 is used to set the gain in the forward path from 0 to 10. Choose the time scale, ms or sec.

   Run the VI. Save all Front Panel data, including the response curve.

8. Change the gain k_a to 1 and run the VI again. Save data.

Effects Test

9. In the *Settings* recessed box set the *Tests* vertical slide control to *Efcts*.

10. Repeat step 5.

11. Enter the correct device number, the values of k_g and k_p, Wait time of 5 seconds or more, Nsampl, Scan Rate (15,000, 5000), and the Channels (2,1,0) in the Channels string control because in this test we are acquiring data from three channels. Also set k_a to 5. Run the VI. This test requires more than a minute to acquire and process the data.

LabVIEW outputs a 15° input angle step, acquires data, and processes data. It calculates the overshoot, and the value of damping (ζ). It also determines the maximum motor speed that occurs at the inflection point of the first position response peak. The three values are stored in an array and the test is repeated eight more times, with the input angle incremented by 5° each time. The three curves are then plotted on an X-Y graph as a function of the input angle. Save the response waveform data.

Analysis

1. Modify this VI to include the measurement of settling time. See step 6 of the Procedure for details. Insert a digital control called *Band* in the Front Panel for user to input the desired band value, and a digital indicator for the measured settling time. Run the VI and measure and save the settling time value. Save data.

2. Use theory to verify the numerical values that LabVIEW determined on the basis of data. This includes the four peaks, POT, damping, the natural and the damped resonant frequencies, settling time, and the system time constant. Justify differences in excess of 15%.

3. Repeat the verification procedure in step 2 for $k_a = 1$.

4. Explain the effect on the position response due to change in gain k_a.

5. Suppose that it was necessary to apply an input angle of 120°. What difficulty may be encountered that is related to the limitation of the DAQ board? Suggest a hardware/software solution to this problem.

6. As suggested earlier, the potentiometer noise may interfere with the settling time measurement. Explain and illustrate how this can occur.

Experiment 16-6: Position Control Servo With Velocity Feedback

The previous experiment examined the characteristics of a position control servo. In its basic form without compensation or any special type of feedback, the behavior of a position control servo is that of an underdamped second-order system. In this experiment the effect of velocity feedback on the system step response is examined.

LabVIEW is used to collect and process measured data. There are three parts to this experiment: simulation, theory, and measurements. The data from the three parts must be within an acceptable margin of error. The emphasis is on the use of the computer as a tool to collect and process the acquired data and to extract meaningful conclusions from the analysis.

Equipment
Analogue Unit 33-110
Mechanical Unit 33-100
Power supply
LabVIEW (version 5.1 or higher)
DAQ board
Extender board

Introduction
The block diagram of the position control system with velocity feedback as shown in Fig. 16-28 is similar to the position control system in Fig. 16-23 of Experiment 16-5. However, the system shown here also uses velocity feedback or derivative feedback, as it is sometimes called. The motor transfer in Eq. 16-31 of the previous experiment is same here.

When we multiply a quantity by s in the s-domain we differentiate in the time domain (TD). Hence, $s\theta_o k_g$ is the generator voltage. A fraction k_v of this voltage is applied to the summing junction at point D in Fig. 16-28. The gain of the summing junction and the potentiometer P1 determine the forward path gain k_a, and P2 provides the fraction k_v of the generator voltage.

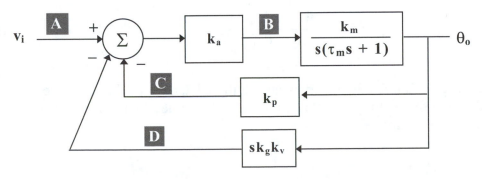

Fig. 16-28 Block Diagram of the Position Control Servo With Velocity Feedback

Fig. 16-29 illustrates the effect that the velocity signal has on the position response. The velocity signal is the derivative (slope) of the position curve multiplied by −1. As the motor shaft advances in a clockwise direction, for example, between t_1 and t_2, the velocity signal acts in the opposite direction (counterclockwise), reaching its maximum value at the point of inflection on the position curve. At t_3 the velocity is zero as the motor momentarily stops and begins to rotate counterclockwise. Here the velocity signal reverses its polarity, providing the drive in the opposite clockwise direction.

The opposing action of the velocity feedback reduces the overshoot, thus increasing the value of the damping ζ.

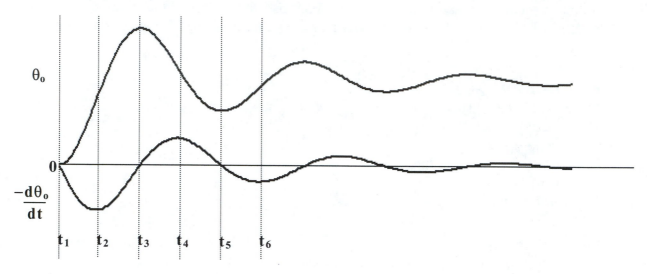

Fig. 16-29 Position and Velocity Feedback Signals

Analysis shows that after obtaining the transfer function of the position control system with velocity feedback and comparing it to the universal second-order transfer function expressed by Eq. 16-3, the expression for ω_n in Eq. 16-34 remains unchanged. However, the new value of damping coefficient that includes the effects of velocity feedback is expressed as follows by Eq. 16-36:

$$\zeta = \frac{1 + k_a k_m k_v k_g}{2\sqrt{k_a k_m k_p \tau_m}} \qquad (16\text{-}36)$$

The value of ζ in Eq. 16-36 includes an additional term $k_a k_m k_v k_g$ in the numerator, clearly indicating that velocity feedback increases the value of damping and consequently reduces percent overshoot.

Software

Open Position Control Servo With Velocity Feedback.vi from the Servo folder in the Book VIs folder. Examine the Front Panel, shown in Fig. 16-30. The *Settings* recessed box includes digital controls that pass values set by the user to the Block Diagram. The *Tests* vertical slide provides three experiments that the user can select.

The *Step Response* test was included in the previous experiment; however, in this experiment velocity feedback is included and its effects are examined in the time domain.

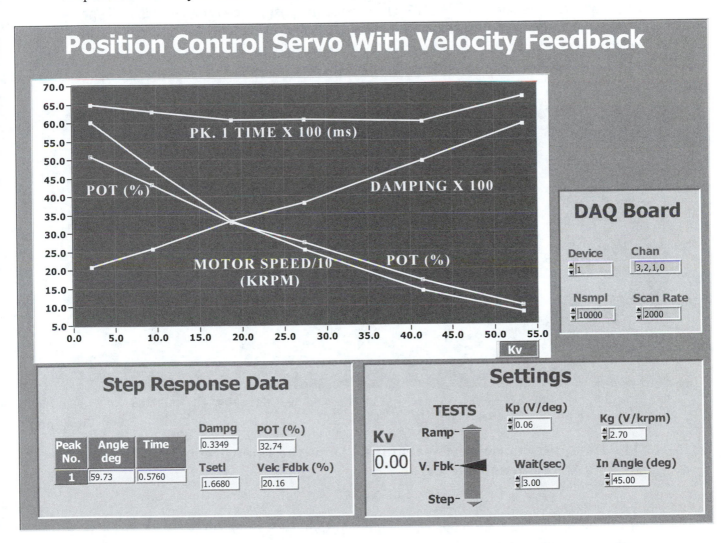

Fig. 16-30 Front Panel of the Position Control Servo With Velocity Feedback

As shown on the X-Y graph in Fig. 16-30, the Velocity Feedback test (V. Fbk) plots percent overshoot (POT), damping ζ, maximum motor speed, and the time of the first position response peak as a function of percent velocity feedback.

The ramp response test (Ramp) is the third test. In this test the ramp waveform generated by LabVIEW is applied to the summing junction of the position control servo with velocity feedback. LabVIEW collects the response data, computes the error, and plots the three waveforms as shown in Fig. 16-31. Because the position control system is a Type 1 system, there is no error when a step input is applied; however, there is an error when a ramp input is applied. Fig. 16-31 shows the three waveforms of a typical ramp response: the input ramp, the lagging motor response ramp, and the error.

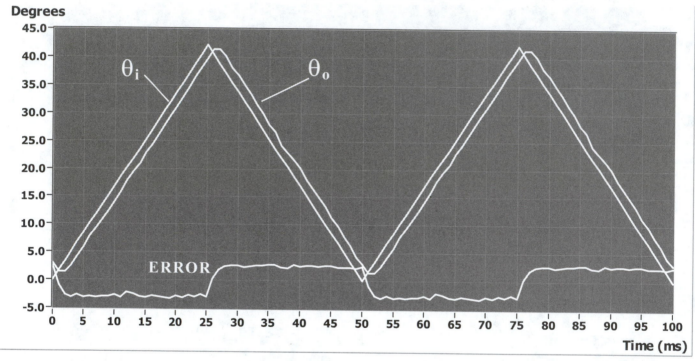

Fig. 16-31 The Ramp Response Test Waveforms Show the Input Ramp, the Lagging Motor Response Ramp, and the Error

The VI uses the user-provided values in the *Settings* recessed box to compute the appropriate response data.

The *DAQ Board* recessed box requires values from the user such as the DAQ board number (device), the analog input channels used, the number of samples to be acquired, and the sampling rate.

Procedure

1. Wire the position control servo as shown in Figures 16-28 and 16-32. The low-pass filter shown below is used to filter potentiometer wiper arm noise. This noise, if left unfiltered, may interfere with the measurement of settling time where the position response voltage level and the signal-to-noise ratio are both low.

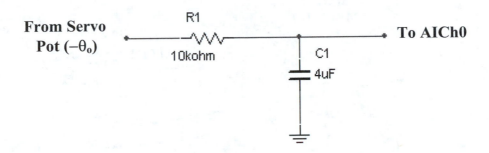

DAC Out Ch0, the D/A converter on the DAQ board, provides the input step representing the desired output position. The user chooses the value of the step in the Front Panel. LabVIEW then converts the angle to voltage and outputs this voltage step at DAC Out Ch0. As shown in Fig. 16-32, position response (output angle) is wired to AICh0, and the input angle to AICh1.

In this experiment the generator signal provides the velocity (or the derivative) feedback. As shown in the wiring diagram of Fig. 16-32, the potentiometer P2 provides a fraction of the generator voltage that is applied to the summing junction. LabVIEW calculates k_v, the percentage of velocity feedback, and displays this value in the Front Panel. *Note:* The filter used in step 1 may be omitted if noise is not a problem.

2. Open Position Control Servo With Velocity Feedback.vi from the Servo folder of Book VIs. Examine the Front Panel and become familiar with the functions of all controls and indicators.

Step Response Test
This test measures the response of the position control servo to a step input. The use of velocity feedback increases damping and reduces the overshoot.

3. Select *Step* from the vertical slide in the *Settings* recessed box.

4. Because some Front Panel controls and indicators are hidden, enter a wrong device number, such as 100, and run the VI. A pop-up window informs you of device error. Click on Stop. At this time, proper controls and indicators required by this test appear in the Performance recessed box.

5. In the *Settings* recessed box input the following values:, k_p, k_g, the Band value, and the input angle. And in the DAQ Board recessed box enter the correct device

number, number of samples (Nsmpl), Scan Rate, and the analog input channels (3,2,1,0). Use Nsmpl and Scan Rate values from the previous experiment. See step 1 of Analysis. Settling time software must be designed by you. You may copy the settling time software from the previous experiment.

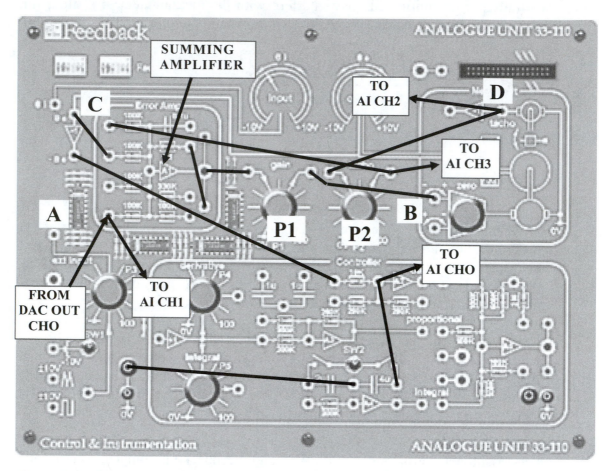

Fig. 16-32 Wiring Diagram for the Position Control Servo With Velocity Feedback

6. Set the gain k_a to 5 and 50% velocity feedback. Run the VI and save data. The value of k_a may be readjusted for best results.

7. Change velocity feedback to 10% and run the VI again. Save data.

8. Vary k_v until the position response is approximately critically damped. Record the k_v value.

Velocity Feedback Test

Because velocity feedback has a significant effect on the time domain step response, it is appropriate to examine this effect in greater detail. Four performance characteristics are plotted as a function of velocity feedback: damping, percent overshoot, time of the first position response peak, and motor speed. Motor speed is maximum because it is measured at the point of inflection of the first position response peak (at t_2 in Fig. 16-29).

9. Select the *V. Fbk* test on the vertical slide in the *Settings* recessed box.

10. Because some Front Panel controls and indicators are hidden, enter a wrong device number, such as 100, and run the VI. A popup window informs you of device error.

    Click on Stop.

11. In the *Settings* recessed box input the following values: k_p, k_g, input angle (45°), and Wait value (3 sec). The *Wait* digital control provides time delay allowing the system to reach the steady state at each test point.

    In the DAQ Board recessed box enter the correct device number, the number of samples (Nsmpl), Scan Rate, and the analog input channels (3,2,1,0). Use Nsmpl and Scan Rate values from the previous experiment, or choose values for best results. Adjust k_a to 5, or choose a value.

12. Run the VI. Set the gain k_v to the value shown by the blinking digital indicator in the Front Panel. You will be prompted at each gain adjustment. LabVIEW acquires, processes, and plots the data on the X-Y graph. Save this data.

Ramp Test

LabVIEW generates and applies a ramp waveform to the summing junction of the position control servo with velocity feedback. The position response is also a ramp, however it lags the input ramp. The difference between the two waves is the error.

13. Select the *Ramp* test from the vertical slide in the *Settings* recessed box.

14. Because some Front Panel controls and indicators are hidden, enter a wrong device number such as 100 and run the VI. A popup window informs you of device error.

    Click on Stop.

15. In the *Settings* recessed box input the following values: k_p, Amplitude (45°), and Frequency (0.5 Hz), and in the DAQ Board recessed box enter the correct device number. Set k_a to 10 on the Analogue board by adjusting the P1 potentiometer to 1.

16. ***Run*** the VI. Immediately the three curves are plotted point by point on waveform chart; the input ramp, the lagging response ramp, and the error (difference between the two curves). During program execution, adjust the velocity feedback constant k_v

by adjusting the potentiometer P2 on the Analogue board for a flat top steady state error waveform. This adjustment reduces POT, making the response curve parallel to the input, and consequently results in a flat top error waveform. The Stop button terminates VI execution.

LabVIEW does extra processing to save the data on a waveform graph located on the right side of the Front Panel. This is because the data cannot be saved on a waveform chart in the usual way (Data Operations>Make Current Value Default).

Save this data for future reference, as you will need it in the Analysis section, by clicking on the *Save* button in the Front Panel. Enter the number of cycles to be saved in the Ncycles digital control.

Analysis

1. Modify this VI to include the measurement of settling time. See step 6 of the Procedure in the previous experiment (Position Control Servo) for details. Use the digital control called *Band* in the Front Panel for user to input the desired band value. The settling time software may be copied from the previous experiment. Run the VI and measure and save the settling time value. Save data.

2. Does the data in step 6 show overshoot? If so, calculate the first peak value in degrees and time. Also calculate ζ, POT, and settling time. Compare these values to the data in step 6.

3. Repeat step 2 of the Analysis for the data in step 7 of the Procedure.

4. Calculate the value of ζ and compare it with the data in step 8 of the Procedure.

5. The data in step 12 of the Procedure includes four curves plotted as a function of percent damping k_v: ζ, POT, peak 1 time (T_1), and motor speed ω_m. Since motor speed $\omega_m = d\theta/dt$, derive the equation for ω_m.

6. Using appropriate theory, calculate ζ, POT, peak 1 time (T_1), and motor speed ω_m at the first, third, and sixth data points on the measured curves. Use the cursor to measure the corresponding values from the data curves. Compare data and theory.

7. Using appropriate theory calculate the steady state error at four selected time points and compare these values with data. Use the cursor to read the data.

Note: Data and theory must not differ by more than 20%. If such large differences occur, retake the data. If after retaking the data the results are still the same, provide theory-based justifications for such discrepancies.

Chapter 17
TCP/IP and the Internet

TCP/IP Protocol

Introduction

A protocol is a set of rules that govern an orderly flow of data between computers on a network. Transmission Control Protocol/ Internet Protocol (TCP/IP) is one such suite of rules that includes several communication protocols. It was developed by the Department of Defense in 1960s and soon became popular among most manufacturers. Because of its popularity and wide usage, it became a standard that is supported by most operating systems including Windows and Macintosh, as well as LabVIEW. TCP/IP can be used on a single network, such as Ethernet, or on the Internet, which links many networks.

IP is in the Network layer and TCP is in the Transport layer of the Open System Interconnect (OSI) model. Because IP is in a lower OSI layer, it has a limited task responsibility and will not guarantee the delivery of data to the destination station.

TCP, on the other hand, is in a higher OSI layer and takes greater care in transporting data by breaking the data down into packets and checking for errors on the transmitted packets. The packets containing errors are retransmitted in order to maintain the integrity of the data received by a remote station. When operating together with TCP, IP still checks for errors.

Client/Server Communication

In the client/server communication diagram shown in Fig. 17-1, a client requests information and the server responds by sending the requested data to the client. The server and the client may be on the same network or they may be on different networks separated by a large distance, communicating over the Internet.

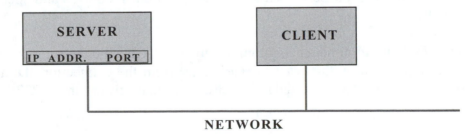

Fig. 17-1 Client/Server Diagram

A client cannot communicate with the server without the address and the port number of the server. Potentially, any computer on a network can be a client or a server, hence it must have a unique 32-bit IP address. For example, the IP address 120.1.20.32 written in decimal format has its binary equivalent of 01111000.00000001.00010100.00100000. Each computer on the network is identified by its unique IP address. The port number identifies a specific port at the IP address; the computer may have several ports.

When a client wishes to make a connection with the server, it must specify the server's IP address and the port number at that address.

TCP/IP Client Model in LabVIEW

In communicating with the server, the client must execute a sequence of steps as follows:
1. Establish connection with the server.
2. Send the server a request for data in the form of a command.
3. Get data from the server.
4. Break connection with the server and report errors, if any occurred.

Fig. 17-2 shows a typical Block Diagram of a TCP/IP client. The following TCP/IP functions play an essential role in communication between the client and the server.

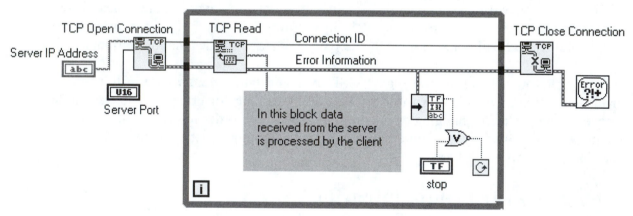

Fig. 17-2 General TCP/IP Client Block Diagram

 The **TCP Open Connection** function (Communication > TCP palette) makes a connection with the server station at the specified IP address and specified port number.

 The **TCP Read** function (Communication > TCP palette) returns number of bytes (specified by the *bytes to read* input) from the connection ID at the *data out* terminal. This is data in string format from the TCP/IP server.

The **TCP Close Connection** function (Communication > TCP palette) closes the connection specified by the connection ID (to server). In the Windows NT operating system all unread data is deleted.

The **connection ID** is a unique reference number that identifies the connection between the server and a printer, or a network connection number. The probe illustration below shows a typical connection number:

Errors

Error data is carried by the error information line. The error data type is a cluster of three data types, as shown in Fig. 17-3a. The Boolean indicator declares if an error has occurred, the long integer shows the code for the error, and the string indicator states where the error occurred.

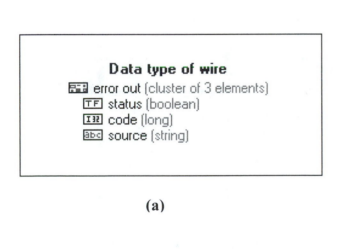

(a)

(b)

Fig. 17-3 (a) Error Data Type, (b) Specific Error Illustration

Fig. 17-3b illustrates an error whose code is 63 that occurred at the TCO Open Connection function in a VI whose name is TCP Client.vi. This type of error could have occurred if the client tried to make a connection with a server that was not running.

TCP/IP Server Model in LabVIEW

As it waits for a request from the client, the LabVIEW server executes the following commands:

1. Initialize the server.
2. Wait for a connection to the client.
3. After the connection is made wait for a command from the client.
4. Execute the received command.
5. Return the data to the client.
6. Close the connection.

Steps 2 through 6 are done inside the While Loop, which continues to listen for a request from the client in an asynchronous arrangement. Once the client's command has been executed, a FALSE is passed to the condition terminal of the While Loop, terminating loop execution and closing the connection to the client. As the last step in the communication sequence, the server is shut down.

Fig. 17-4 shows a typical Block Diagram of a TCP/IP server. The following TCP/IP functions play an essential role in communication between the client and the server.

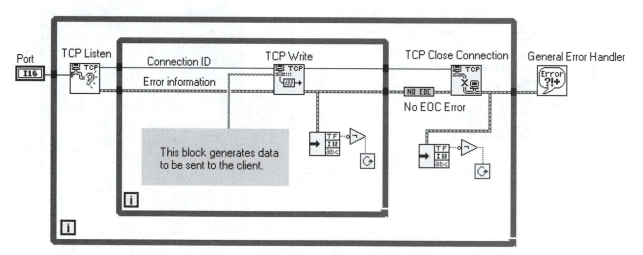

Fig. 17-4 General TCP/IP Server Block Diagram

 The **TCP Listen** function (Communication > TCP palette) waits for a connection at the specified port. When the connection is made by the client, the inner While Loop generates data and sends it to the client.

 The **TCP Write** function (Communication > TCP palette) writes a data string to the specified connection ID.

The **TCP Close Connection** function (Communication > TCP palette) closes the connection specified by the connection ID (to the client). In the Windows NT operating system all unread data is deleted.

In Fig. 17-4 the task of the outer While Loop is to detect a connection to the client. When a client makes a connection, the inner While Loop generates and sends data to the client at the Connection ID.

Error signals occur on the Error Information line. If the client terminates the connection, the error "TRUE" will terminate the execution of the inner While Loop and pass the error "TRUE" to the outer While Loop. When the client closes the connection, a TRUE is reported on the Error Information line from the TCP Listen function. When the error cluster is unbundled, the TRUE condition (indicating that an error occurred) is inverted, as shown in Fig. 17-4, and applied to the condition terminal, thus terminating the execution of the inner While Loop and passing the error condition to the outer While Loop. Because the outer While Loop is waiting for the connection from the client, it is essential that its continuous execution not be terminated. This is accomplished by the NO EOC function that resets the TRUE to a FALSE. Hence, the outer While Loop will continue running until an error other than the client terminating the connection occurs, at which time it will terminate execution and the General Error Handler will report errors.

Experiment 17-1: TCP/IP
Introduction
In this experiment, TCP/IP functions are used to build a TCP/P server and a TCP/IP client. The client will request data and the server will generate and send the requested data to the client. What is demonstrated here is the feasibility of exchanging data on a network such as Ethernet, using the TCP/IP protocol.

Procedure
1. Build the Client.vi Front Panel and Block Diagram as shown in Fig. 17-5 and save it as Client.vi.

2. Build the Server.vi Front Panel and Block Diagram as shown in Fig. 17-6 and save it as Server.vi.

3. Determine the IP address for your computer and enter it in the Front Panel string control of Server.vi. Choose the server's port number and enter it in the digital control of the Front Panels of Server.vi and Client.vi.

4. This experiment is done by a team of two workstations. The station designated as the server must open Server.vi and run it. The station designated as the client must open Client.vi and run it. Server.vi must run first. The server can now choose the waveform to be sent from the Function menu and also change the parameters of the waveform. The client should receive the updated data.

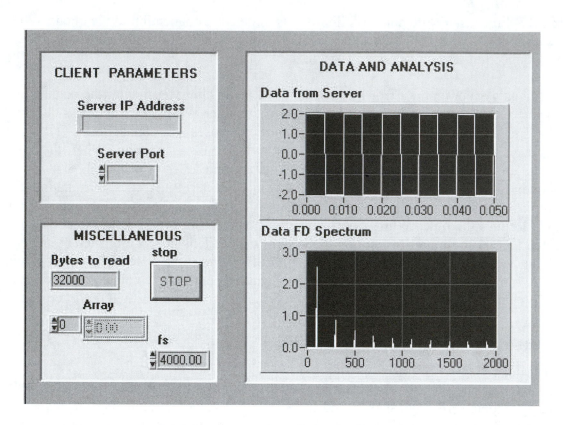

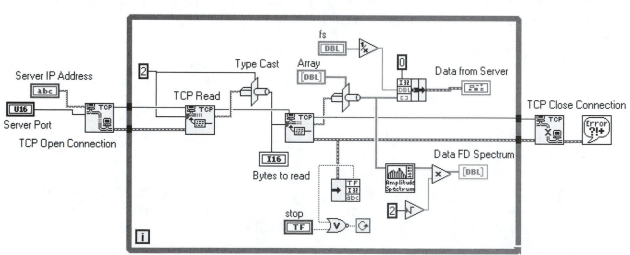

Fig. 17-5 TCP/IP Client Front Panel and Block Diagram

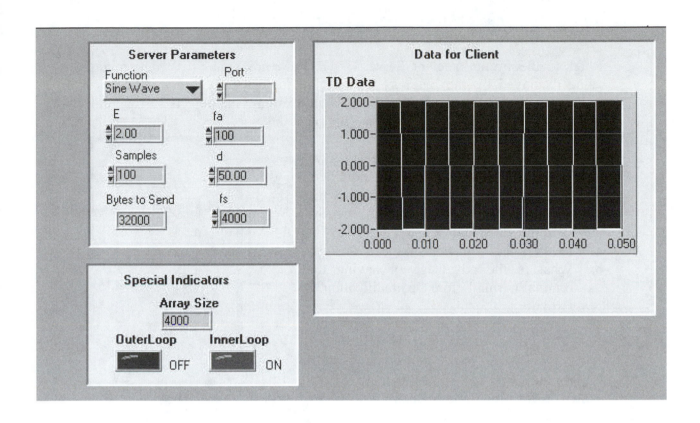

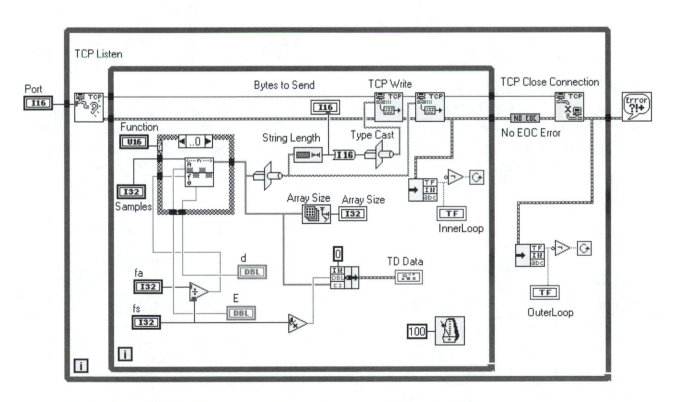

Fig. 17-6 TCP/IP Server Front Panel and Block Diagram

Analysis

1. Refer to the Front Panel of Server.vi. Explain how LabVIEW determines the value displayed by the *Bytes to Send* digital indicator. Devise a simple test in LabVIEW to check your answer. You should examine the objects surrounding the *Bytes to Send* object in the Block Diagram to get the clue.

2. Modify the server's Block Diagram as shown in the illustration. Run the Server and Client VIs with and without the modification while observing the Boolean indicators for the inner and the outer loops.
   a. Explain the behavior of the indicators.
   b. What is the advantage of having the original wiring before the modification? Explain.

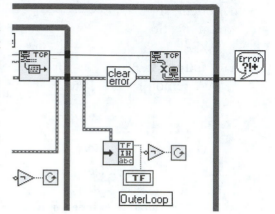

3. Modify the client's Block Diagram as shown in the illustration. Determine what else must be modified to make the Server and Client VIs work. Note: the required changes are very small.

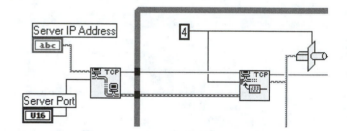

4. Modify Server.vi and Client.vi as follows:
   This time the client generates a waveform (a square wave or a sine wave) and sends it to the server. The server receives the waveform from the client, displays it on a waveform graph, and saves it to a spreadsheet. The user must have a Front Panel option in the server as to whether to save the data to the spreadsheet or not to save it.

LabVIEW on the Internet

Introduction

A communication model between a remote client and a server is shown in Fig. 17-7. The client uses a browser to access the server and process a request. The request is processed by the Common Gate Interface (CGI) program at the server's station and is returned to the client. Fig. 17-7 shows the specific steps that must be executed during such communication.

A. Client Connects to Server

A remote client uses a browser, such as Netscape Navigator or Internet Explorer, and the server's IP address to make a connection with the server and open the server's home page. The client clicks on the link that represents the client's request on the server's home page.

B. Server Passes the Request to CGI

The CGI called CGICTRL.vi processes the client's request. The server communicates with the CGI through the use of the environmental variables, which contain the parameter values of the client's request. CGICTRL.vi uses the parameter values to run the CGI application called Thermo.vi in our experiment.

C. CGI Returns Data

CGICTRL.vi returns the data to the server, and the server forwards the data to the client. Specifically, the server opens the second web page that contains the Front Panel of the CGI application Thermo.vi, which that remote client can view on the browser's screen.

D. Next Request

The remote client sees the Front Panel of Thermo.vi, where temperature data is being acquired and displayed on a chart in real time. The temperature data reflects the temperature in a laboratory. The Front Panel also includes several hyperlink objects that the client can operate and set the cooling fan's trip point. Every time the client clicks on a hyperlink, a new request is issued and the server will process it.

The reader should be familiar with the following definitions:

Universal Resource Locator (URL) – A standard method for describing various resources that are available on the Internet. A client can use URL VIs to request a CGI application from the G Web server. URL VIs can be found in the Internet Toolkit subpalette (Functions>Internet Toolkit>URL Client VIs).

Common Gate Interface (CGI) – A standard for communicating with information servers. The term CGI can be used to describe a CGI application or an interface to the CGI application. CGI VIs that can be used to build a CGI application can be found in the Internet Toolkit subpalette (Functions>Internet Toolkit>CGI VIs).

Hyper Text Markup Language (HTML) – HTML is a computer language that is used to build web pages. A web page has embedded HTML code that a browser interprets to display the web page. HTML uses tags to break up the document into sections, which perform specific tasks.

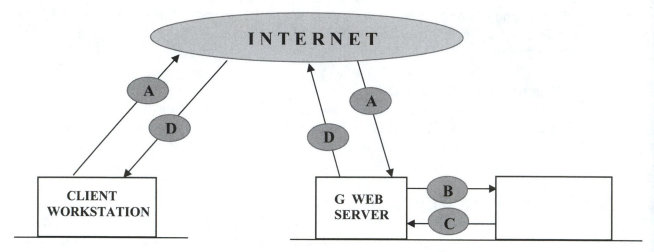

Fig. 17-7 Communication Model Between the Client and the Server

SOFTWARE

The Browser

Once the server is running on the host's workstation, the remote client opens a browser, such as Netscape Navigator or Internet Explorer, and enters the server's IP address, such as 127.0.0.1. At this time the client is ready to communicate with the server over the Internet. In this experiment it is possible to achieve the exchange of information between the client and the server in one of three configurations:

Client and server are connected by a network such as Ethernet
Client and server are far apart and are connected by Internet
Client and server are on the same workstation

The client runs the browser and the browser opens the server's first web page (home page). It contains a single link called "Temperature Control" as shown in Fig. 17-8.

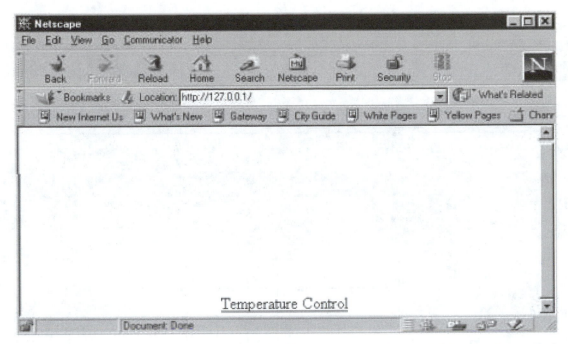

Fig. 17-8 Server's Home Page with the Temperature Control Link in Browser's Window

Server's Second Web Page

When the client clicks on the Temperature Control link, a second web page, shown in Fig. 17-9, opens and appears in the browser's window.

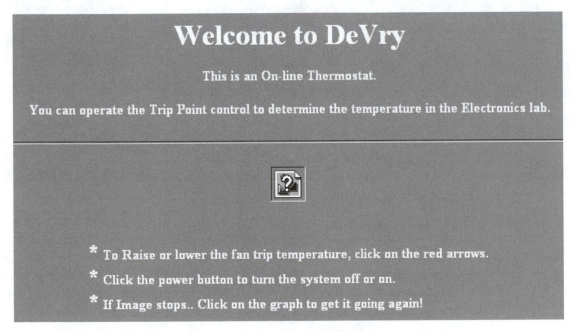

Fig. 17-9 Server's Second Web Page with a Message for the Client

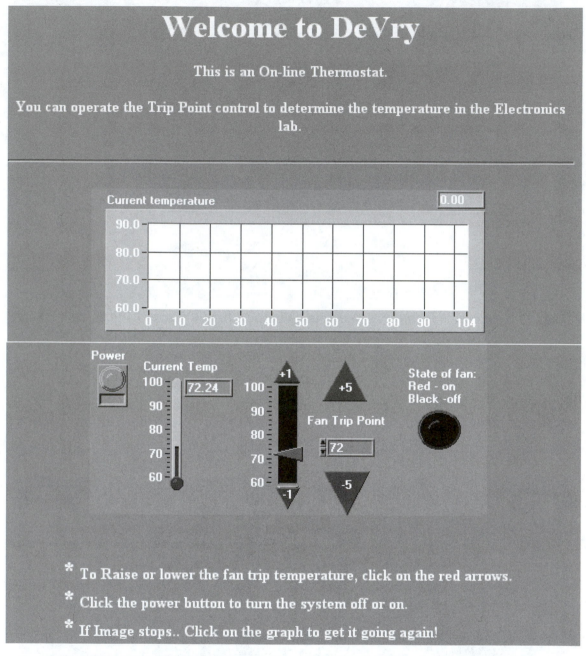

Fig. 17-10 Server's Second Web Page Updated a Few Seconds Later

This web page includes a brief welcoming message and instructions on operating the Front Panel objects of the LabVIEW's temperature control VI. This window persists for several seconds, allowing the client to read the instructions and then make changes as shown in Fig. 17-10. It now displays the Front Panel of the temperature control VI called thermo.vi that is running on the server.

It also includes five hyperlinks, +1, −1, +5, −5, and a Power switch. The remote client can operate each hyperlink.

HTTP Server

In order to process a remote client's request, the HTTP server must be running. An example of the server window is shown in Fig. 17-11. The black bar displays on the left side current data and time, and on the right side the elapsed time and the number of processed requests. Displayed below the black bar are the server's address, port number, and "Running" (the current status of the server).

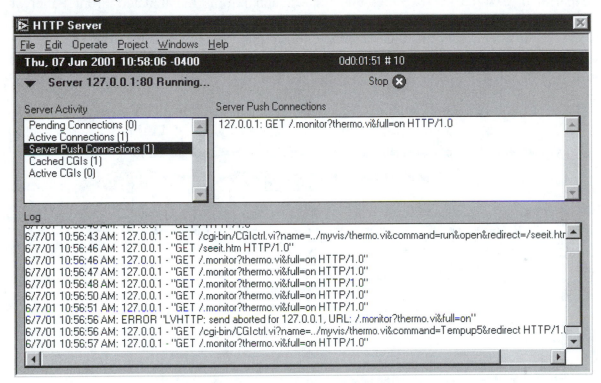

Fig. 17-11 The Window of the HTTP Server

The Server Activity section displays

Pending Connections - Connections accepted by the server but not yet processed.

Active Connections - Displays the number of open connections

Server Push Connections - Displays the number of server push connections. When this line is selected, as shown in Fig. 17-11, the display includes the address and VI name (thermo.vi) whose Front Panel is returned to the client once a second.

Cached VIs - Displays the number of CGI VIs currently in memory and their activity status.

Active CGIs - Indicates the number of CGI VIs processing requests at the present time.

Log indicates the most recent requests sent by the client and errors.

Common Gate Interface

The Common Gate Interface (CGI) is a VI written in G language using CGI VIs, such as Keyed Index Array.vi, that are located in Internet Developer's Toolkit for G. A CGI VI is an interface between the HTTP server and a CGI application. The server communicates with the CGI VI by means of environmental variables, which include the parameter values of the client's request.

Suppose the client clicks, for example, on the " +5" hyperlink. This represents a request from the client to increase the fan trip point by 5° F in the CGI application (thermo.vi in our experiment). The HTTP server passes this request to the CGI VI (CGIctrl.vi in our experiment). CGIctrl.vi executes this request and returns the updated Front Panel of thermo.vi to the client's browser. This is where server push animation is very important. Without this capability the client is unable to see the updated Front Panel that shows the fan trip point indicator increased by 5. The G server as well as Netscape Navigator support server push animation.

The CGI VI must comply with the following format shown in the Block Diagram of Fig. 17-12.

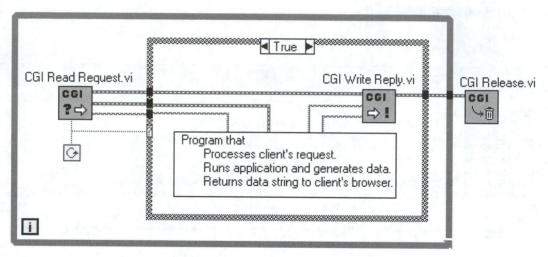

Fig. 17-12 General Block Diagram of the CGI Interface

The following CGI VIs are absolutely essential and must be included in the Block Diagram:

 CGI READ REQUEST.VI – Detects the client's URL

 CGI WRITE REPLY.VI – Responds to the client's request and returns the data to the client's browser.

 CGI RELEASE.VI – Informs the HTTP server that the CGI has completed execution and can be unloaded from memory.

In addition, the While Loop must include code to process the client's request and return data to the client's browser.

CGIctrl.vi

CGIctrl.vi is the interface designed for our experiment. It includes, of course, the three CGI VIs described earlier and additional code as shown in Fig. 17-13.

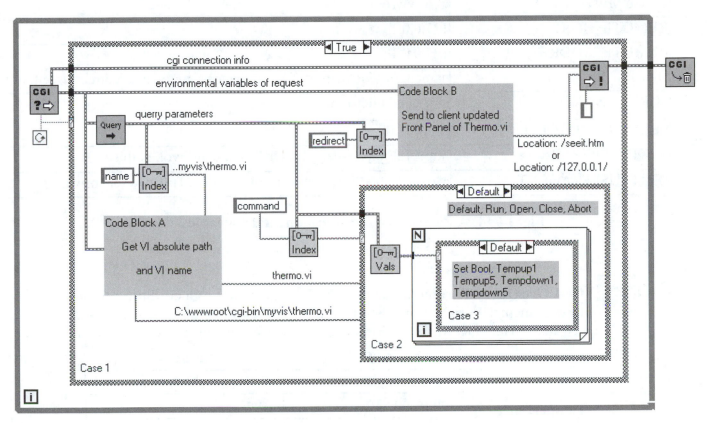

Fig. 17-13 Block Diagram of CGIctrl.vi

The Common Gate Interface is the interface between the server and our CGI application, thermo.vi.

CGI Read Request.vi passes a TRUE to the Condition Terminal of the While Loop if the request from a client is a valid request. If the request is valid, the True frame of Case 1 will then be executed. This frame includes the necessary process client's request. If CGI Read Request.vi passes a FALSE to the condition terminal, the execution of the While Loop will be terminated.

CGI Read Request.vi returns an Environmental Variables cluster that includes two one-dimensional arrays: the keys array and the values array. Fig. 17-14 shows a partial Environmental Variables list. The keys are listed in the left column and their corresponding values, in the right column.

Of particular interest is the key called the Query String. Its value is a URL that includes a relative path to the CGI application thermo.vi and to the parameters of the client's request.

Keys	Values
SERVER_NAME	127.0.0.1
SERVER_PORT	80
DOCUMENT_ROOT	C:/wwwroot
REMOTE_HOST	127.0.0.1
REMOTE_ADDR	127.0.0.1
QUERY_STRING	name=../myvis/thermo.vi&command=run&open&redirect=/seeit.htm
SCRIPT_NAME	/cgi-bin/CGIctrl.vi
SERVER_PROTOCOL	HTTP/1.0
HTTP_REFERER	http://127.0.0.1/seeit.htm URL link that invoked this CGI
HTTP_USER_AGENT	Mozilla/4.06 [en]C-gatewaynet (Win98; I) (Browser software)

Fig. 17-14 A Partial List of Environmental Variables

Query → **CGI Get Query Parameters.vi** in Fig. 17-13 extracts the Query String from the Environmental Variables and returns a cluster of two one-dimensional arrays of keys and values. The Query String's keys and their corresponding values are shown in Fig. 17-15.

	Keys	Values
Temperature Control link	name	…/myvis/thermo.vi
	command	run
	open	
	redirect	/seeit.htm
Hyperlink Tempup1	name	…/myvis/thermo.vi
	command	Tempup1
	redirect	
Hyperlink Tempup5	name	…/myvis/thermo.vi
	command	Tempup5
	redirect	
Hyperlink Power Switch	name	…/myvis/thermo.vi
	command	SetBool,Power,toggle
	redirect	

Fig. 17-15 Keys and Values of the Query String

This table shows that each set of keys and values corresponds to a specific request by the client. For example, when the client clicks for the first time on the "Temperature Control" link, the key/value combination of name/run is the request from the client to open and run the thermo.vi CGI application, and the redirect/seeit.htm opens the second web page and presents the client with the Front Panel of thermo.vi. The Front Panel shows the temperature data being acquired and displayed on the waveform graph in real time.

As the client clicks on the +5, −5, +1, −1, and Power switch hyperlinks, the Query String command values change. For example, the command value Tempup5 results when the client clicks on the +5 hyperlink. CGIctrl.vi attends to this command and increases the fan trip point by 5°. When the client clicks on the Power switch, CGIctrl.vi uses the "SetBool, Power, toggle" command to stop execution of thermo.vi.

 The **Keyed Array Index** CGI VI used in Fig. 17-13 returns the value of the command that corresponds to the specified input key.

Code Block A decodes the name of the CGI application and its absolute path. The nested Case structures 2 and 3 execute the client's request represented by command values such as run, Tempup5, SetBool, Power, toggle, and so on. They receive as inputs the name of the CGI application (thermo.vi), its absolute path, and the Query String.

Code Block B updates the Front Panel of thermo.vi, which is viewed by the client in the browser's window. As mentioned earlier, the HTTP server as well as Netscape Navigator support server push animation, but Internet Explorer does not. In the event that the user wishes to use Internet Explorer, the refresh command is added to the embedded second web page HTML code. The refresh command will update the Front Panel temperature graph in thermo.vi. The client will notice the blinking effect that is due to the updating operation.

CGI WRITE REPLY.VI – Responds to the client's request and returns data to the client's browser. As shown in Fig. 17-13, it receives as inputs the CGI connection info from CGI Read Request.vi and the header generated by Code Block B. As shown in Fig. 17-16, CGI Connection Info is a cluster of three elements that are the name of the interface, its type, and the number of the session, and the header references the server's second web page that contains the Front Panel of thermo.vi.

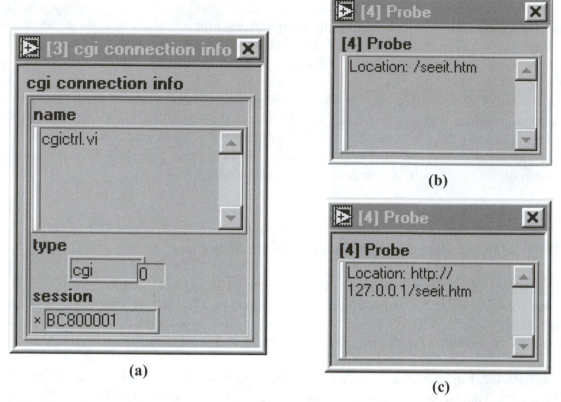

Fig. 17-16 (a) Connection Info Cluster from CGI Read Request.vi, (b) Header for the CGI Connection Info After Client Clicks for the First Time on Temperature Control Link in Server's Home Page, (c) Header for the CGI Connection Info After Client Clicks on a Hyperlink in Server's Second Web Page

Thermo.vi

This is the CGI application LabVIEW program. Fig. 17-17 shows the Front Panel and the Block Diagram of thermo.vi. The temperature is displayed in real time on the waveform chart and on the thermometer indicator. The hyperlinks labeled as +1, −1, +5, −5, and Power control allow the client to operate from the browser's window. When the client clicks on +5, for example, the Vertical Pointer Slide and its digital display will increment by 5 degrees. This is an indication that CGIctrl.vi has responded to the client's request. The LED indicator labeled as State of Fan will turn red when the temperature exceeds the fan's trip point value.

In the Block Diagram AI Sample Channel.vi returns one temperature data point from channel 0 of DAQ Board No. 1 every 0.2 seconds. This data value is converted to degrees F and compared against the fan trip point value that is under the client's control. If the actual temperature has exceeded the fan trip point value, the TRUE Case structure is executed and AO Update Channel.vi will output +5 V to turn the fan ON for cooling. If the actual temperature is less than the fan's trip point value, AO Update Channel.vi will output 0 V in the FALSE frame of the Case structure to turn the fan OFF.

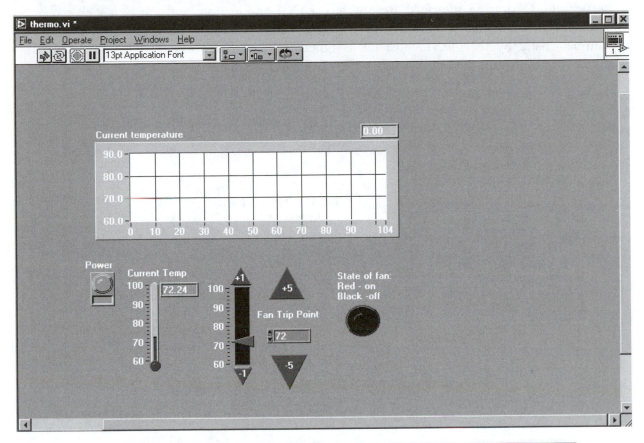

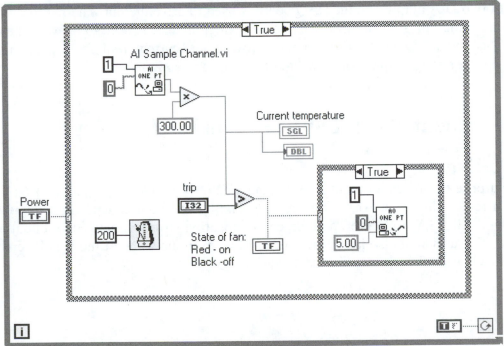

Fig. 17-17 The Front Panel and Block Diagram of Thermo.vi

Hardware

The hardware setup on the server is shown in Fig. 17-18. It includes the running HTTP server, the CGIctrl.vi interface, the thermo.vi CGI application, and the two web pages. The DAQ board must be configured by running the wdaqconf utility, and the device number, used by thermo.vi, must be assigned. As usual, the extender board provides access to all pins on the DAQ board, which is physically inside the PC.

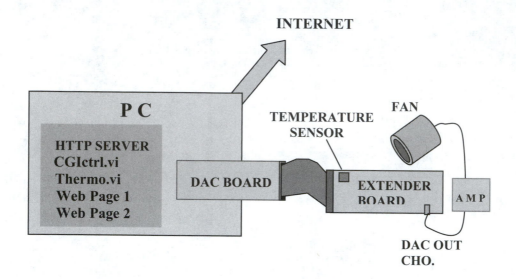

Fig. 17-18 The Hardware Setup on the Server

One of two D/A outputs on the DAQ board is used to drive the fan. Thermo.vi outputs +5 V to drive the fan. An amplifier such as a Darlington chip, is used to provide the current drive, as the DAQ board is unable to provide the current required by the fan. The temperature sensor chip is connected to one of the analog input channels.

Experiment 17-2: LabVIEW on the Internet

Objective

The purpose of this experiment is to demonstrate how a client can open a VI on a server from a remote location. Also its intent is to familiarize the user with the Common Gate Interface VIs, G Web Server, and the communication structure between a client and the server. The user will also have an opportunity to examine some of the HTML embedded code, such as the image map for the hyperlinks.

Parts

Hardware

DAQ board (such as LabPC$^+$ or MIO-16E-10)
Extender board for the DAQ board
Fan
Darlington to provide current drive for the fan

Software

LabVIEW (version 5.1 or higher)
Internet Developer's Toolkit for G
wwwroot folder (provided with this book)
Browser such as Netscape Navigator or Internet Explorer

Procedure

1. Configure the hardware as shown in Fig. 17-18. Connect the temperature sensor to AI Ch0 and the fan to D/A Ch0. Thermo.vi assumes that the device number for your DAQ board is 1. If it is not, change the device number in thermo.vi to your device number.

2. Load the wwwroot folder and the Internet Developer's Toolkit for G in the same folder on the server's hard drive.

3. Open LabVIEW, switch to New VI, and from the Front Panel run HTTP server (on the server's station) by choosing Project > Internet Toolkit > Start HTTP Server. When the server's window opens, it will appear as shown in Fig. 17-11. Record the server's IP address.

4. Open the browser on the client station; in the browser's location window, enter the address you recorded in step 3 and hit the Enter key on the keyboard.

5. When the browser opens the server's home page, you will notice a link labeled Temperature Control. Click on this link.

6. After a few seconds, the server's second web page opens with the Front Panel of thermo.vi and real-time temperature data acquisition displayed on the waveform chart. Operate the links to change the setting of the fan's trip point. The Power switch toggles when you click on it, turning thermo .vi ON and OFF.

Chapter 18
GPIB Instrument Control

Introduction

The GPIB (General Purpose Interface Bus) interface has its origin in the HPIB (Hewlett Packard Interface Bus) developed by the Hewlett-Packard Corporation in the 1970's for interfacing their instruments. It gained widespread popularity and soon became a de facto industry standard. In 1975 the IEEE standards group published the standard IEEE-488 for the GPIB interface. Industry uses this standard to this day as a parallel interface for various applications in instrument control.

The GPIB interface is a parallel 24 pin conductor bus. It includes eight data lines for control messages that are often ASCII encoded, and various management and handshake lines. Handshaking is used to transfer messages between the PC and the instrument being controlled.

Fig. 18-1 shows a typical connection between the PC with the GPIB interface board and the software for controlling the operation of an instrument. Not every instrument has a GPIB capability. This capability is designed into the instrument by the manufacturer, and a GPIB connector is provided on the instrument housing. The cable used to connect the instrument must also be GPIB compatible and meet the IEEE-488 protocol.

As shown in Fig. 18-1, the operation of several instruments may be controlled. Consequently, each instrument must have a unique address between 0 and 30, with address 0 usually being reserved for the GPIB interface board. National Instrument Corporation has an array of GPIB interface boards with different capabilities. As shown in Fig. 18-1, the AT-GPIB interface board is used in this paper. It has a maximum data transfer rate of 1 Mbps.

The NI-488.2 is the driver software for the GPIB interface. GPIB devices can be Talkers, Listeners, or Controllers. Listeners receive messages while Talkers send messages. The Controller, usually the PC where the GPIB board and the NI.488.2 software are installed, manages the flow of commands or messages on the GPIB bus. The communication between the PC and the instrument is message based. ASCII encoded message strings are transferred using handshaking.

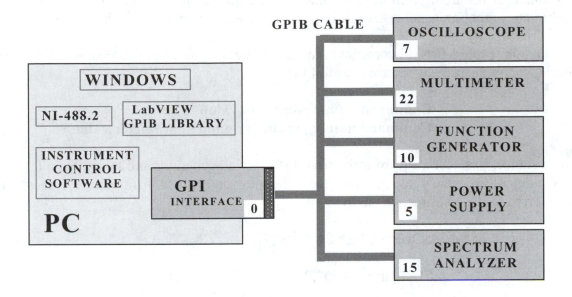

Fig. 18-1 GPIB Instrument Control Configuration. Each Instrument and the GPIB Interface Has a Unique Address

Functions>Instrument I/O>GPIB is the location of available GPIB VIs for building an instrument control VI, a VI that is custom tailored to meet the user's exact needs. On a small scale this is a driver.

The most used GPIB VIs, shown below, are the GPIB Read and the GPIB Write VIs. The former reads a data string from the instrument with the specified address and the latter writes a control string to the instrument with the specified address.

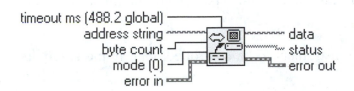

The **GPIB Read** VI reads the specified number of bytes from an instrument with the specified address string.

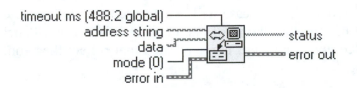

The **GPIB Write** VI writes the data string to an instrument with the specified address string.

GPIB Instrument Commands

Five instruments are used in the experiment, described in this chapter: the power supply, multimeter, function generator, oscilloscope, and spectrum analyzer. If the instrument has GPIB capability then the manual for that instrument must include a list of GPIB commands. Table I lists all commands for the instruments used in this chapter. A typical format for an instrument control is as follows:

Command name: command value
Command name: parameter name: parameter value

For instance, to select a square wave from the function generator and to set its voltage to 5 V and its frequency to 500 Hz, the following commands from Table I must be transferred to the instrument at address 10:

SOUR:FUNC: SHAP SQU
SOUR:VOLT 5.0
SOUR:FREQ 500.0

In a similar fashion, commands are issued to control the operation of other instruments.

Instrument Control Template

The GPIB instrument control VI need not be a commercial driver. In most applications a custom driver that meets the needs of the user is more efficient and faster.

The user must know the commands for the instrument that he wants to control. If the instrument is GPIB compatible then the commands can be found in the manual for that instrument.

Fig. 18-2 shows a general-purpose instrument control template that can be adapted to almost any GPIB instrument. It consists of two Sequence structures inside a While Loop. As long as a TRUE is applied to its condition terminal, the code inside the While Loop will be repeatedly executed. The user must click on the STOP button in the Front Panel to terminate execution.

The Sequence 1 structure selects the instrument to be controlled. One frame is required for each instrument.

The Sequence 2 structure generates the necessary commands to turn the instrument, select one of its outputs or parameters, and adjust the selected parameter to the desired value. Also included in this frame is code that detects whether the user changed any of the Front Panel controls. If not, the empty FALSE case will be executed and no message will be sent. However, if the user changed the setting of one or more of the Front Panel Controls, the TRUE case is executed and commands are generated that reflect the Front Panel changes made by the operator.

Table I   Instrument Control Commands for HP Instruments

INSTRUMENT	COMMAND NAME	COMMAND VALUE	ACTION
POWER SUPPLY HP E3631A **Address 5**	OUTP	ON	POWER ON
		OFF	POWER OFF
	INST:SEL	P6V	SETS OUTPUT TYPE TO +6 V
		P25V	SETS OUTPUT TYPE TO +25 V
		N25V	SETS OUTPUT TYPE TO –25 V
	VOLT	NUMERIC	SETS OUTPUT VOLTAGE:USER ASSIGNED
	CURR	NUMERIC	SETS OUTPUT CURRENT:USER ASSIGNED
FUNCTION GENERATOR HP 33120A **Address 10**	SOUR:FUNC:SHAP	SIN	SETS OUTPUT TYPE TO SINE WAVE
		SQU	SETS OUTPUT TYPE TO SQUARE WAVE
		TRI	SETS OUTPUT TYPE TO A TRIANGULAR WAVE
		RAMP	SETS OUTPUT TYPE TO RAMP WAVE
	SOUR:FREQ	NUMERIC	SETS OUTPUT FREQUENCY VALUE
	SOUR:VOLT	NUMERIC	SETS OUTPUT VOLTAGE VALUE
MULTIMETER HP34401A **Address 22**	CONF:	VOLT:DC	SETS THE METER TO READ DC VOLTAGE
		CURR:DC	SETS THE METER TO READ DC CURRENT
		VOLT:AC	SETS THE METER TO READ AC VOLTAGE
		CURR:AC	SETS THE METER TO READ AC CURRENT
		RES	SETS THE METER TO READ RESISTANCE
		FREQ	SETS THE METER TO READ FREQUENCY
		CONT	SETS THE METER TO TEST CONTINUITY
		DIOD	SETS THE METER TO TEST DIODE
OSCILLOSCOPE Address 7	READ?	NONE	REFRESHES THE SCREEN
	WAV:DATA?	NONE	ACQUIRES THE DATA AS A STRING

The Front Panel control "A" in Fig. 18-2 is compared to the local variable "B" (the "B" indicator is inside the True Case frame). If they are the same, nothing happens because the empty True Case is executed. But if the operator changed the value of "A" in the Front Panel, the values of "A" and "B" are now different. Consequently, the True Case is executed and appropriate commands are dispatched to change the instrument setting to the new value of "A" and the value of "B" in the True Case is set to that of "A."

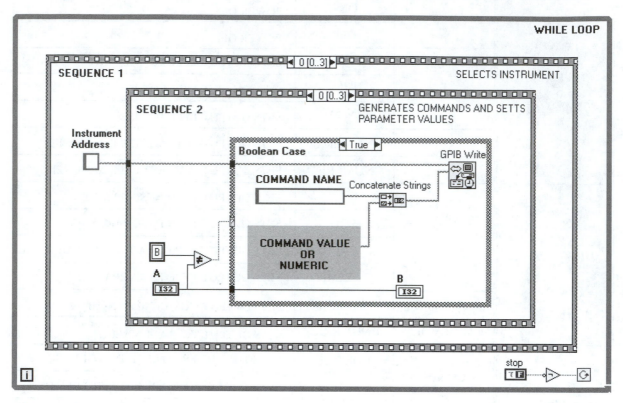

Fig. 18-2 GPIB Instrument Control Template

This is illustrated in Fig. 18.3. Suppose that the Freq. digital control in the Front Panel is set to 500 Hz and the VI is executed. The value Frequency 2 is zero by default. The result of the comparison of Freq. and Frequency 2 yields FALSE at the output of the Not Equal To function, and the True Case is executed, storing 500 Hz in the Frequency 2 indicator. The command is also sent to change the frequency of the Function Generator at address 10 to 500 Hz. This process is repeated each time the operator changes the value of the frequency.

Before transmission, all commands must be converted to strings because strings provide the most flexible format. They can represent alphabetical characters, numeric characters, and almost any other character on the keyboard. The transmitted command strings are usually ASCII encoded.

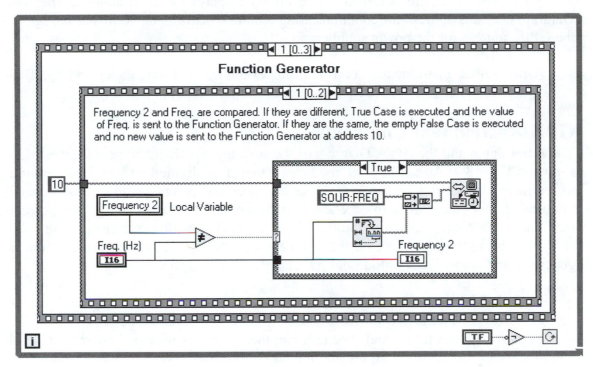

Function Generator

Frequency 2 and Freq. are compared. If they are different, True Case is executed and the value of Freq. is sent to the Function Generator. If they are the same, the empty False Case is executed and no new value is sent to the Function Generator at address 10.

Local Variable

Frequency 2

Freq. (Hz)

SOUR:FREQ

Frequency 2

Fig. 18-3 The True Case is Executed When the User Changes the Value of Freq. Digital Control in the Front Panel.

Experiment 18-1: GPIB Instrument Control

Objective

In this experiment the user is acquainted with the basic functionality of the GPIB instrument control. This includes the software and the hardware configuration.

Procedure

Initial preparation for this experiment includes the installation and the configuration of the GPIB interface board. The instruments to be controlled are then connected to the GPIB interface by the GPIB cable, as shown in Fig. 18-1.

The instruments that you wish to control may be different than those used here. In order to make GPIB Instrument Control.vi work for your instruments, you must modify the commands in the Block Diagram. To do that, obtain the commands for your instrument from the instrument's manual and enter the new commands in the Sequence structure. Each frame in the Sequence 1 structure is assigned to one instrument, and all frames of Sequence 2 structure apply to that instrument. For example, in Fig. 18-4 Frame 0 of Sequence 1 is assigned to the power supply and frames 0 to 3 of Sequence 2 generate commands for the power supply.

When a GPIB instrument is turned ON, it briefly displays its GPIB address. The Block Diagram uses this address for communicating with that instrument. Note that in Fig. 18-4 the GPIB address for the power supply is 5.

1. Open GPIB Instrument Control.vi and examine the Front Panel and the Block Diagram. The following section provides information about this software.

GPIB Instrument Control.vi

The Front Panel and the Block Diagram of the software, GPIB Instrument Control.vi, are shown in Fig. 18-4. As can be seen, the Front Panel is subdivided into sections pertaining to the instruments being controlled. There are three instrument sections: power supply, multimeter, and the function generator. Each section includes controls for selecting instrument functions and for adjusting the selected function parameter value. For example, the user may select on the vertical slide the square wave from the function generator and adjust its amplitude and frequency on the adjacent digital controls.

In the Block Diagram of Fig. 18-4, the software is inside the While Loop that provides the user with an interactive environment. The loop will execute as long as TRUE is applied to its condition terminal. The user can then adjust Front Panel controls to set the instruments to the desired outputs and adjust their values. To terminate execution, the user must click on the STOP button in the Front Panel.

Due to the repetitive nature of the instrument control software, Fig. 18-4 shows the Block Diagram details that apply to the power supply in Frame 0 of the Sequence 1 structure. The four frames of Sequence 2 (inside Sequence 1) structure include the generation of appropriate commands to control the operation of the power supply.

Frame 0 of Sequence 2 checks the state of the power switch. If the user changed the state, Case 2 generates the command OUTP followed by ON or OFF from Case 1 and sends it to the power supply at address 5.

Frame 1 of Sequence 2 checks the state of the power supply output type (6 V, +25 V, or −25 V). If the operator changed the state, Case 4 generates the command INST:SEL followed by P6 or P25 or N25 from Case 3 and sends it to the power supply at address 5.

Frame 2 of Sequence 2 checks the state of the output voltage. If the user changed the voltage value, then Case 5 generates the command VOLT followed by the new value of voltage from the Front Panel digital control Voltage. It then sends this command to the power supply at address 5. The decimal form of voltage is converted to a string by the To Fractional function.

Frame 3 of Sequence 2 checks the state of the current setting. If the user changed the setting, then Case 6 generates the command CURR followed by the new value of current from the digital control Current. It then sends this command to the power supply at address 5.

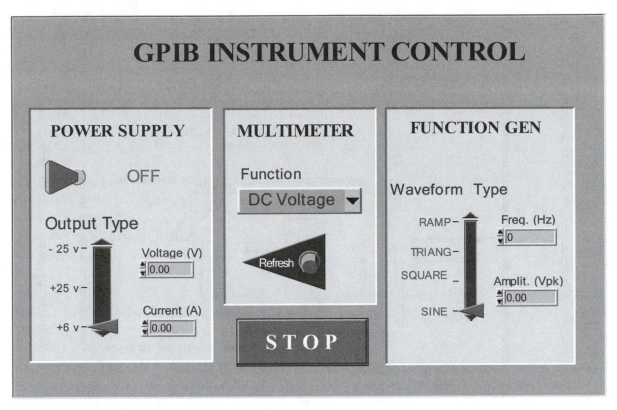

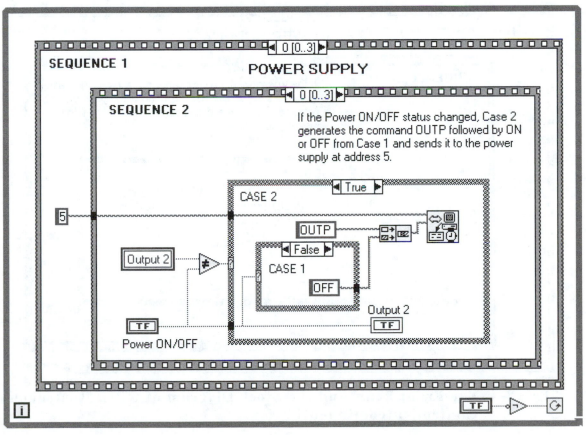

Fig. 18-4 The Front Panel and the Block Diagram of GPIB Instrument Control.vi

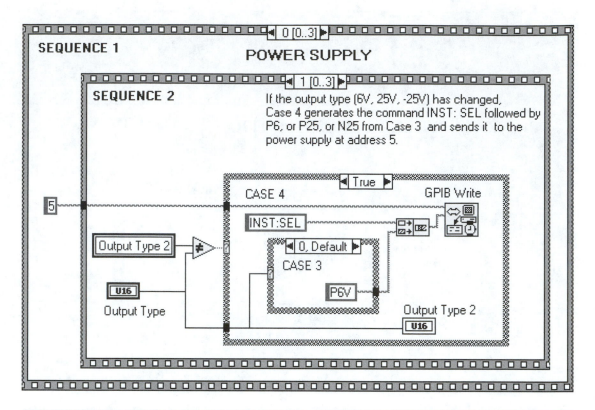

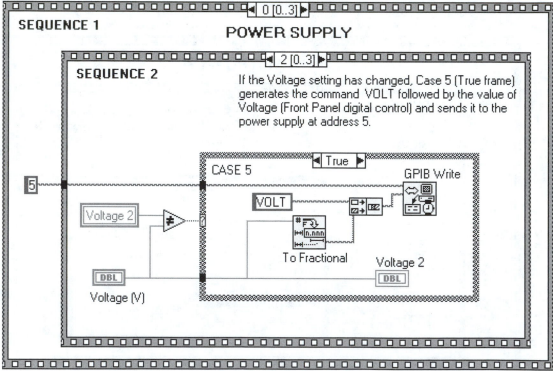

Fig. 18-4 The Front Panel and the Block Diagram of GPIB Instrument Control.vi (continued)

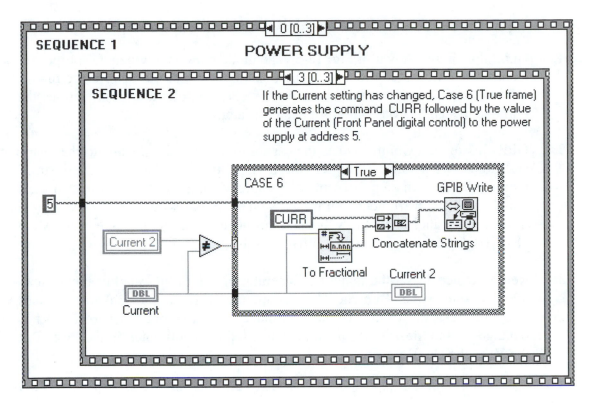

Fig. 18-4 The Front Panel and the Block Diagram of GPIB Instrument Control.vi (continued)

2. ***Run*** GPIB Instrument Control.vi.

3. **Power Supply**
    Select +6 V from the Output Type vertical slide in the Front Panel and enter voltage values such as 2, 3, 4, and so on in the Voltage digital control. Observe the power supply to see these changes occur.

4. **Function Generator and Multimeter**
    Select Sine on the Waveform Type vertical slide in the Front Panel. Set the frequency to 500 Hz and the amplitude to 5 V. Connect the multimeter to the output of the Function Generator. Set the Function control to AC Voltage. Note the measured value on the multimeter. Now set the Front Panel Function control to Frequency and note the reading on the multimeter instrument.

Analysis

1. Build the circuit shown in Fig. 18-5. The diode may be any general-purpose diode. Connect the Multimeter to V_o and the Function Generator to V_i. Before running the VI and making measurements you must make the software modifications that are outlined in step 2. You will return to step 1 later.

2. GPIB Instrument Control.vi has the capability to send commands to the instrument and adjust its setting. However, there is no provision to read the instrument's measured value. In step 1 we must measure V_o and display its value in the Front Panel of GPIB Instrument Control.vi. Modify the VI so that it displays two values in the Front Panel: the RMS value and the DC or the average value of the V_o in Fig. 18-5 [Hint: Consider using the GPIB Read function].

3. Return to step 1. Set the Function Generator on the Front Panel of the VI to a 4 V, 500 Hz sine wave. Run the VI. Record the RMS and the average values. To measure the RMS value, set the Multimeter on the Front Panel of the VI to AC Voltage, and to measure the average value, set the Multimeter on the Front Panel of the VI to the DC Voltage.

4. Repeat step 3 for a square wave and then for a ramp. Record two values (Front Panel display) for each waveform.

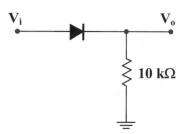

Fig. 18-5 Test Circuit for the Analysis Section

5. Using calculus, derive the RMS and the average values for the three V_o waveforms (sine, square, and ramp). Use the derived equations to calculate the six values and compare them to the measured values.

6. Is it possible for the calculated and measured RMS values to differ by a large amount? Explain what could be a major source of error. [Hint: Consider the true RMS voltmeters and the older types whose scale depends on the calibration.]

Experiment 18-2: GPIB Oscilloscope

Objective

As mentioned earlier, the remote operation of any instrument that has a GPIB interface (a GPIB connector, usually in back of the instrument) can be controlled by GPIB commands. The oscilloscope used in this experiment is a GPIB instrument. The user will become familiar with specific GPIB commands that control its operation.

Procedure

1. Initial preparation for this experiment includes the installation and the configuration of the GPIB interface board. The oscilloscope to be controlled is then connected to the GPIB interface by the GPIB cable as shown in Fig. 18-1.

The oscilloscope used in this experiment is HP-54600B, and the LabVIEW program Oscilloscope.vi is designed for this oscilloscope. In order to make GPIB Instrument Control.vi work for your oscillocope, you must modify the commands in the Block Diagram. To do that, obtain the commands for your instrument from the instrument's manual and enter the new commands in the Sequence structure. The commands are strings such as those shown below. The "Channel 1: Range?" shown below is a command requesting the oscilloscope to send back the voltage range value for channel 1.

The voltage range for channel 1 is received, divided into 8 divisions, and displayed on the Ch1 V/Div digital indicator in the Front Panel as shown below:

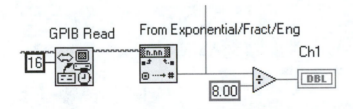

This is done also for channel 2. Similarly, in response to the request for time range,

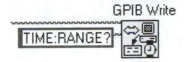

the oscilloscope sends back the value for the time range. The range is divided into 10 divisions and displayed on the Front Panel digital indicator Time/Div as shown below.

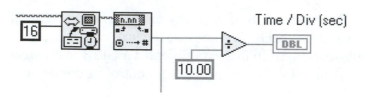

Thus, if the oscilloscope that you are using is different, you must modify all commands in the Block Diagram of Oscilloscope.vi to make the VI work for your instrument.

2. Open Oscilloscope.vi (Book VIs>GPIB>Oscilloscope.vi). The Front Panel is shown in Fig. 18-6.

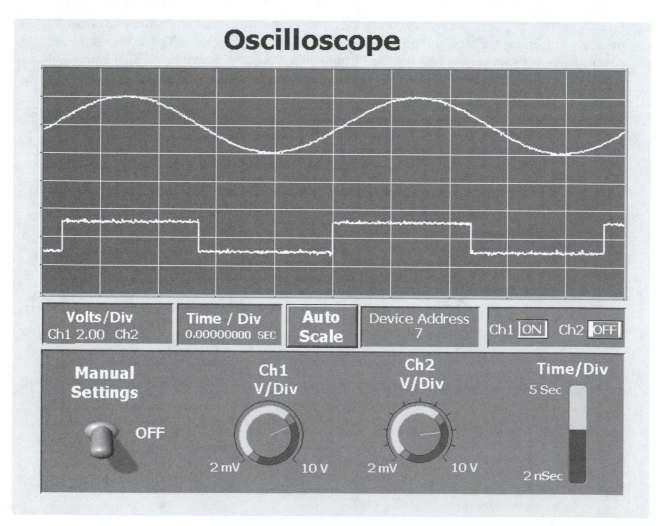

Fig. 18-6   The Front Panel of Oscilloscope.vi

The oscilloscope's device number, or the address for the oscilloscope used in this experiment, is 7. The Auto Scale button provides automatic scaling for the time base and Volts/Div for both channels. To active auto scaling, click on this button. The Manual Settings button, when activated by clicking on it, allows the user to set Volts/Div and time manually.

Examine the Block Diagram and study the software organization. See how scaling both the auto and manual settings is accomplished in the software.

3.   Connect to Ch1 of the oscilloscope instrument a sinusoidal waveform whose amplitude is 4 Vpk and whose frequency is 1 kHz.

4.   Connect a 2 kHz TTL square waveform to Ch2 of the oscilloscope instrument.

5.   Operate the Front Panel controls to display the waveforms in auto mode. Check the peak voltage and the time period of the waves.

6.   Display the waveforms in manual mode and set the appropriate voltage and time scales. Check the peak voltage and the time period of the waves.

Experiment 18-3: Bode Plotter

Objective

The purpose of this experiment is familiarize the user with the GPIB equipment that measures and plots the Bode magnitude and phase plots. The user will also become familiar with the LabVIEW software and specific GPIB commands that make such plots possible. In this experiment the emphasis is on computerized data acquisition and data processing.

Introduction

The Test Configuration for the Bode Plotter is shown in Fig. 18-7. The GPIB interface in the PC provides the interface or the gateway for GPIB commands and data. The sine wave generator (on the Function Generator) and the AC voltmeter (on the Multimeter) are connected to the GPIB interface by the GPIB cable.

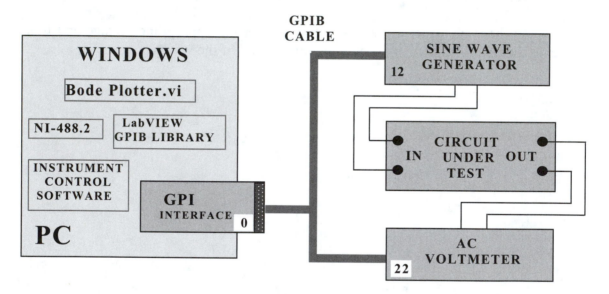

Fig. 18-7 Bode Test Configuration. The Circuit Under Test is Driven by a Sinusoidal Waveform. The Output is Measured by the AC Voltmeter. GPIB Commands and Data are Exchanged Over the GPIB Cable.

LabVIEW sends commands to increment the frequency of the sine wave generator and thus vary or sweep the signal frequency applied to the circuit under test (CUT). The output of the CUT that is measured by the AC voltmeter, is returned to LabVIEW for additional processing.

The CUT is one of two filters: a low-pass filter (LPF), shown in Fig. 18-8a, whose transfer function is $G(s) = 20000/(s + 5000)$, or a high pass filter, shown in Fig. 18-8b, whose transfer function is $G(s) = 5s/(s + 8000)$.

Note: Only one circuit can be tested at a time in Fig. 18-7.

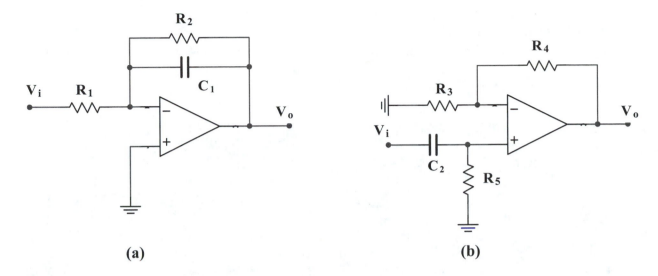

Fig. 18-8 The Circuit Under Test: (a) Low-Pass Filter, (b) High-Pass Filter

Before performing the test on the Bode Plotter, follow the procedure below to determine the parameter values for both filters.

Design Procedure

1. **Simulation** on Workbench or Multisim: Using the given transfer function run the simulation test for each filter. From the frequency domain response extract the values of the 3 dB frequency and the DC gain for each circuit.

2. **Transfer function:** Derive transfer functions for each filter in terms of Rs and Cs. Express the magnitude $M(\omega)$ and phase $\varphi(\omega)$ for each filter in terms of Rs and Cs. The transfer function $G(s) = V_o(s)/V_i(s)$ in the LaPlace's s-domain. The magnitude $M(\omega) = |G(j\omega)|$ and the phase $\varphi(\omega)$ is the angle of $G(j\omega)$.

3. **Design equations:** Using the results from steps 1 and 3, determine the design equations. From the design equations determine the parameter values of the Rs and Cs in Fig. 18-8. There will be more unknowns than equations, making it necessary to pick some values.

The section that follows provides a brief description of the LabVIEW program Bode Plotter.vi, which is used to generate the Bode plots.

Software

Fig. 18-9 shows the Front Panel of Bode Plotter.vi, the LabVIEW software. The graph displays include the magnitude, the phase, and the error (the difference between the data and the theory) plotted as a function of frequency.

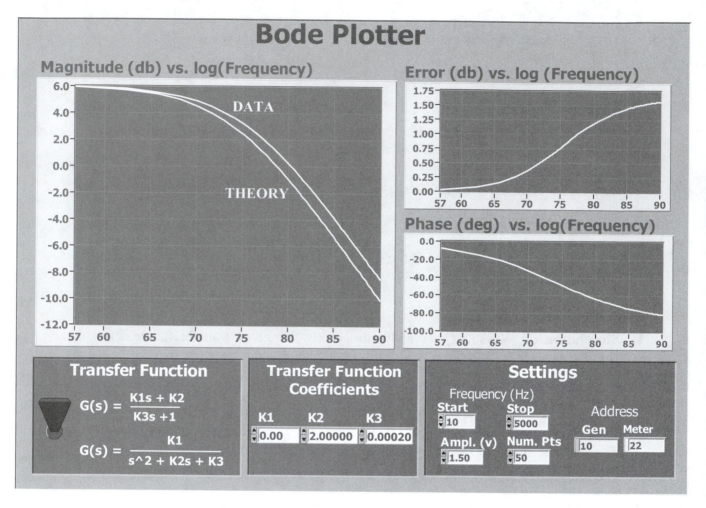

Fig. 18-9   The Front Panel of Bode Plotter.vi

The Transfer Function section allows the user to select a first-order system or a second-order system by means of a switch. The Transfer Function switch selects the first-order system or the second-order system. In the illustration the switch is set to select the first-order system.

The user must enter the values of K1, K2, and K3 in the Transfer Function Coefficients section.

In the Settings section the user must enter the Start and Stop frequencies, the amplitude of the sine wave applied to the CUT, the number of points to be measured, and the GPIB addresses of the Function Generator and the Multimeter. Once running, Bode Plotter.vi will acquire the number of specified data points and process the data. Processing includes the plotting of the magnitude, the phase, and the error curves as a function of frequency.

The VI also plots the magnitude curve based on theory and the values of K1, K2, and K3.

Procedure

1. Connect the equipment as shown in Fig. 18-7. The station must be configured for the GPIB operation. The equipment used in this experiment is the Function Generator E-4431B and the Hewlett Packard Multimeter 34401A.

   Note: If the equipment that you are using is made by another manufacturer, then before proceeding, additional work must be done. First, the commands in the Block Diagram for Bode Plotter.vi must be modified. To do that, you must look up the GPIB commands for the Function Generator and the multimeter in the manuals. For example, in Frame 0 (Block Diagram of Bode Plotter.vi), the first command shown below

   sets the Multimeter to read AC voltage, and the second command sets the Function Generator to output a sine wave (SIN). Some commands are concatenated by a value as shown in the next illustration.

   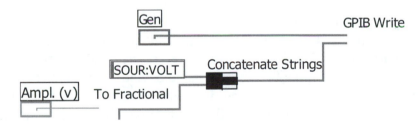

   The command SOUR:VOLT must be followed by the amplitude value set by the user in the Front Panel. As usual, all GPIB commands must be expressed as strings. You can proceed to the next step after all Block Diagram modification is complete.

2. Using the parameter values that have been previously determined, build the two filters.

3. Open Bode Plotter.vi (Book VIs>GPIB>Bode Plotter.vi). Examine the Front Panel and become familiar with all controls and indicators.

4. Connect the LPF circuit as shown in Fig. 18-7.

5. Run the VI and acquire the real data in graph form. A typical run for the LPF is shown in Fig. 18-9.

6. Repeat steps 4 and 5 for the high pass filter.

Analysis

1. Plot the exact Bode plots for each filter based on theory. Use LabVIEW to generate such plots (Formula Node may be used for this purpose). Other equation plotting software may also be used. A Bode plot must include $M(\omega)$ in dB versus $\log(\omega)$, and $\varphi(\omega)$ versus $\log(\omega)$.

2. Present here the following results in a format that is suitable for comparison:

    Simulation data from step 1 of Design Procedure
    Real data from steps 5 and 6 of Procedure
    Theory based data from step 1 of Analysis

    We shall assume that the data in question represents the magnitude plot versus frequency. Ignore the theory based response that is presented on the same graph in step 5 of Procedure. This may be used for comparison purposes with the results obtained in step 1 of Analysis.

3. **Test Results and Conclusion.**
    Compare the real data with that of the simulation and the theory based response. The comparison may be done at several frequencies and the deviation or the error noted. In a practical situation the acceptable deviation or the error is determined on the basis of subsystem specifications. For example, suppose that you are a designer working for company X and a vendor places an order for a subsystem to be designed by you. After all testing is complete you may have all or some of the results of step 2 above. It will be clear if your subsystem design is acceptable or not because the vendor will, no doubt, specify tolerances.

    In the absence of specifications and tolerances, let's assume that the deviations noted are due to component tolerances, as resistors and capacitors have tolerances. Measure the actual values of all passive components used and calculate deviations from the nominal values. Using these deviations, repeat step 1 above and obtain a new response curve. Does the real OP amp have any effect on the response?

    Make conclusions based on suggested analysis.

Experiment 18-4: The Spectrum Analyzer

Objective

The purpose of this experiment is to familiarize the user with the GPIB equipment that measures and plots the spectrum of a signal. The user will also become familiar with the LabVIEW software and specific GPIB commands that make frequency domain response possible.

Equipment and Test Setup

As shown in Fig. 18-10, signal generator output is a carrier amplitude modulated by one frequency, generally referred to as tone modulated AM wave. The spectrum analyzer takes this wave and produces the frequency domain response or the spectrum of the AM wave, which consists of the carrier spectral line at the center of the response and two sidebands.

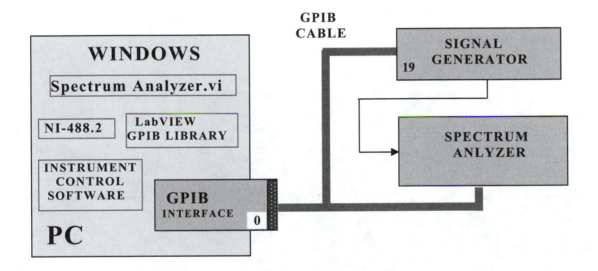

Fig. 18-10 Spectrum Analyzer Test Configuration. The Spectrum Analyzer Input is a Tone Modulated AM Wave. GPIB Commands and Data are Exchanged Over the GPIB Cable.

LabVIEW Software

Fig. 18-11 shows the Front Panel of Spectrum Analyzer.vi. The recessed box section on the right includes the controls that provide various parameter values that are used by the VI's Block Diagram. Here the user chooses the type of waveform from the menu ring, the carrier parameters that include its frequency and its level, and the modulating signal parameters that include the frequency and the AM depth or the modulating index. The RF switch is used to turn OFF the AM wave, and the Preset control resets the signal generator.

In the lower section, the user may adjust the width of the sweep for a better display, set here for 100 kHz. This section also includes digital indicators that display the AM wave bandwidth, the carrier power, and the total sideband power.

The GPIB device addresses 18 and 19 are for the equipment used in this experiment. These addresses make possible the communication of data and commands between the equipment and the LabVIEW software.

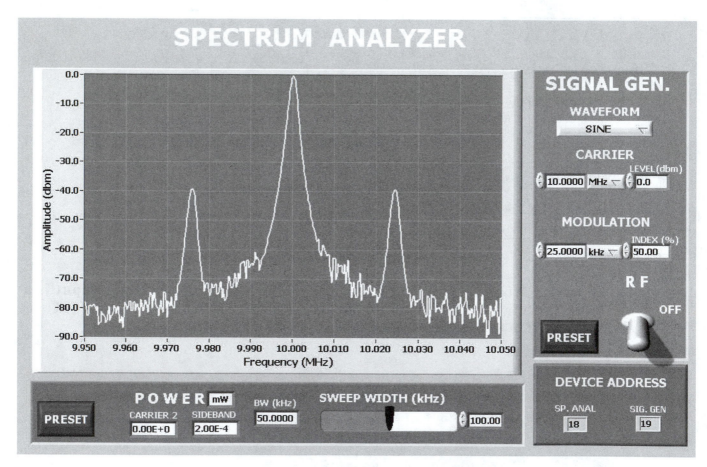

Fig. 18-11   The Front Panel of Spectrum Analyzer.vi. The Display Shows the
Spectrum of a 10 MHz Carrier Amplitude Modulated by a 25 kHz Signal.

Procedure

1. Connect the equipment as shown in Fig. 18-10. The station must be configured for the GPIB operation. The equipment used in this experiment is from a specific manufacturer. If the equipment that you are using is from a different manufacturer, then the comments in step 1 of the Procedure in the preceding experiment apply here as well. This means that the Block Diagram GPIB commands must be modified to suit your equipment and new GPIB addresses for your equipment must be entered in the Front Panel. If all that has been done, then you may proceed to the next step.

2. Open Spectrum Analyzer.vi (Book VIs>GPIB>Spectrum Analyzer.vi) and be familiar with the functionality of all Front Panel controls and indicators.

3. Decide on the parameter values for the AM wave and enter these values in the Front Panel.

4. Run the VI and save the spectrum display.

5. While the equipment is set up, you may need more data. See step 2 of Analysis.

Analysis

1. Calculate the AM wave bandwidth, the carrier power, and the total sideband power. Compare these values to those in the Front Panel display. The calculated and the measured values should be almost exactly the same.

2. Devise a simple algorithm in LabVIEW to examine the format or the manner in which the spectrum analyzer sends data to Spectrum Analyzer.vi program. You may place probes at strategic places in the Block Diagram to view the data.

3. Modify the VI to include the time domain wave that corresponds to the frequency domain response acquired. Fig. 18-12 illustrates the two responses.

Challenge Design Problem

Develop a LabVIEW VI to accomplish the following:

A station in the AM broadcast band has been approved by the FCC to transmit radio programs at 1 MHz. The modulation consists of the following frequencies: 200, 400, 600, 800, 1000, and 1500 Hz. All modulation components have the same level. You may choose the values of the modulation and the carrier levels.

Your LabVIEW VI must display the TD and the FD responses. A GPIB link that controls the modulator and the spectrum analyzer must be used, since both are actual equipment. If you wish, you may synthesize the modulation waveform in software. This waveform is then used to drive the signal generator's carrier.

The LabVIEW VI must display the AM wave bandwidth, the carrier power, and the sideband power. You must apply appropriate theory to verify these values.

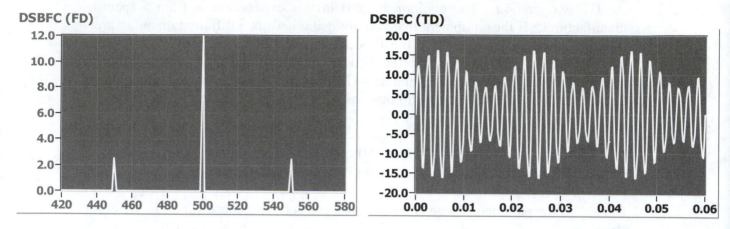

Fig. 18-12   Frequency Domain and Time Domain Responses for an Amplitude Modulated 500 Hz Carrier by a 50 Hz Tone

Index